SHARON

SHARON

The Life of a Leader

— BY HIS SON —

GILAD SHARON

Translated by Mitch Ginsburg

HARPER

An Imprint of HarperCollins*Publishers*
www.harpercollins.com

HarperCollins books may be purchased for educational, business, or sales promotional use. For information, please write: Special Markets Department, HarperCollins Publishers, 10 East 53rd Street, New York, NY 10022.

Frontispiece: Courtesy of Ziv Koren

Unless otherwise noted, all illustrations are from the Sharon family archive.

Maps by Nick Springer, Springer Cartographics LLC

FIRST EDITION

Library of Congress Cataloging-in-Publication Data has been applied for.

ISBN: 978-0-06-172150-2

11 12 13 14 15   OV/RRD   10 9 8 7 6 5 4 3 2 1

To my beloved wife and children,
my brother, my mother, those who preceded us,
and especially for you, Dad—
you are the hero of this book,
the hero of our lives.

CONTENTS

The index is available online
at www.harpercollins.com/books/sharon.

ACKNOWLEDGMENTS

Thank you to my wife, Inbal, without whom there would be no book. The same goes for my wonderful editor, Claire Wachtel. Thanks to Marit Danon, whose hard work is exceeded only by her sensitivity; my brother Omri; Meirav Catriel Shtarker; Shira Hadad; Arnon Milchan; Reuven Adler; Dr. Avigdor Klagsbald; Dubi Weissglass; Roni Schayak; Lior Shilat; and Jane Friedman. At HarperCollins I must thank my publisher, Jonathan Burnham; Kathy Schneider; Tina Andreadis; Joanna Pinsker; Brenda Segel; John Jusino; Leah Carlson-Stanisic; Lucy Albanese; Richard Ljoenes; Archie Ferguson; Maureen Healy-Murray; and, of course, Elizabeth Perrella.

AUTHOR'S NOTE

Over the course of nearly sixty years my father has been on the front line of all major national events in Israel, and he has always played an active and central role. There is no matter of national importance that he has not addressed, influenced, or left his mark upon—from national security and the Israel Defense Forces, where he set new standards and stretched the army's capacity to record heights, to the continued development of agriculture, industry, and foreign relations, among many others.

His fingerprints can be found all across the length and width of this country—in the form of over one hundred blooming settlements in the Galilee, the Golan Heights, Samaria, Judea, the Negev, and the Arava.

Even at a young age my father had a strongly developed sense of history. Over the years he meticulously kept material, including letters, notes, maps, records of conversations, articles, speeches, and detailed notebooks. In the writing of this book I used all this authentic material. In addition, I read many books and articles that described the events covered in the book. I spoke to people who had accompanied him on his journey since childhood, during the long years in the army and in the public and political spheres, including the prime minister's office. I met with President George W. Bush in the Oval Office of the White House and with Prime Minister Tony Blair in Jerusalem, two people who played major roles in the diplomatic events described in this book. Ever since I was a young boy, I was a close confidant of my father's, and I believe I know him well. I personally witnessed many of the events described in this book and heard from him about those that occurred before I was born. I tried to bring to this book all I had learned and absorbed during those many years. I hope I was successful.

Gilad Sharon
Shikmim Farm
August 2011

SHARON

My father with my brother Gur in the Sinai,
August 15, 1967. (Eli Chen, courtesy of IDF
and Defense Establishment Archive)

THE EVE OF ROSH HASHANAH CLUB

On the eve of Rosh Hashanah 1967, when I was eleven months old, my older brother Gur was killed in an accident with a gun. He was almost eleven, and it was without a doubt the event that most affected me, my family, and our home.

I am told I was present, as was my then-three-year-old brother, Omri. I do not know what he saw or what he remembers, but there is no doubt in my mind that he, too, was profoundly touched by what happened.

Deep down inside, in my core, there is a seed of sadness, and it is rooted in that event.

But my nights of crying in bed and years of living with this open wound were dwarfed by my parents' suffering.

The sorrow, the longing, the feeling that it could have been different; the loss of a whole life, full of promise, of a special boy who, as I learned over the years, was a gifted horseback rider, had a sharp sense of humor, and was a natural leader among his friends.

In my mind, I relate the astounding fact that my father did not do harm to himself to the fact that he still had two children and a wife. Otherwise, how much suffering can a person take—and for what? I say this despite his strength and fortitude, qualities my father possesses in huge measure.

His keen sense of responsibility, which extended well beyond his family, also helped him a great deal. But his attuned sensitivity and acute intelligence in no way eased his struggle to stay alive and maintain his sanity. On the contrary, I think those attributes were a burden under those circumstances.

From left to right: Me, my brother Omri,
my father, and Gur, 1967.

As he once poignantly remarked: "The pain's intensity is not diminished by the years; it's only the intervals between the stabbings that grow longer."

My father returned from the Six-Day War victorious and covered in glory, embraced by an adoring nation. I believe it was then that the enduring cries "Arik: King of Israel!" began. And suddenly my parents found themselves in unfamiliar territory.

My father was no longer anonymous; a large and loving crowd surrounded him. It was almost impossible to go out into the streets or to the movies. Nonetheless, my mother, Lili, always said that he returned from the war altered as in David Atid's beautiful song— *You learn to return to old habits, but your face, my boy, remains altered . . . Stand in the gate, my girl, stand in the gate, I'm coming back to you on dirt roads. I was a boy, girl, I was a boy, now smile at me, familiar and beautiful.* She would sing along with the radio, and grow sad.

All the deaths and bereavements, the injuries and the pain, affected him. He is tough, but as I said, also sensitive.

After the war, my father had the opportunity to take Gur on a tour of all the areas suddenly accessible to Israelis. The photographs are still in our home. But then, as in a Greek tragedy, disaster struck.

Since then, year after year, at exactly 10 a.m. on the eve of Rosh Hashanah, our closest friends arrive at the cemetery in Kiryat Shaul, outside Tel Aviv. We do not notify them; they simply come.

The story of our family and friends is reflected in the small group that gathers each year. The passing of time is mirrored in this group, too; each year someone else is no longer present.

The regulars included my grandmother Vera, who passed away in 1988, and Moshe "Kokla" Levin and his beautiful and devoted wife, Lea. Kokla, a hulking, strong soldier, was drafted into the Paratroopers Brigade back when my father was the regiment commander. He immediately stood out as a particularly brave warrior and later became a charismatic and daring officer. Chosen to help assemble what would quickly become Israel's top army commando unit, he distinguished himself yet again, and a while later, after his discharge from the army, he was recruited to the Mossad, where he eventually rose to head of the Operations Division. Kokla was a nonconformist; he never flattered anyone and was known for being tough toward subordinates and commanders alike. With us he was gentle and full of humor. During those first years, when my parents had yet to rise from the crushing weight of the tragedy, he used to take us to the beach with our boxer, Ogi. He came to the house immediately after the accident, silent as always, and remained.

Upon returning from reviewing plans with one of the prime ministers, he once told his wife: "I have respect for people, but Arik is the only one I have fearful respect for." He said this despite the fact that my father was not in any high position at that time. That changed nothing for Kokla.

Uri Dan was always there. In 1954, when Uri was dispatched as an Israel Defense Forces (IDF) correspondent for the military newspaper to cover the paratroopers' reprisal raids, my father sent him to take a parachuting course. Uri, an extremely driven and determined journalist and international author, addressed all matters with equal tenacity. He withstood the challenges of the course and afterward joined the paratroopers on their cross-border missions, parachuting deep into the Sinai during the 1956 Sinai War and later accompanying my father during other wars, photographing and reporting throughout. In due course he covered conflicts in Vietnam, Cyprus, Kosovo, and more. For over fifty years he was a close friend of my father's.

With his white beard and ebullient blue eyes, Reb Shlomo Maydanchik, the head of the Israeli Lubavitch village Kfar Chabad, presided over the ceremony. A proud train conductor, he never parted with the Israel Railways hat he wore over his skullcap. He would rise for work before the first light of day, and after clocking out devote himself to his religious endeavors. He knew everyone and everyone knew him—respected him, too, because he always cared first and foremost about the concerns of Lubavitch.

Maydanchik had worked as a train conductor in Russia before he came to Israel, before he met the Lubavitchers, and he "returned" to the faith of his fathers. For years I tried to question him about his adventures in the inns along the endless train tracks in Russia, but he refused to say a word. As far as he was concerned, he was born the moment he met the followers of the Rebbe.

And even among those devout followers, Maydanchik stood out because of his ardency. He wholeheartedly believed that the Lubavitcher Rebbe was the Messiah and refused to accept his death. "Don't you think you're taking this a bit too far?" I'd ask him, but the words never seemed to register.

The Rebbe was our tie to Maydanchik. After Gur's death he sent a letter to my father: "and then he [my father] was famously known as a commander and protector of our holy country and its residents, with noble attributes, and the Blessed One has smiled upon him and given us triumph and success through his actions, leading to an unimaginable victory." For Maydanchik, who saw the letter, it sufficed. He did not need more of an invitation than that. He fell in line like a soldier.

Motti Levy, "our Motti," was always in attendance. Motti had been my father's army driver since 1967, accompanying him throughout the War of Attrition, the fight against terror in Gaza, and the Yom Kippur War. Motti held every possible driver's license and did not let anyone else sit behind the wheel of my father's vehicles, caring for him as though he were his own father.

Badly wounded during the Yom Kippur War, Motti was forced to lie in a full body cast for fourteen months. My mother would bring him his favorite dishes, especially her bean soup. Once he had recuperated, he came to work on the farm with my parents, but shortly thereafter left-wing extremists from Kibbutz Kerem Shalom, protesting my father's political views, held a violent demonstration on our farm and broke Motti's leg again. He was forced to return to bed for another eight months.

Later, while my father served in different ministerial positions in the Israeli government, Motti managed the farm.

Other members included my mother's older sister, Olga; her husband, Shimon, a survivor of the death camps, and their sons; my mother's sister Yaffa; and "Uncle" Saffi, whose relation to us is hard to explain but, after years of acceptance and mutual recognition, remains set in stone.

Saffi Pohachevsky, the son of farmers from Rishon LeZion, had good hands. He would arrive on the farm every winter to help us with our grapes. "You prune a grapevine when it no longer drips if you cut it and when it is not yet dripping," he said, explaining to me the small window of opportunity.

I was young and he was old, a grandfather, and yet he would scale the roofs to reach the grapevines. "Come on, Saffi, leave it to me," I'd say to him, but he'd already have scrambled up. "Do as I do," he'd say, leaping over a tall iron gate with the acrobatic spring of a Romanian gymnast. I did not even dream of trying to follow suit.

Sarah Mankotta, our friend and neighbor from Tzahala in Tel Aviv, came too, as well as Dov Sidon, an old friend of my father's, and Ariella Kisch, a close friend of my mother's and her partner in grief—Ariella lost her daughter during that same period. Danny Tzur, Gur's classmate in Tzahala, continues to come after more than forty years of memorial services, along with Gabi and Yael Cohen, our close friend Reuven Adler, and more.

Since the 1970s, when my parents began taking my brother and me to the memorials, I have not missed one. Before the service we would water the small flower bed and clean the area around it with a rake. In those days a large cypress tree stood alongside the burial plot, and we would hide the watering can and the rake in its branches.

Margalit, Gur's mother, is buried there, and beside her in the small plot is Gur. My mother's sister, Margalit, known as Gali, was my father's first wife. She was killed in a car accident in 1962, when Gur was five.

She had been my father's sweetheart since he was a youth, and her loss was a terrible blow, but worst of all for him was its effect on Gur, who had been by his mother's side day and night, especially when my father was away in the army.

Telling his child about the death of his mother was an unbearable ordeal for my father, and I cannot recall when or in what manner, but he once conveyed the difficulty to me. The conversation has always haunted him, as Gur refused to accept the news.

Two headstones stand in the small square plot. One is round and black and carved from basalt stone with feminine lines. My father brought it from the Galilee region.

Beside it, a marble stone, small, rounded, and childlike.

It was said then: "She took him to her."

By ten everyone would be there, and at exactly the top of the hour Maydanchik would begin the service. My brother and I stood on either side of my father, Omri to his left and I to his right, unrehearsed and yet fixed spots.

There were years when tears did not pool in my father's eyes, and only their shade changed. There were years when tears crept into the corner of the eye, and there were years with tears. Not crying, but tears. And there was one year in which torrential rain came down and drenched everyone.

Even at an early age, I had the feeling that I was supporting my father. When I was a young boy and we went somewhere, I felt as if I were protecting him, despite the objective fact that he was big and strong and I was small and young. Not because he needed protection, not remotely. He had been in combat, at times face-to-face, and had been injured several times. Even after his discharge from the army, I saw him defend the farm on his own against trespassers and thieves, utilizing his signature backhand slap, which I had not seen anywhere else.

And still, although he never requested it, I felt the need to protect him.

The memorials, where I watched him suffer, were particularly difficult for me.

My mother, her sisters, and another friend stood under the shade of the large cypress tree that's no longer there; diagonally to my right, between my mother and me, stood Maydanchik.

Behind us, all our friends. Prayers were said, psalms read. Intoning from the pocket-size prayer books he handed out, Maydanchik called out a name and made sure the person knew where we were reading from.

My father recited the Kaddish confidently, the pain evident in his eyes rather than in his voice.

———————

At the close of the ceremony, everyone lingered and chatted as Maydanchik collected the prayer books and skullcaps he had brought. My father shook everybody's hand warmly. I'd water the flowers again, and everyone would file out toward the parking lot, where Maydanchik handed out citrus honey from Kfar Chabad for Rosh Hashanah.

But the real attractions were in Uncle Saffi's trunk.

He'd pop it open in the cemetery parking lot, revealing homemade delicacies: dried figs he had picked on our farm, dried apricots, raisins made from our grapes, fresh vegetables, pickles, frozen mango slices, half-frozen juices, which usually suited the heat of the late-summer season, and homemade wines and liqueurs. Everyone walked toward his buffet on wheels, crowding around his car.

Dear Daddy,

Daddy, we were very glad to get your letters, postcards and telegrams. When I saw the postcard you sent, at first I thought it was the Empire State Building. But after Lili read it to me I understood it was the hotel you were staying at. I'm getting along pretty well with Omri. This week I was at the orthodontist and he checked my teeth and even I can see that the situation is getting better. Lili and I were on the show "Northern Command Band" and I have to say the band is terrific, the best in the IDF I think. (The songs were not so great but the skits were wonderful.) My boy-scout troop didn't meet today, so I had time to write to you. ([I] also [had time] before, but you must have been a kid and sometimes you're just not "in the mood.")

Daddy, I pumped up my bicycle tire but got a flat. I have not swapped books yet (because the teacher's sick).

I miss you very much but I can handle it.

Grandma told me that the mare's fat and healthy, but I have not seen her since you left.

P.S.

Daddy do you know that Gilad is already really laughing these days, he laughs non-stop till Lili worries he will get a stomachache from laughing so much.

Your son who loves you very much,
Gur March 4, 1967

Gur, six and a half years old, May 1963.

PART I

1928–1948

Samuil, Vera, and my father, circa early 1930s.

ONE

My father's mother was never known as "Grandma." She wouldn't have it; to her "it sounded old." Instead we just called her by her given name, Vyera.

My father's deep appreciation and respect notwithstanding, his relationship with his mother was nothing like my relationship with my parents, nor my relationship with my children. Her feelings for her son, while undoubtedly deep, never took physical expression. She did not hug or kiss him, did not display her affection. I imagine that Vera herself was not shown much tenderness. She was born in 1900, one of eight brothers and sisters, the sole Jewish family in Galevencici, a small village in the Mogilev region, in what is now Belarus, along the Dnieper River.

Somewhere along her father's line one of the men was awarded a swath of forest as recognition for his exemplary service in the czar's army, a rather rare commendation for a Jew in those days. As a result, her father worked as a forester. On the Hebrew holidays a Jewish family or two would occasionally come to stay to celebrate with them.

Mordechai Schneeroff, my father's maternal grandfather, was said to be a man of great physical strength. That description seems to sit well with his work as a forester and the fact that he was able to protect his family in a region where, for hundreds of years, attacks on Jews had been somewhat of a sanctified tradition for the locals, and they were the only Jewish family in the village.

No one dared lay a hand on the Schneeroffs. One day Mordechai heard that some villagers were planning to break into his barn and steal from him. He asked the family to lock him in the barn, where he waited for the thieves. When they showed up, he gave them a solid beating and the lesson was learned.

The Schneeroff family, circa 1910. Vera is in the middle, between her parents, Mordecai and Tehila.

There's nothing left of the family in that small village, which even at its most populous never exceeded a few hundred souls, but neither has it changed radically over the years. As prime minister, my father received pictures of the village along with a listing of the families from his mother's day. The pictures revealed that little had changed—apparently there was still no electricity, no running water, no sewage system, no paved roads; the houses were still made of wood and heated by coal. Nonetheless, the Schneeroffs secured an education for all of their children. Some of them became doctors, one an engineer, and all were well settled in their different locales—Israel, Russia, Turkey, France, and the United States. Perhaps this sheds some light on Vera's own educational methodology and on the importance she placed on studies. Vera, a brave, downright fearless woman, used to describe how, when she was in medical school in Tbilisi, she once noticed a mouse moving under the tablecloth while she studied with her friend. Realizing that her friend would scare easily, she clapped her hand over the mouse and held it, without her friend noticing, until they were done studying, at which point she got up and took care of the rodent.

———

Samuil ("Shmuel" in Hebrew), my father's father, was born in 1896 in the town of Brest in the southwest part of what is today Belarus. Jewish life thrived in the city, but with the outbreak of World War I and the expulsion of the Jews from all parts of southwest Russia, the family journeyed to Baku, on the shores of the Caspian Sea, in what is today Azerbaijan. That is where he completed his studies in a Russian high school, studying French, German, and Latin. At home his father taught him Hebrew and the Bible, and imbued in him a strong love of Zionism. Samuil's father was a religious man who had been ordained as a rabbi. He himself was not observant, but he knew the religion and its customs well. Still today on Passover we sing the songs in the manner in which they were sung at Samuil's house. As for Vera, I'm not sure she knew all of the customs and laws, but she most certainly knew how to catch a chicken and convert it into chicken soup for the Sabbath. While Samuil was an ardent Zionist, Vera had no exposure to Zionism in her small village.

With the Communist Party victorious and the Red Army approaching, Vera and Samuil, who met, so the story goes, on a Tbilisi train, realized that they had to flee. Zionist activists were being arrested, and both of them were realists. They harbored no illusions as to their fate, and so at the last moment, as the Red Army closed in around them, they fled to the Black Sea city of Batumi and from there boarded a ship to the land of Israel. For Samuil, who had completed his studies, was a certified agronomist, and spoke Hebrew well from home, it was the realization of a dream.

His father had already lived in the land of Israel for a period and taught Hebrew there, and had always wanted to return with his family. So for Samuil, settling in Israel and working the land was the pinnacle of his aspirations. But for Vera, it meant the end of her medical school studies during her fourth year, with a short time remaining until she would be a doctor. Even though she thought she might continue her medical studies at the American University in Beirut, she never had the opportunity to do so, an eternally sore subject with her. When she came to Israel, she had no idea what she might find, and in truth she did not find much. She also knew no Hebrew, a language that, even after she had spent close to seven decades in Israel, she still spoke with a thick Russian accent.

*Vera, my father, and his sister Dita, with their cow
Tikvah ("Hope") in their village, circa 1930.*

They arrived in the land of Israel, and about one year later, in November 1922, settled in Kfar Malal, a cooperative agricultural community. Only after living in a tent for a year and a half in the village did Vera and Samuil move into a shack they had built by hand themselves. At first, it had two rooms, one for Vera and Samuil and one for the cow and the mule. Later on, a barn was built for the animals.

Vera bore two children—Arik, my father; and his sister, Yehudit ("Dita"), older by two years—in the village, in the shack she and Samuil had built.

———

M y father would say that he was the most well-mannered child in the village. It happened only once, he said, that "I didn't get up when a woman entered the room, and that never happened again." He would not specify what type of punishment he received for this grave crime, but from experience I can say that I never once saw him remain seated when a woman entered a room or came to the table, even when he was prime minister. The discipline worked; it withstood the test of time.

Vera's disciplinary measures left their mark on my father for years. Whenever in her opinion he acted improperly, she would say, "I'll talk to you later," and that seemingly simple sentence would bother him immensely—he hated the uncertainty of not knowing when she would approach him or how, and it seems he preferred to take his punishment on the spot and not be left hanging.

Samuil, like Vera, was extraordinarily stubborn, and when he believed in something, he was willing to be obstinate, come what may.

———

D uring those years, and in fact well into the 1970s, socialism was the dominant doctrine of the land. There was very little patience for anyone who showed flashes of ideological independence, particularly in the early years. Vera and Samuil lived in a village that was part of the Socialist Workers movement, an ideology that did not suit them, not least because they were unwilling to be told how to think or what to do. But there were no other options available to them.

Samuil, for instance, decided to plant clementine and mango trees in addition to oranges and lemons, even though the board of the village commune had decided on the latter two. His intransigence brought sanctions—his produce was not sold with the rest of the village's, but marketed separately.

Already in the 1930s, Samuil had pioneered the cultivation of avocados in Israel, calling them the fruit of the future. Until recently several avocado trees remained on their plot in Kfar Malal, but back then no one had heard of an avocado, and his decision was greeted with perplexity and ridicule.

At that time the family's clementine grove was one of the first in the country and a marvel for all fruit growers, and its produce helped the family through the lean years of World War II.

Samuil claimed, rightly as it turned out, that potatoes could be planted not only in the spring as had previously been done and as the board of the village had determined, but also in the fall. As an agronomist, he took it as a matter of principle that he practice what he preached.

Occasionally, in the evenings, political party hacks would come to the village and dance with the weary, fervently idealistic village members.

My grandmother Vera was the only woman in the Socialist Workers movement whose bottom was never pinched by Herzfeld. He was a central figure in the powerful workers union, and he dealt mainly with new settlements. At times, after a night of dancing, Herzfeld was able to convince village members to relinquish some of their land for a new settlement—in this case, Ramot Ha-Shavim. In exchange, they were promised a new home. Samuil believed, once again rightly, that the village farms needed larger plots than were deemed necessary by the party. He was blessed with foresight. In the future, he said, children would be born, and each family would need more space so that they could make a living from their plot. Nonetheless, the village board decided to relinquish some of the community's land. Although Vera and

Samuil voiced their opposition to this decision, the village went ahead and allowed a parcel of their land to be fenced off for the new community. Samuil was not at home that night, but Vera set out anyway, alone, with her two children sleeping in the shack. She crossed the few miles to the fence at a quick clip and snipped the wire, an act that could lead to arrest. Although an investigation was launched, Vera had made her position to the contractor plain: Do not stretch a wire over our land, or it will be cut again. Later on, Vera said that the only thing she worried about that night was the safety of her two sleeping children. The family continued living in the shack for several more years.

This kind of nonconformism alienated the family from others in the village and even led to a workers' union hearing, where the village board tried to have Vera and Samuil ousted from the village and the union. On a visit to the union as prime minister, my father was given the seventy-seven-year-old court dossier.

11 August 1924

To the Agriculture Center!
We hereby come before you with the absolute appeal that the trial of Comrade Scheinerman be carried out in a principled manner, without any compromises or concessions. To us it is clear that there are but two choices here: either to oust them from the workers' union or to demote and expel the board in a whirlwind of intrigue. We recognize no other option.

Board members,
Y. Ben Yehuda
A. Shapira
Y. Tapuchi
Y. Bogin

The verdict arrived some two months later.

2 October 1924

To the Village Board
Honorable Comrades

   . . . During our last session we once again discussed the matter and reached the following conclusions:

1. Under no circumstances can we abide the expulsion of Comrade Scheinerman from the village. Expulsion is the harshest, most severe punishment we can serve and despite the complexity of the current predicament we have yet to reach the conclusion that said punishment is necessary.
2. We recognize that Comrade Scheinerman was utterly undisciplined and therefore we find cause to serve him warning and also to punish him with the negation of his active and passive right to vote on village matters for the period of six months.
3. We further find that the board did not exhaust all measures in the pursuit of peace and the resolution of said misunderstanding, measures that would have been helpful in this situation. We favorably recognize the actions of the board but it is incumbent upon us to note that it failed to grasp how to keep matters peaceful in this regard.

   In the coming days members of the Agricultural Central Committee will visit you, offering a more detailed account of the committee's decision.

With Greetings from Your Comrades,

The effect of the village's ostracism reached the point of absurdity when my father, age five, fell off his donkey and split his chin open. Rather than rushing him to Kfar Malal's medical clinic, which was nearby, Vera ran with him in her arms, bleeding, for several miles to her friend Dr. Fogel's clinic in Kfar Saba. My father recalled that he saw the light in the windows of the village's clinic as she ran, but that she continued past it, in the dark, a dangerous time to be on the road back then. When they finally arrived, Vera pounded on the iron gate. Dr. Fogel and her husband, a Ukrainian, appeared with a lantern in hand, and by its kerosene light my father saw his mother drenched in his blood.

Members of the village of Kfar Malal, circa 1920s.
(Courtesy of the Kfar Malal Archives)

The members of the village were strong, stubborn people, and they held on to their land with tenacity. They lived under constant threat of Arab terror. But still, there were quite a few eccentric people who added color to the hard life in the village. One member, for example, was in the habit of taking Vera's copy of the daily paper, until they reached an understanding whereby Vera would get the paper and leave him the previous day's paper in the box, which he could feel free to steal. That satisfied the needs of both sides.

Another member, in charge of the communal milk container, used to measure the amount and the quality of the milk each farmer produced. For years his cows produced the most, until one day he was too sick to report to work and someone replaced him; suddenly his cows' productivity declined sharply.

Another member was once seen leaving our barn with a sack of grain on his back. "What exactly are you doing?" he was asked, and he answered with a question, "Is this yours? So then, take it," as though the whole episode had been nothing more than a misunderstanding.

There was also a woman who stood outside the fence to our property, clothed in a filthy jute bag. This woman liked Vera and was aware of the ostracism but also of Samuil's recognized knowledge of agricultural matters, and she would holler, "Scheinerman, when do the potatoes get planted?"

Another village member, due to his honesty and rectitude, was chosen to be in charge of measuring and marking members' plots. This member limped, as one of his legs was shorter than the other. When it became known that his plot was larger than all of the others, he responded by saying, "I measured mine with the longer leg."

While Vera worked hard, led a Spartan life, and rarely left the village, Samuil, who also toiled, was prone toward inquisitiveness by nature, and was a worldlier individual. His work as a manager of orchards took him far and wide, and he spent many days and some nights away from the village. Although Israel is a small country, in those days the means of transportation were limited and poorly developed, and travel was never entirely safe due to Arab terrorism. Samuil survived two Palestinian attacks, one while on his mare on the way back to the village and one when ambushed by a sniper who had purposely tampered with a water pipe.

The life of an uncompromising idealist is difficult, particularly in those days, in a village, for one who spoke several languages, played music, and drew artistically. Samuil was a violinist, and new immigrants from the village of Ramot HaShavim flocked to his shack, instruments in hand. He also had a good drawing hand; to this day we have his pencil sketches and his aquarelles, the ones that survived.

Drawing by Samuil, Rome, June 1949.

However, there was a great divide between their rich cultural life and their poor material means.

There was friction between the life Samuil chose as a soil-tilling pioneer and his interest in worldly affairs, involving science, politics, economics, defense, and matters pertaining to the Jewish people at large. He read the daily papers, and his radio, courtesy of his sister Sana's husband, was one of the first in the area.

His neighbors in the village did not share his love of culture, and his devotion to the arts was seen as odd. Even his clementines were packaged in an aesthetic way, with several green leaves left on one fruit, an act that, at best, seemed strange to others. The conflicts between him and his neighbors were unavoidable and enduring: in his will he asked that no one from the village eulogize him and that he be led to his grave in his own pickup truck, not one of the village's.

He was voted secretary of the country's agronomists association and helped breed goats. Vera and Samuil had a small breeding station and a fine billy goat, who was matched with the region's nanny goats. My father once told us that when he was still a teenager, a woman brought her nanny goat to their stud and then had complaints about the process. "It's too fast, it's impossible, I know," she had said. And since he was from the "groom's" side, my dad responded with a smile, "Ma'am, it's not what you know, it just is what it is."

Samuil was sixty when he died, still young, beset by the kind of heart problems that are easily treated today. He was an ardent Zionist, a firm believer that non-Jews had full rights "in" the land but only Jews had rights "over the land." My father learned that from his father, and I in turn learned it from him at an early age.

Years after my grandfather passed away, from the vantage point of distance, my father regretted not getting to spend more time with him.

———

Samuil, the ideologue who believed manual labor was virtuous in and of itself, was not a healthy man; on the other hand, Vera, whose Zionist beliefs grew with her years in the country, was hale and hearty. "I was strong as steel," she told my father in an unusually candid conversation late in life.

The truth is, this was hardly a revelation to us.

If from Samuil my father received his Zionism, his love of the land, his concern for every Jew the world over and for Jews as a nation, as well as his love for art and his sensitivity, then from Vera he received his strength, his toughness, and his ability to withstand any hardship. Vera, a practical woman, always contended that you could have a dispute with anyone, which indeed happened to them often enough in the village, but that you should never get to a point where you no longer spoke with the other side; in other words, the channels of communication should always be left open. She was more moderate than Samuil in this matter.

Beyond the demanding farm work and the raising of the children, Vera also carried with her a heavy longing—for her family,

dispersed around the world, for Russian culture, and quite possibly for the landscape of her childhood, painted in her mind in
green and gold in summer and a frozen shade of white in winter.
She was a devoted correspondent with her far-flung relatives, and
every once in a while she would shut herself up in a room to write,
in what was referred to as Letter Day. A sorrow would descend
on her on those days, and the rest of the family would just let her
be—it was best that way for all parties concerned.

Because of her slanting eyes, high cheekbones, small size, and
strength, both physical and predominantly mental, she always
seemed to me a remnant of a pogrom, perhaps a descendant of
Genghis Khan. Every time there was some mention of her an-
cestry, I'd make galloping noises for my father by drumming on
the table. Everyone in the house knew what that sound meant:
Mongolian horsemen, many, many of them, galloping on their
short horses across the Russian plain. Short, strong, and determined, they ride with their slanted eyes narrowed against the
wind. Nothing deters them, nothing stops them. Between their
saddle and their horse's back they store a piece of meat, softened
by the friction and the horse's sweat. All this came to mind when
I saw my beloved grandmother.

*My father with his dog and the dagger that he received from
Samuil on his thirteenth birthday, circa early 1940s.*

TWO

During my father's early years, Samuil spent many days working away from the village, at times unable to return at night. My father's youth was spent working the land. He had chores on the farm, helped with guard duty at night, went to high school in Tel Aviv, a bus trip that took an hour each way, and was a member of the youth movement of the Haganah, the underground Jewish defense organization during the British Mandate.

At age seventeen he graduated from high school and completed his matriculation exams. By then he was very active in the Haganah and had even completed the two-month squad leader course in the isolated kibbutz of Ruhama (near where we live today, on the Shikmim Farm) while still working on the farm in Kfar Malal.

Samuil wrote to my seventeen-year-old father during his squad leader training course in Ruhama:

July 19, 1945

Dear Arik,

We received your first letter. It reached us within five days. Although you were terrifically succinct, it was still a comfort to receive it.

There is nothing new at home. We're now finishing the irrigation of the clementines. They're showing a lot of fruit and we have to watch over them because they're the only hope for this year.

What have you learnt in your new environment? Have you noticed differences in climate and soil there as opposed to in our village?

. . . When do you think you'll be able to come visit? We eagerly
await your next letter and do not fulfill your promise—not to write.

We all send our warmest regards
And kiss you

Yours, Dad

P.S. Write us some details about the kibbutz, what are their main
crops.

At the time my father thought he might join the ranks of the Palmach, the elite arm of the Haganah. Then one day, while they were working side by side in the orchard, Samuil told him that anything he decided to do in life would be acceptable, but he had to promise one thing—"Never, never participate in turning over Jews to non-Jews. You must promise me that you will never do that." This was during the Saison (from the French *saison de chasse*, "hunting season"), when the Palmach, the leading force in the Haganah, turned in members of the Irgun and the Lehi (or "Stern Gang") to the British.

Although the Palmach was not mentioned by name, the message was clear. My father did not join their ranks and instead became a covert member of the Haganah, working aboveboard as a policeman in the British Jewish Police and secretly as a member of the underground, which was becoming more and more like an army as the War of Independence drew near.

The message of never surrendering a Jew stuck with my father, and he staunchly opposed the extradition of Jews. Let them sit in jail in Israel if they are guilty, he said, but under no circumstances should we participate in extradition.

In the summer of 1947, my father received a letter. On the envelope was the following: "Ariel, Labor Youth Camp, Kfar Vitkin."

Kfar Malal, August 8, 1947,

Ariel, my dear!

I bear good news: We received a letter yesterday from the Hebrew University in Jerusalem—you've been accepted. We were very happy and I believe you will be, too. At long last a man must also look out for his own best interest. Up until now you have worked for the farm and for all of us. Two years have passed with you working day and night. Know this, too: lest you say—now is not the time for abstract studies—there will always be, even in the best of times in this country, when our sweetest dreams are fulfilled, at all times, our hands, those of the youth and those of the elderly, will be full of work for peace, and we will be forced to split our time and our strength toward labor on several fronts.

Yours with abundant love,
Dad

Despite their joy over his acceptance to the department of agriculture, war drew ever nearer, and it wiped out any chance of his enrolling in university.

He left home at nineteen, and at the close of that year, 1947, the Haganah was rendered a full-fledged army and my father a full-time soldier. At home, the time my father spent with his father, the kind of time that these days would be called quality time, was all too often thwarted by hard physical labor, financial difficulties, and incessant security problems. After leaving the farm he threw himself headlong into his army service, an intensive period that included the War of Independence, where he was badly wounded in the battle of Latrun; his period leading the commando Unit 101;

and his time with the Paratroopers Brigade, which carried out
the lion's share of Israel's retaliatory attacks and spearheaded the
Sinai Campaign. Two months after the end of Operation Kadesh
(the Sinai War), and four days after Gur was born, Samuil died,
on December 31, 1956.

My father with his sten gun, February 1948. (Courtesy
of the IDF and Defense Establishment Archive)

   The day before Samuil's death, my father went to visit him in
the hospital. He looked weak and tired, and as my father helped
him shift his position in bed, Samuil said to him, "It's too bad I'm
going to die, you still need my help in so many things." At the
time it sounded strange to my father. He was young, strong, and
successful. He had been the commander of the retaliatory cross-

border raids; he had brought about a major shift in the capacities of the IDF and in its belief in itself. He was bursting with vigor and self-confidence. And yet years later he said that his father had been right. He had felt his absence.

Although my father was but a twenty-eight-year-old officer, his abilities and accomplishments had caught the attention of Prime Minister David Ben-Gurion, who had taken a liking to him over the previous three years.

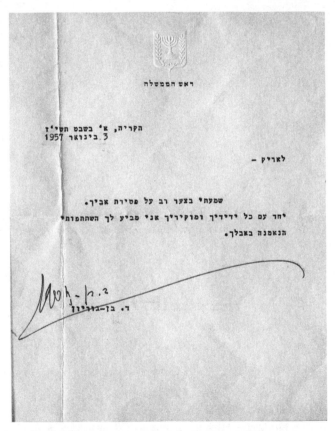

Condolence letter from Prime Minister Ben-Gurion,
dated 3 January 1957, upon the death of Samuil.

The Prime Minister

The Kiriya, 3 January 1957

To Arik,
> *It pained me to learn of your father's death.*
> *Along with all of your friends and those who hold you in high regard I express to you my loyal condolences for your loss.*
>
> *D. Ben-Gurion.*

In 1951, after my father had suffered for two years through intermittent bouts of malaria, the doctors recommended that he spend some time in a radically different climatic zone and rid himself of the disease. Aside from the cross-border raids, the ensuing trip was his first time out of the country. During the eleven weeks he spent in Paris, London, and the United States, traveling and staying with friends and family, he remained in touch with his family in Kfar Malal, a correspondence apparently characterized by detailed letters on the part of Samuil and brief responses from my father. Samuil wrote the following in my father's traveling diary on the day he set out:

To Arik:
> *A) Bring close hearts that are far and near.*
> *B) Proudly fly the flag of Israel and the IDF.*
> *C) Learn all and from everything. May your eyes be alert to every detail, because the world is made of details.*
> *D) Write daily and do not rely on your memory.*
> *E) Know what to respond . . .*
> *F) Don't forget—when all is said and done the best place in the world is—home.*
>
> *Go in peace, return in peace.*
> *Dad.*

Kfar Malal 11.19.51

Arik my dear,

Your first letter, which we awaited with great anticipation, arrived several moments ago. You have no idea how happy we were. Perhaps just to stoke our frustration they delivered the mail today at three in the afternoon (!) and you can imagine how many times mom ran to the gate to check the post office box. I am sure that once you see how pleasant it is to receive letters from home—you will then keep the commandment of sending them.

. . . I am happy to hear that for the period of 2–3 days you will be "a gentleman," that is, you will be dressed according to the European standard.

Over here it just started to rain today, rather well. All this time there's been wonderful, hot weather ("clear skies") so we continued to irrigate.

So all is well, I am reading a lot and both of us—await letters from you.

Yours with great love from me and from mom,
Shmuel.

Kfar Malal, 10 December 1951

Hello there Arik my dear,

After a "cold war" that lasted more than four months and after all nature of trickery employed by the ruling group for the past 30 years—our list has emerged victorious, and I as of now am head of the "Yvool," the orchard committee, and in essence am in charge of all the village's affairs. I sign in the name of the village. This has influenced the aforementioned group as much as the cross influenced Mephisto (see the opera Faust*).*

See many places and meet many people of character. Write down many conversations and interesting phenomena. Not everyone can do this. In point of fact I am now reading the Hebrew translation of One World *by Wendell Willkie [the former candidate for the presidency of the United States]. The book made waves around the world—he circled the globe, flying 31,000 miles in 160 hours, but the entire book leaves the impression of a tourist skittering along, and despite the vast potential laid before this particular tourist—it has the depth of an omelet.*

Your next letters must be laden with details!
Yours with love, Dad.

Happy Holiday to all,

Kfar Malal, 12.28.51

Hello there Arik my dear!
    Here all is fine, we're healthy and during the winter days we rest more and read a lot. By the way I am in competition for reading now with mom and she is even ahead of me, and only when she falls asleep do I get the book and carry on.
    I am busy with work, especially public works. One of the managers of Tnuva Export [the Israeli exporter of citrus fruits] approached me and asked if I would be willing to accompany the first shipment of oranges that is to be sent to Russia in two to three weeks' time. I responded that I would be willing to look into it as long as they could guarantee that I would be let out of that bewitched land and allowed to return home in peace.
    As for your desire to study—I applaud you. In terms of choosing a profession—think carefully. After reading your letters and witnessing your ability to take in with great acuity what is happening around you—maybe you should look into studying an interesting and very alive profession—journalism? Columbia University has a special and famous department, this is in essence a general-literary development and this could be of aid to you if you remain in your current

profession. What say you? And now, if you like, I turn to the more prosaic: Mom asks that you send a package, (not a scoop) of rice and laundry soap—and starch 1 kg and that is all.

> *I kiss you and all those dear to me.*
> *Dad.*

During the brief period of several months beginning in the fall of 1952, my father enjoyed the student life. While my father was studying at the Hebrew University, Samuil sent him the following letter:

Kfar Malal
Friday, 3.6.53

Shalom to you, Arik!
. . . How are things, studies, health, the weather in our capital etc. etc. Stalin is dead, they announced today, and it seems, he will be dead for a long time.
. . . Best to you from me and Vera and warm regards to Margalit. Hope to meet up next Shabbat eve.

> *Yours, Shmuel*

Best of luck on exams.

Vera and Samuil, as opposed to many others in Israel, harbored no illusions about the murderous regime of Joseph Stalin. In Israel many of Stalin's idiotic followers mourned his death, declaring that the "sun of all nations had set." In Kfar Malal, Vera and Samuil always referred to him derisively in Russian as the "Gruzinski kham," the Georgian ignoramus, the crude Georgian . . . and so, as can be seen from the previous letter, the news of his death did not elicit any great sorrow from them.

Aᴅnd so, what remnants did my father have of his father? Shards of memories that he would relay to us: his father drying the dishes while his mother washed, and the two of them after supper speaking gently (while my father was in his bed at the end of the room, pretending to sleep) in the little shack; Samuil at the violin along with friends playing chamber music; my father riding with his father to the market with a truck full of watermelons that they had grown and picked with great effort, the day hot, my father sitting on the edge of the truck with a club in hand, guarding the merchandise, his father unable to sell the watermelons, occasionally a crude merchant approaching and splitting a watermelon by throwing it to the ground, examining its flesh, and laughing contemptuously, and in the end the merchandise being sold for pennies and the two of them returning home. And there was also the time when the two climbed up on the roof of the barn during a stormy night to make sure that the roof tiles would not blow away, and most of the tiles were saved, as was the hay. Afterward Samuil told my father of another storm, years earlier, that had raged on a night he was not at home and how Vera saved the roof by tying it all by herself to the wagon—the two of us did all right, but it wasn't up to the job Vera did, he said.

Loading clementines near the house in Kfar Malal, fall 1944.

Once, during one of the short periods when my father worked on Shikmim Farm, sometime during the 1970s, I remember him driving an open tractor, cultivating the fields. My brother and I were kids, and we each sat on a wheel wing, facing each other, while my father sat on the operator's seat. It was hot, the field was big, and there was still plenty of work ahead of us. When he noticed our signs of fatigue, he stopped and told us how when he was a child and he and his father were out sowing the watermelon fields, it was hot, it was hard for him, and his father would stop for a moment and tell him "not to look ahead at all the work we have still to do, but back at all we have already done." Encouraged by what had been done, he'd find the strength to go on, and so did we.

Our father, Omri, and I on the tractor,
circa mid-1970s.

The social ostracism and the absence of physical displays of affection in the house forged in my father the need for warmth and a great appreciation for all expressions of emotional tenderness. Any such acts on the part of his soldiers left indelible marks on him; when one of them covered him with a jacket or a scratchy military blanket during a break in the action or after a mission, hushing people and saying, "Arik's tired, let him sleep," the voices would drift into his consciousness as he fell asleep, and he would feel embraced by the familial warmth of his soldiers.

The affection women showed him over the years in my opinion filled the void of his childhood, and despite his daring, his vigor, and his downright stormy nature at times, he was always gentle and kind with women.

Childhood memories have stayed with my father forever, such as the one of the package his uncle Solomon, Vera's brother, sent him—a jar laden with carefully arranged smoked fish, or of the sweet strips of apricot fruit leather. Packages in general were a rare occurrence at the time in the moshav, which perhaps explains why on one occasion a member of the moshav came running to my grandparents' shack from the Public House, where the mail arrived, and told them excitedly that a package had come for them, promptly detailing its contents. For years I attributed his love of food to a similar source—the austere years of his childhood.

I n the aftermath of the War of Independence the tensions in the village subsided. The small village had suffered many deaths and injuries, and that put the old struggles into perspective. My father would tell us that the first time he ever saw the inside of the village clinic was only after he was badly wounded at Latrun and needed to have his bandages changed. The small village cemetery—like other cemeteries in the battlefields of the War of Independence—is filled with the graves of the village young who died in battle. My father's military service as an officer and the manner in which he excelled brought Vera and Samuil great pride.

Great was my desire to see my children, my son Ariel and my daughter Yehudit, continue our work and settle on the land into which we poured the best of our strength, however I could not resist when my son Ariel was called to the flag. With all the integrity, warmth, devotion and love for the homeland that he inherited from my father Rabbi Mordechai

Scheinerman, he served the IDF and even earned the right to spill his blood for his people and his land. My son Ariel—my pride, and I do not wish to deter him from the path he has chosen. Rise and be successful!

S. Sharon 8.14.56.

Vera and Samuil, May 1956.

Those were the words Samuil wrote in his last will and testament.

After Samuil's passing, the tension between Vera and the rest of the village further subsided, and for many consecutive years Vera was elected to the board of the village and was widely admired for her hard work, her stubbornness, and the exemplary farm she ran. Vera, then past eighty-five, told my father, who wrote down her words:

I always tried to make sure that the farm would be in order so that everyone would know and see that everything was cultivated. Scheinerman is with us or not—the farm is cultivated.

When I came to the land of Israel I had no idea how many difficulties we would face—I thought I'd bring six children into the world in order to build up the country.

When Samuil passed away it was all on me, I was so busy. A type of ambition formed that all of it would be organized. So no one would say that the Scheinerman place was in disarray.

Vera was in the habit of telling my father that the animosity had died down over the years. "Don't write about Kfar Malal," she'd say, "no need to look for hatred from one hundred years ago." But toward the end of her life she told him candidly that the hatred had been passed down, not just to the second generation but also to the third. The two times that my father was elected prime minister, he won by an immense margin. On both occasions I checked the voting records from the village and found that he had not secured a majority there. This did not surprise me. Our family always had some friends here and there in the village, but on the whole, as a single body, it remained hostile.

Years after Vera's death, I prevented a neighbor from building a fence to usurp a small piece of our land in the village. Thinking perhaps it was an insult, he said, "You're exactly like your grandmother." I was overjoyed, taking the statement as a great honor. When she was around eighty years old, my father took away the firearm that she kept at the ready. She was not pleased, but at least she was able to hold on to the heavy club that she kept under the bed to her final day. "What are these stains on the club, Veruchka?" I'd ask her. "Nothing," she'd say in her Russian accent, "they are nothing, Zolutka" (golden one). But I knew that those stains had not come into existence because Vera used the club for mashing tomatoes; she simply wasn't a big talker.

With us she softened—maybe on account of Gur's death, "my Gurele," as she called him, and maybe that is just what happens with grandchildren—but with others she remained as tough as ever.

One day, over at our house, already an older woman, she sat down next to the phone and called information. She got the number and placed the call. To our surprise we heard her say in a very relaxed tone of voice, "If you ever tie your horse to my fence again, I will let it go free," and then put the receiver down. The horse was never tied to her fence again.

Her idioms are with us still; it's enough for one of us to start one, with or without the accent, and any one of us can finish it.

"Whoever brings news also take news," she would say of gossipers who came by.

"Explain to him so he understands," she would say, without limiting the means of clarification.

When she was in her eighties and Omri in his early twenties, she asked him, "Did you hear about the disease?" Feigning ignorance, he said, "What disease?" "AIDS, you didn't hear of it?" "I'm careful, Vera," he said. "What do you mean careful?" "I only go out with girls from good homes." He was being facetious: that expression, too, was one of her own, and she used it to mean educated and cultured people, the kind with whom she would let my father play as a boy. "The world is in a panic, and you make jokes," she responded.

"Vera's roast chicken" is what my kids, born after she died, call the fantastic dish we often have, a recipe that has stayed in the family. When she was in the hospital toward the end of her life, my father sat by her side, and she stroked his hand and he stroked hers, an expression of affection and tenderness that may well have been unprecedented. Sixty at the time, my father felt he had come full circle, that some internal emotional wall had finally fallen, and he was relieved. It's not that she didn't love. She loved very much: "The essence of my life is my family—Dita, Arik and Lily, Omri and Gilad, they are the essence of my life," she said near the end. She simply loved differently.

My father with Vera and me, outside our home on the farm, September 1984. (Courtesy of Professor Boleslav "Bolek" Goldman)

PART II

1948—1973

My father, in front, reviewing his platoon, April 1948.
(Courtesy of the IDF)

THREE

pparently all it took was a trip or two to the village kindergarten. After that, my father decided to stay at home. He'd play with his dog, Shpitz, his duck, and the little donkey that he received at age five. This was classified information when I was a child. My parents didn't want us getting any ideas into our heads as we trudged to kindergarten.

He did not enjoy his experiences at his elementary school, either—the Aharonovitz School for the Children of the Workers in the Sharon region. As prime minister, some sixty years after he graduated, my father was asked to send a congratulatory letter in honor of the school's seventieth anniversary. The request sat on his desk unanswered. Several weeks later, Marit Danon, whose official title was Prime Minister's Secretary, said, "Mr. Prime Minister, this is getting awkward, we have to give them an answer, perhaps just write something nice." My father snapped, "You don't understand, there was nothing nice about that place." The greeting was never sent.

One of the traumas my father experienced in school occurred at perhaps the beginning of the second grade, during the class play. A makeshift stage was assembled, and curtains fashioned out of white sheets. Each child was given a little piece of cardboard, in the shape of a blackboard, a pencil, and so on. My father was the chalk. He had one line. He was supposed to say something like, "The chalk is also happy to go back to school." The white sheets parted, and an audience of attentive parents, including Vera, waited expectantly. The chalk's turn came, and suddenly a long and painful silence descended on the room. The line was gone from his head. "We won't let him be in any more plays," the children grumbled. Vera waited . . . it was useless—the chalk could not recall his line. Finally the teacher wrote out the sentence for him, and he was able to recite it.

My father (top left) in high school, circa 1940s.
(Courtesy of the Hefler family, Haifa)

The unfortunate chalk incident is why my father never again spoke before an audience without notes in his pocket. Neither the size of the crowd nor his familiarity with the material had any bearing on the decision. The knowledge that he had the notes had a calming effect.

My father was not a particularly popular child on the village. He was a daydreamer, sitting on our family's plot of land in Kfar Malal and gazing at the ridgeline of Samaria in the distance, imagining all sorts of fantasies and heroics, with himself in the starring role.

Nor did his military career start out with immediate success.

He finished squad leader training at Kibbutz Ruhama with the dubious title of "squad leader under supervision." "When the war started," he used to tell us, "all my supervisors were gone." He thought he had done well during the course. "We'd do platoon drills where one squad attacked Tel Nagila, another squad attacked Tel Hasi, and the third one remained in the rear, in reserve," he would say with a smile. (The hills were several kilometers apart, too great a distance for one platoon to cover.) Again and again we heard the tale of their platoon commander who bungled the navigation during the course. The cadets were supposed to head back to Ruhama for the night, but the platoon commander marched them long and hard in the direction of a distant light. The long trek led them to a tractor, plowing Kibbutz Dorot's fields, far west of Ruhama and rather close to the farmland my parents bought less than thirty years later.

My father's military success was earned under fire. It was in those situations that he remained calm. He began to notice that the soldiers in the platoon stayed close to him. His leadership ability flowed from his soldiers. He instilled confidence in the men under his command because he knew that they trusted and relied on him, and they trusted and relied on him because of the confidence he instilled. (During the Yom Kippur War, twenty-five years later, soldiers would cling to his shirt, needing to touch him amid the madness. Under terrifying barrages of fire, they were both soothed and emboldened by the sound of his calm, unwavering voice over the radio.)

At the end of one mission he replaced the platoon sergeant. "I ousted him," he would say, laughing. A short while later, after a mission in the village of Bir Ades, he replaced the platoon leader.

The village of Bir Ades, which no longer exists, was near Kfar Malal. During the War of Independence, Iraqi troops established positions within the village. My father, a twenty-year-old sergeant, was chosen to lead the troops. The force assembled in the synagogue in Magdiel, a village close to Kfar Malal. Outside the rain came pouring down. He told us many times how he had proudly briefed the soldiers, telling them how he would lead them through the orchards, the fields, and the wadis to the rear of the village. The infantry force was joined by a platoon of heavy machine guns, under the command of a man who would become a friend— Talik, who years later became Major General Yisrael Tal, the developer of the Merkava tank. That was the night they first met.

The force set out, through the mud and the steady rain. The machine gunners were barely able to march through the sludge. At some point it became clear that part of the force had splintered off. "I'll get them," my father said. He realized that they had not seen, through the heavy rain, that after walking due east the company had turned sharply toward the rear of the village. He hurried after them through the thick mud and found them. With the force back together, they arrived, undetected, at the outskirts of the village. Talik's gunners set up positions and opened fire. The forces within the village responded in kind. "We shot, they shot, and the order to get up and charge never came. I gave it, and that's how I wound up ousting the platoon commander," my father would say, laughing.

M ay 24, 1948: The Thirty-second Regiment of the Alexandroni Brigade makes its way by bus from the Sharon region to Kibbutz Hulda, the staging ground for the attack on Latrun, on the road to Jerusalem. The bus passes through Tel Aviv, and my father sees signs calling for the public to enlist in the armed forces. He's puzzled. The country's been at war for half a year, and he hadn't thought that there was anyone who had not yet joined up. Who are these placards for? he asks himself. But the cafés in the city were full. They traveled on. Near Bilu Junction, close to the Arab village of Akir, a roadblock stands in their way. The soldiers try to persuade the villagers to let them pass, but they're denied access. The rickety convoy of buses turns around and reaches Hulda via a different route. As my father recalled in an article he wrote for *Yedioth Ahronoth*, on April 29, 1998, fifty years later:

An olive grove near Hulda. My platoon and I were passing the heat of the day, thinking prebattle thoughts, blending with the water-smoothed stones and the earth, feeling part and parcel of the land: a rooted feeling, a feeling of a homeland, of belonging, of ownership.

Suddenly a convoy of trucks stopped next to us and unloaded new, foreign-looking recruits. They looked slightly pale, and were wearing sleeveless sweaters, gray pants, and striped shirts. A stream of languages filled the air; names like Herschel and Yazek, Jan and Maitek, Petr and Yonzi, were thrown around. They clashed with the backdrop of olives, rocks, and yellowing grains.

They'd come to us from Europe's death camps, through blocked borders, on illegal ships, and then again to internment camps, British ones this time, and from there, back on the boats, straight to the front.

I watched them. They stripped, their backs white. They tried to find uniforms that fit, and fought the straps of their battle jackets as the commanders that had just now met them helped them suit up. They did this in silence, as though they had made their peace with fate. Not one of them cried out: "Let us at least breathe the free air after the awful years we endured." As if they'd come to the conclusion that this was yet another final battle for the future of the Jewish people. They surely did not know that among the Jewish community, far too many clung to their villages, defending them, despite Ben-Gurion's appeals. And that at that time, many had not yet enlisted at all. And no small number of the "moneyed aristocrats" of the time had sent their sons overseas to keep them out of harm's way in a time of war.

These recruits were known as the Gachal [an acronym for overseas draft], and they were sometimes called Gachleitzim, with a measure of derision. They were not commemorated in song or spoken of around the bonfires. They were not models to be emulated. No one waited for them at home, eager for their tales. They did not have homes. People from a different world, with foreign experiences. Young people like us, and yet they were a thousand years older than us.

[At the cemetery] on Mount Herzl in Jerusalem, in a mass grave where the remains of fifty-two warriors from our platoon, the fallen from a single battle, are interred, among the dead of B Company, the Thirty-second Regiment, Alexandroni Brigade, are four unknown soldiers. For fifty years now, each time I've passed by the monument I've stopped and wondered, Who were they? Where were they from? Do any of their family members remain? To this day, has anyone ever looked for them? Is someone looking for them still? And I have no answer. No one does.

My parents at the Latrun Memorial, 1998. My father is touching the name of Mordecai Dochemener, whom he refers to in his letter to Gulliver more than fifty years earlier. Above his name is that of Moshe Duvdevani, who was MIA for fifty years. (Courtesy of Zalman Enav)

The mission is a nighttime strike against the Arab forces in Latrun. The goal: open the corridor to Jerusalem, which has been cut off from the rest of the country by the Arabs. The discussions in regiment headquarters continue into the night, and the force sets out late. A strong smell rises off the fields of chickpeas. My father spearheads the lead platoon, which leads the battalion into battle.

May 26: Morning comes, and with it a summery haze. My father and his platoon arrive at the wadi beneath their target, the shallow channel that now runs east of the road that links Kibbutz Nachshon and Latrun Junction. The haze has burned off, and my father and his platoon are exposed. Machine-gun fire, heavy and accurate, rips through the force. The offensive is broken. In the summer of 1973, my father said this in an interview:

I commanded our force's lead platoon. We were to capture the village of Latrun and the monastery. We set out very late. We encountered murderous fire. Many fell wounded. I had to circle back around. I entered a gully and from there tried to take up the offensive again. It was hopeless. In the meantime morning had risen. The fire was awful. In my platoon I had fifteen dead, eleven wounded, and five POWs. Only four members of the platoon were in one piece at the end of that day of blood.

Throughout the day, alongside the vineyard, at the foot of the monastery, we took awful fire. The entire offensive was broken. It was a crushing defeat. We clung to the sides of the gully. In the afternoon, a sudden silence came over the field; I thought the fire had stopped. I gave the order to prepare to charge. The wounded readied themselves, too. Anyone who could move was ready to charge. I looked back and saw that I had misinterpreted the sudden silence. The entire mountainside behind us was covered with Arab villagers. They butchered our wounded, the ones left in the field by other units. We had no means of communication with the other units in the battle. We were lost. I was badly wounded. My arm was in a cast from a previous injury. And here, in this operation, I had been severely wounded in the stomach and thigh. I was unable to move. I knew with full certainty that if we stayed in the field, or if we attempted to evacuate the wounded, not one member of our platoon would remain alive. The villagers approached. We no longer had the power to deter them.

All around me—the dead and the wounded. All friends, all from the Sharon region, most from a single village. People you grew up with. Here they were, right in front of you, in this awful field, close to death, and there was nothing you could do for them. They were lost.

This was the most difficult order I ever gave in my life. There were others, of varying magnitudes, of different degrees of responsibility, but none was as grave as that one. I looked at my wounded. I knew I was seeing them for the last time. I knew they would be butchered. I gave the order. For the first and last time in my life as a commander, I gave the order: retreat, retreat and leave the wounded in the field.

There was no choice. I had to save the few that were still alive. I lay there, tormented by pain. The few who were able to move, passed me by. "Should we leave you here, too?" Yes, me too. I saw the eyes of those who fled. They contained shock and sorrow, immense pain. That look accompanies me to this day, always.

Welcome to the end of the road, I told myself. I don't know how, I don't when, but I started to move. I dragged my body centimeter by centimeter. The rocky terraces in the field had to be crossed. I knew that I did not have sufficient strength for that. Another centimeter, and another one. And the sounds of the pillage and the slaughter being perpetrated by the villagers reverberated through my eardrums. There was this one guy there, also wounded, also crawling, also progressing with great difficulty, also with little chance of surviving. I did not know him. He was sixteen. He had been sent my way on the eve of the mission, and I had not yet exchanged a word with him. His face had been smashed, the fields were on fire, everything smoldered, the burns on my feet were killing me. Now, too, we did not speak. When we reached the first terrace, without saying a word, he pulled me up, carried me with what remained of his strength. And again at the second terrace, and the third. We progressed together.

I owe him my life. A straightforward debt, piercing, overt. His name is Yaakov [Bogin]. He spent a long time recuperating in the

hospital. I visited him often. I got out before he did and managed to
participate in the battles in Operation Dani.

Every day, on his way from the farm to Jerusalem, my father
drove past that wadi. He took us there when we were young,
showed us where he was lying wounded and where he crawled.
I've never passed by there and not thought of him in the awful
heat, injured, bleeding, issuing the command to retreat, barely
able to crawl over the stone terraces.

My father's experiences at Latrun left an indelible mark—the
wounded left behind, the fate of those who were mutilated and
killed by the Arabs. As a commander of the Paratroopers, he instilled in his soldiers an ironclad rule: No soldiers could be left
in the field. That oath, later adopted and held dear by the entire
army, can be traced back to that brutally hot day in May 1948
when my father was left wounded on the wadi beneath the monastery at Latrun.

My father was part of the first group of soldiers to be treated at Tel
Hashomer Hospital. There he endeared himself to a young nurse.

October 18, 1949

Arik,

    Surely you are quite surprised about the nature of this letter, but
without bothering you any more than necessary, filling your head with
vain words, I shall express my wish to you sparingly: I would be
endlessly thankful to you if you would be so kind as to pay me a
visit here, in the place where I am "buried." If you choose to fulfill
my wish, ask for Hut 28, my quarters, or Hut 14, which is where I
work. If this is not easily accomplished, do not relent and do continue
to seek me out in all sorts of other places.

    Arik, do not be lazy and do fulfill my wishes. I am waiting.

N.

Kfar Malal, November 5, 1949

Hello N,

　　I received your letter and was not surprised by its nature. I am generally less innocent than I look. It is not a good idea, in my opinion, to be too grateful about the prospect of a visit of mine, not every visit bears the desired results and at times we are disappointed. At any rate, if I am in the vicinity, I will come by to see you. Know that I have come home only for Shabbat and even that may not happen much in the near future. Send regards from me to Ruchaleh the redhead from Kfar Vitkin, one of my closest friends, there are others there, but I do not recall their names. I was among those who had the honor of being laid up in the hospital around the time of its opening roughly one year ago. Not a bad place and there is no need to be "buried" there, as you write. You can have yourself a rather good time, we did at any rate when we were there.

　　Regarding the end of the letter—it would not be wise to hang too many hopes on matters that appear rather unclear, especially for a girl like you who surely has everything yet ahead of her.

See you,
Arik

Even as prime minister the letters continued to come:

June 15, 2001

Hey Arik!

　　Out of the blue—just wanted to let you know that if and when you would like to escape all the "festivities" for a few moments, the wars and the other depressing stuff—I am here, and so is the coffee! So it would be nice if you were to come. That said—I am sure the coffee on your farm is no less good and one can even hug a goat or lamb on the way—so I'm sure I could find my way over if you invite me! I'm sporty, spirited, a bit infantile apparently, sharp witted, funny, and divorced. I am not too sure this letter will reach you, quite likely it will get stuck with some secretary or other and find its way to the garbage, but I am hoping anyway. So what do you say, Arik, are you coming? I believe Lili would be happy.

Bye—
R.

My father participated in Operation Dani, one of the later missions of the war. He was promoted to battalion intelligence officer and later company commander. He was in his early twenties, but his letters reveal a young-old man whose youth had been taken from him. Most of his friends had been killed or injured. His platoon was ravaged, his company torn to shreds. The casualties and their families were a sensitive subject for him already then:

March 4, 1950

Hello Gulliver,

I was asked by the IDF archives (Warriors section) to write about our people who fell. However, I managed to write only about some of them and am leaving for a week. While I am gone, please write about Mordechi Dochemener, Yaako Zelberstein, Shalom Udi, Shlomo Glidaei, Elyakim Zamer. If you have their photographs, attach them.

Pay attention: Those people have no relatives in Israel, and without us, no one will perpetuate their memories.

See you,
Arik
["Gulliver," Yithok Ben Manachem, a childhood friend of my father's and a brave soldier, was killed in the Sea of Galilee operation in 1955.]

In July of 1948, only two months after being wounded, he requested a transfer to a commando unit. The request was granted.

In 1949 he was made commander of the Alexandroni Brigade's reconnaissance unit and later of the Golani Brigade's recon unit. Avraham Yoffe, the brigade commander, was his superior officer, and the two forged a close friendship that lasted for decades. From there he continued on to the battalion commander course in Tzrifin, led at the time by Yitzhak Rabin. Binyamin Gibli, the IDF's chief intelligence officer, was taking the brigade commander course at the same time in Tzrifin. He heard about my father from Avraham Yoffe and approached him with a proposal. "He sat down on one of those military cots," my father said, "and I thought he was going to offer me the position of brigade intelligence officer, and I planned to take it. Instead he offered me the post of chief intelligence officer of the entire Central Command. I said yes without showing any sign of enthusiasm. It was as though someone had grabbed me by the lapels and pulled me up." During the twenty-eight years of his military service, there were probably only three officers who helped him advance: Gibli, Yoffe, and, some fifteen years later, Rabin, who pulled him out of the deep freeze he had been placed in by others.

He served as chief intelligence officer of the Central Command under Major General Zvi Ayalon, who, in his late thirties, was called "the old man" by his staff officers. At the time the army held one of its biggest field maneuvers, pitting the Central Command against the Southern Command, as led by Major General Moshe Dayan. Ayalon loved assessments, and throughout the drill he huddled with his officers and went over data. The IDF chief of staff, Lieutenant General Yigael Yadin, was attending the drill, and he put an end to that. "The situation's already changed one hundred times," he barked. My father took note: Keep the assessments and the orders short, clear, and relevant.

Toward the end of the exercise, the Central Command went on the offensive, infiltrating deep into the Southern Command's territory. The route was long and difficult. My father knew the terrain better than the others, and he led the brigade. Later it was said that the chief intelligence officer was needed alongside the commander,

not leading a brigade. This was probably correct, but already then
it was clear that providing details and data for the commander was
not for my father. (He does not recall seeing the brigade commander
that night. That taught him an important lesson: The real commander is the one who is in the crucial place and can make decisions
based on firsthand knowledge. This lesson was driven home during
the Six-Day War and even more so during the Yom Kippur War.)

My father's next position was as chief intelligence officer for
the Northern Command. When Dayan was appointed commander of the Northern Front, he told my father, "As far as I'm
concerned, a good intelligence officer is someone who knows the
terrain better than me." Those were welcome words to my father.
He made Dayan, who loved fruit almost as much as he loved rob-
bing antiquities, a fruit map of the region, marking the different
orchards and groves, with particular attention devoted to figs,
Dayan's favorite.

At the time two Israeli soldiers were being held captive in Jordan.
Dayan was of the mind that if the IDF took a few prisoners, an exchange might be possible. "See if you can do anything about this,"
Dayan told my father, in his typical way giving an order that was
not an order: if it worked, great; if it failed, he was not responsible.

My father didn't need things spelled out. Nor did he ask for his orders in writing. He took a pickup truck and called one of the officers
under his command, Shlomo Hefer (Gruber), and they headed in
the direction of Kibbutz Maoz Chayim near the Jordanian border.

The two men stood on the Israeli side of the border, beneath
an acacia tree, alongside the ruined Sheikh Hussein Bridge, and
looked toward the Jordanian police station, which was barely visible on the eastern side of the river. While plotting how he might
capture some of the Jordanians in the police station, my father
saw a Jordanian sergeant and three soldiers approaching the
bridge from the south. "Maybe I won't need a plan at all," he
thought to himself.

He walked down to the bank and called out to the armed police
officers. Once they drew near, he scrambled up onto the ruined

bridge and approached them. His pistol was in his pocket, hidden from view, and all four soldiers were armed. They came forward without hesitation. "Some cows have gone missing," he said. This was not an unusual occurrence; horses and cattle were stolen and smuggled across the river all the time by Arabs. The men began to talk amiably, and my father invited them to come sit under the shade of the acacia tree on the Israeli side of the river and continue their conversation. He realized that four men might be hard to handle, so he asked the sergeant to send one of his men back across to check with the local police about the theft. At that stage, after the friendly conversation, my father felt bad about what he had to do. However, when the sergeant sent two of his men, he knew he had to act. As soon as the two men disappeared into the reeds on the far side of the river, he got to his feet. The sergeant, who perhaps became suspicious, rose too, as did the man under his command. My father smacked the sergeant hard and grabbed him by the holster that he wore across his torso. The Jordanians were disarmed, handcuffed, and blindfolded, and taken back to headquarters in Nazareth. Dayan was not there. My father left him a casual note: "The mission is accomplished, the prisoners are in the cellar." Shortly thereafter the exchange was made, and the Israeli and Jordanian prisoners returned to their respective homes.

Upon release, the sergeant told my father, "Your army is not orderly."

"How would you know?" my father said. "You were in the cellar the whole time."

"I heard the reveille whistle blow more than once," said the Jordanian, a student of British military discipline.

Dayan gave my father the silver-plated pistol that the sergeant had received as a gift from the king and recommended to Chief of Staff Yigael Yadin that he be awarded a citation for bravery.

He was not awarded the citation. He did, however, receive this note from his commander, Moshe Dayan:

September 5, 1952

To: Major Ariel Sharon
    You have my great esteem for the operation of yours yesterday, the capture of the two legionnaires. Even old warhorses perk their ears at the profound wisdom and daring of this solo operation.
    There is no one like you guys!

Moshe Dayan—General
OC Northern Command

My father was discharged from the army in the fall of 1952 and enrolled in the university.

This was his second attempt at entering college. Back in 1947, before the war, he had planned on studying agriculture and had already been accepted to the Hebrew University. Five straight years in uniform left him less certain that this was what he wanted to do. He considered studying agriculture at the University of Colorado. He considered the benefits of law school. In the end, he decided to enroll at the Hebrew University in Jerusalem and study history of the Middle East. He was twenty-four at the time. The feeling that his youth had passed him by had already set in, and would accompany him forever.

He was always curious about student life. When I was in university, he constantly asked me what it was like. I know he regretted having missed that period. Campus life was something he brought up many times during our conversations at home. To him, the student life was the pinnacle of freedom; to be free of responsibility was something he had known for only several months during his entire life, if at all. War and military service had snatched his youth away after high school, and schoolwork and chores on the farm before that.

One night, some time after high school and before I was drafted into the army, I sat outside in the yard with some friends and we sang at the top of our lungs. My father was home that night. He later told me that to him it seemed like the end of my youth, the looming draft date, and tears welled in his eyes. It touched him deeply.

With Gali (Margalit), Jerusalem, 1955.

He was not sure he should continue in the army, despite his awareness of his talent for military matters. He knew how seriously Vera and Samuil took the matter of education, especially in light of Vera's own truncated schooling. University was the logical next step.

Margalit and my father got married on March 29, 1953. They went to the office of an army rabbi, an acquaintance of my father's, and he married them. Later they sent a telegram to Vera and Samuil, to Margalit's parents, and to Aunt Sana in New York. My wife, Inbal, and I were married in a similar manner. It seemed natural to us. After the ceremony, we called our parents, since by then telephones were more readily available.

Despite the feeling of freedom, the joy of living without the yoke of responsibility on his shoulders, ten days after his wedding, in the wake of the murder of Jews in the Sharon region, my father wrote a letter to David "Dado" Elazar, the Operations Officer of the Central Command.

April 8, 1953

Hello Dado!
If your plans include a reprisal raid in response to the most recent murders, then I am willing to do it. I can round up a few people, excellent people, for this mission, so that the operation would have a non-military feel—I envision a small force that can carry out this type of mission with a greater degree of success than we've been used to seeing of late.

Since I am quite certain that this type of action is necessary and since I am a son of the region, I am keen to carry this out and I know the area well. I will execute this successfully.

I am willing, despite the fact that I am rather immersed in my studies, to leave Jerusalem for a few days in order to carry out an appropriate mission. I believe the general is likely to support this suggestion of mine.

I await your response in this matter.

Because my father was a battalion commander in reserves, he was not wholly detached from the military. He was twenty-five years old and newly married, a student, but still national matters trumped all other considerations.

The need to respond to terror attacks against Jews was something he believed in wholeheartedly. I'm not sure that he remembers writing the letter to Dado. We found it among his papers, and it is, I suppose, the first seed of what later grew to be the 101 Unit, the commando unit under his command, which then grew into the Paratroopers Brigade and eventually became a towering, historically significant accomplishment of his military career.

Hurva Synagogue in the Old City of Jerusalem, circa 1930, before it was destroyed by the Jordanian Legion in 1948.

FOUR

Some see the murder of Rabbi Avraham Shlomo Zalman Tzoref in Jerusalem in 1851 as the first nationalistically motivated murder of Jews by Arabs—the birth of Palestinian terror.

Previous murders had been deemed a result of religious or financial conflicts, or whatever other motives Arab Muslims have had to murder Jews ever since the founding of Islam.

Rabbi Tzoref's murder, perhaps a milestone in Jewish-Arab relations, began with the Ottoman authorities' granting permission to rebuild the synagogue that was known as the Hurva [ruins], or the Hurva of Rabbi Judah the Pious, in the Jewish Quarter of the Old City. A Jewish center, synagogue, yeshiva, and rabbinic court were all established there. The development was not welcomed by the Arabs, who killed Tzoref. There was also some talk about a 150-year-old debt Jews owed to Arabs and about payments that the Arab leaders had received early in the restoration and which had since dried up. Rabbi Zalman Tzoref's story stuck with my father for years, and as prime minister, on July 21, 2002, he passed Cabinet Decision 2268: "We decide [unanimously] to work towards the construction and restoration of the Synagogue of Rabbi Judah the Pious, 'the Hurva.' "

The more Jews immigrated to Israel, the greater the Arab violence against Jews grew. Arab nationalism in the land of Israel, and with it Arab violence against Jews, was a side effect of Zionism, the ideology of Jewish nationalism. There is no evidence that the Arabs of the land of Israel had their own nationalist aspirations before the rise of Zionism. In fact, if they had any nationalist aspirations at all, they were always as a part of a greater Arab nation.

A deciding moment in the historic development of Zionism was what is known as the Balfour Declaration, issued by British foreign minister Arthur James Balfour at the urging of Chaim Weizmann. This declaration promised the Jews a national homeland in the land of Israel.

This of course prompted a hostile response from the Palestinians. However, in 1919 Faisal bin Hussein (Faisal I) of the Hashemite family signed an agreement with Weizmann, the head of the Zionist Confederation, accepting the principles of the Balfour Declaration.

Article 4 of the agreement states: "All necessary measures will be taken to encourage and stimulate immigration of Jews into Palestine as quickly as possible to settle Jewish immigration upon the land through closer settlement and intensive cultivation of the soil." After accepting the Jewish right to settle in the land of Israel, they paid lip service to the Arabs of Israel, saying all Jewish development must preserve the right of the Arab peasants (*fellahin*) in Israel. Thus, the Hashemites were willing to accept the existence of a Jewish entity in the land of Israel, and just like the rest of the Arab leaders, they were not really interested in the Palestinians.

Nonetheless, after the Balfour Declaration and on into the 1920s, Arab violence grew, became more organized, and took on a distinct anti-Zionist ideology.

In 1920 Arabs attacked Jewish settlements in the north of the country, including Metulla, Ayelet Ha'shahar, Degania, Menachemia, and Tel Hai. On March 1, 1920, eight Jewish guards, among them Zionist activist Joseph Trumpeldor, were killed in Tel Hai. The settlement of Bnei Yehuda, on the eastern side of the Kinneret, was abandoned.

On April 4, 1920, in Jerusalem, during Passover, an Arab mob left the Nebi Musa Mosque east of Jerusalem and attacked the residents of the Old City. The mob, whipped into a frenzy by religious leaders, spent four days running wild in the Jewish neighborhood, killing six and injuring some two hundred people. The British were nowhere to be seen, although they knew in advance of the attack. Ze'ev Jabotinsky, who organized the Defense of the Jewish Quarter, was arrested, and he and his men were sentenced to long prison terms. After a short period of time, though, they were released.

The link between Muslim clergy and mass incitement, the fanning of the flames of hate, is quite strong. Many of the marauding attacks against Jews began with sermons in the mosques, a phenomenon that is ubiquitous still. These days the mosque is used as a place to both recruit terrorists and organize terror cells. In a 2007 article in the Israeli daily *Maariv,* a jailed Hamas terrorist, Muamar Shahruri, told Yoram Schweitzer how the terrorist Abed al-Basset Udeh was recruited at the close of a prayer service in a mosque in Tulkarm. Udeh later blew himself up in Netanya's Park Hotel on Passover Eve, 2002.

The 1920 attacks signaled the beginning of organized Palestinian terror. It was in that year also that Izz ad-Din al-Qassam, a Syrian-born radical Muslim clergyman, fled to the north of Israel from Syria, where he was wanted by the French colonial government for calling for armed resistance. In Israel, in both Haifa and the Galilee region, he found work as a clergyman and began inciting the Arabs to resist both the Jewish and British presence in the land. He advocated for violent struggle that would oust the British and eliminate the Jews, making way for Greater Syria, an Arab state that would include the land of Israel.

In the late 1920s al-Qassam assembled a clandestine organization called the Black Hand, using his cover as a man of God to hide the organization's activities. Nine Jews were murdered, the last of whom was a sergeant in the British Police. This elicited a British reaction, an operation that ended with the death of al-Qassam and several of his men. This despicable murderer was revered among the Arabs of Israel and seen as a role model. The terror he inflicted was conducted under the cloak of religion, both in terms of ideology and organization, and this made a mark on the later generations. He was also socially active, helping the lower classes and recruiting from among them. This is the exact formula embraced, some sixty years hence, by Hamas—religious and social activity, under which it recruits for terror activities; the authorization to kill in the name of God, with senior clerics giving their blessings to suicide bombings. In fact al-Qassam's influence was so widespread that Hamas has called its terror groups the Izz ad-Din al-Qassam Brigades, and its rockets, which it fires from Gaza into Israel, are Qassams—also named after the murderer.

Nonetheless the man who influenced Arab terror in Israel and who inculcated the notion among the Palestinians that the Jews had no right to a national home in Israel was Amin al-Husseini. It was he who led the charge against granting Jews so much as a sliver of land in Israel.

His unwillingness to compromise was the result of several factors: a true hatred of the Jews, his belief in Pan-Arabism, his fanatical religious beliefs, and his battle for the supremacy of his family over the other elite Arab families, especially their archenemies the Nashashibis. This internal struggle led on many occasions to radicalization merely for the sake of standing out and proving that the Husseinis were the true leaders in the battle against the Jews and the British. Amin al-Husseini was the primary inciter of the Passover riots in 1920. He was tried in absentia and sentenced to a long prison term, but fled from justice. Nonetheless, in 1921, as part of an attempt to appease the Arabs, the British themselves appointed him to the position of Mufti of Jerusalem, the highest Muslim position in the land and the keeper of the holy places.

In March of that year the Arab workers' union, headed by Amin al-Husseini's uncle, asked Winston Churchill, the colonial minister at the time, during a visit to Palestine if he could nullify the Balfour Declaration, put an end to Jewish immigration, and establish an Arab government in the land. Churchill refused, saying, "It is manifestly right that the Jews, who are scattered all over the world, should have a national centre and a National Home where some of them may be reunited. And where else could that be but in this land of Palestine, with which for more than three thousand years they have been intimately and profoundly associated?"

In May 1921 riots flared up in Jaffa, Petach Tikva, Kfar Saba, and other settlements in the Sharon Plains. Altogether, fifty Jews were killed or injured. One of the villages attacked was Kfar Malal, where my father was born in 1928. The village was abandoned. Arabs ransacked it entirely. A year later it was reestablished, and it was around that time that my grandparents, my father's parents, joined. The trauma of the attack and the ransacking of the village stuck with the villagers for years, and the tension that terror created was always very palpable in Kfar Malal.

The readiness for and likelihood of an outburst of terror at any moment were a constant for my father and his parents, and me, and in fact for every Jew in Israel, but there was no fear in my father's house in Kfar Malal, which may explain why he used to tell us over and over again when we were children, "You don't have to be afraid, only to be careful."

In the eyes of my father, as a child and later as a young teenager, the village was well fortified; all the village members did regular guard duty, and he was quite confident it would never fall.

In November 1921 Arabs attacked the Jewish Quarter of the Old City of Jerusalem.

In August 1929 another wave of terror occurred. The violence began right after the conclusion of the mufti Amin al-Husseini's Friday sermon at the al-Aqsa Mosque, and a marauding crowd of Arabs swooped down on the Old City, brandishing knives. Nineteen Jews were killed in Jerusalem, and a synagogue was burned and destroyed, along with much other Jewish property.

The terror continued the next day in Hebron, where sixty-seven Jews were killed and the ancient community in the city was destroyed. The murders were ghastly, with many of the dead showing signs of terrible brutality, evidence of a deep and profound hatred of Jews by people who had lived by their sides in the same city for hundreds of years. Their property, of course, was pillaged. The Jewish community in Gaza was also destroyed, and Jews in Safed, Beit Shean, Motza, Tel Aviv, and Haifa were also terrorized, with the dead including women, children, and the elderly. All told, 113 Jews were killed, and some 340 were wounded.

From 1936 to 1939, what's known as the Great Arab Revolt, during which Arab terrorists killed four hundred Jews, raged through the land. The events can be explained, reasons given, chronologies detailed, and I will explicate briefly, but the pattern does not change, despite the shifting pretext. The bottom line is always the same, Arabs murdering and maiming Jews, marauding and pillaging. Sometimes the excuse is the desire to chase the Jews out of the country and prevent them from establishing a national home; another time it's rejecting the recommendation to divide the land of Israel by the Peel Commission, which was founded to investigate the sources of the violence; another time it's in opposition to the UN's decision to enact the partition plan; another time it's anger and revenge for Arabs' losing in the War of Independence, which they started in order to annihilate the young Jewish state; another time it's the desire to return to the borders that existed before the war of 1967, another war that the Arabs instigated, once again to try to wipe Israel off the map; and another time it's because what they were offered in 2000 and 2001 by the prime minister at that time, Ehud Barak—and he offered very close to 100 percent of their demands—was not enough; and yet another time because they wanted Barak's offer back, but my father and the citizens of Israel were not prepared to extend it again.

(The Palestinians blame the Jews for all their problems and failures. They don't seem to analyze their own actions; perhaps their choices dragged them down and they are to be blamed. If they had accepted the UN Partition Plan of 1947, they could have had a state the same age as Israel. Their all-or-nothing approach proved time and again to be devastating for them.)

In 1936 the Arabs wanted to bar the Jews from attaining a state, put an end to Jewish immigration, prohibit the purchase of land by Jews, hinder the Jewish economy by striking for long periods of time, end the British Mandate, and establish an Arab state. Jewish villages and neighborhoods were attacked, roads rendered unsafe for travel. Cars and buses were ambushed, farmers and villagers preyed upon, their fields and silos burned, their orchards uprooted. The Peel Commission, established by the British in 1936 to ascertain the reason for the outburst of violence, handed in their recommendations in the summer of 1937. They submitted that the best course of action would be to partition the country along the following lines: two countries for two peoples, with the Jewish share comprising a mere 17 percent of the land west of the Jordan River, including the strip of beach north of Tel Aviv and the Galilee. The Arab state was to include Judea and Samaria (the West Bank), the lowlands, the southern stretch of coast, Gaza, and the Negev. Jaffa, the international airport near Tel Aviv, Jerusalem, and Bethlehem were to be left in international hands. The same went for mixed cities, like Haifa, Tiberias, and Safed.

The Jews agreed in principle to this offer, and the Jewish community's leadership was authorized to negotiate the size of the land apportioned to them. The Arabs rejected it out of hand and immediately went back to their campaign of violence and terror, which abated only close to two years later. My father was nearly eight years old then, and the truth is that even before that, there was not much in the way of quiet for the Jews in the land of Israel. When he was five years old, he told us, he and Vera took a bus from Tel Aviv to Jerusalem to see Dr. Ticho, the famous eye doctor, and spent the entire journey looking out the window, hoping to catch sight of Abu-Gilda, the infamous one-eyed Arab bandit.

On November 29, 1947, the UN voted to partition the land into two states, Arab and Jewish. Although the Jews were offered a small, noncontiguous sliver of land, they accepted; the Arabs resolutely refused the offer.

My father, by chance, was at home that night in Kfar Malal with his parents, listening to the vote on the radio, and not with his Haganah unit. He told us that the villagers burst out into the main road and started to dance once the plan had been approved. The following day, November 30, a terror attack was launched against Jews, signaling the start of the War of Independence. A

PEEL COMMISSION
MAP, 1937

FRENCH
MANDATE
OF SYRIA

Acre

Haifa

JEWISH STATE

Sea of
Galilee

Nazareth

Mediterranean
Sea

Jenin

Nablus

Jordan River

Tel Aviv

Ramla
MANDATED
TERRITORY

Jerusalem

Bethlehem

Gaza

Hebron

Dead
Sea

Khan Yunis

Beersheba

ARAB STATE

BRITISH MANDATE
OF TRANSJORDAN

EGYPT

N

0   40 km

band of Arab terrorists opened fire and then threw grenades at a bus near Petach Tikva.

Arab bands chose the roads as a focal point of their wild attacks: they didn't require much in the way of manpower, and it was a technique that could inflict many casualties.

The early 1950s were marked by the arrival of what was known as the *fedayun*, bands of Palestinians that crossed the border into the young state of Israel and stole and sabotaged agricultural equipment. Before long, they turned to terror. During that time hundreds of Jewish civilians were killed by these *fedayun*, mostly near the border regions.

The murder of Yochanan Nahari serves as a classic example of the type of attacks the Palestinians perpetrated at the time. Yochanan was a member of Kibbutz Tzura'a, situated near the biblical village of Tzura'a, the home of Samson, near the Jordanian border. He was twenty-three, and was the kibbutz's shepherd. On a spring day, March 17, 1950, he set out with his sheep toward the area of the monastery Dir Rafaat, about a mile and a half from the kibbutz.

We, too, have been raising sheep for close to forty years. I grew up with them, and I can well understand the joy of heading out with the flock on a spring day, the hills coated in green, especially near Tzura'a, a region covered with cyclamens and anemones. It's the best time of year for sheep and shepherd alike, especially for those who love these animals that have been grazing on these hills since the time of the Bible.

A band of *fedayun* beat and stabbed Yochanan to death and made off with the kibbutz's flock for Hebron, which was then in Jordanian hands—the murder of a Jew, based on the old hate, followed by the looting of a Jew: the ideal Palestinian terrorist attack.

On August 23, 1950, Tamar Oren and her boyfriend, Binyamin, or "Bnaya" as he was known, sat on a mound within the orange groves of the village of Yarkona. Bnaya hugged Tamar and asked her to marry him. *Fedayun* bullets killed Tamar and went through the hand he was hugging with. She was only nineteen, a

UN PARTITION PLAN, 1947

LEBANON

SYRIA

Mediterranean Sea

Haifa

ARAB STATE

Sea of Galilee

JEWISH STATE

Jordan River

Tel Aviv

Jaffa

ARAB STATE

Jerusalem

INTERNATIONAL ZONE

Dead Sea

Beersheba

JEWISH STATE

TRANSJORDAN

EGYPT

N

0   40 km

twin to her brother Uri and the younger sister of Amos, who was in my father's elementary school class in Kfar Malal.

Leah Pastinger was born in Budapest, Hungary. As a child she spent a lot of her time figure skating. She survived the Holocaust and managed to make it to Israel in 1948. Eighteen years old, she worked at a Jerusalem law office to help her newly reunited family financially. On December 6, 1951, she was kidnapped on the way back to her parents' house by a band of Arabs, who robbed, raped, and murdered her. Her body, bearing awful marks of violence, was only found weeks later.

I could carry on and list hundreds of instances where Jews, who wished no one any harm, who wanted simply to live tranquil lives, were murdered. Such was life in Israel at the time. The young Israel Defense Forces, which only five years earlier had simultaneously defeated the armies of Syria, Lebanon, Iraq, Jordan, Saudi Arabia, and Egypt as well as the Palestinian forces, was by 1953 thin on manpower and weak in performance. Many had left the army after the war and tended to their own affairs. But the terror attacks only mounted, so much so that in 1953 alone there were some three thousand attacks. The situation was unbearable.

Before 1967, in a country whose total size was roughly 8,000 square miles, smaller than the state of New Jersey, the line between tactical and strategic measures is frequently blurred. In a country so small, where a single car accident on the main road from Tel Aviv to Haifa or Jerusalem to Tel Aviv could bring the entire state to a standstill, in a country lacking any meaningful distinction between front line and home front, terror, even as perpetrated by small bands and individuals, is a strategic threat, capable of grinding everyday life to a halt. The unanswered terror attacks seeded widespread panic, particularly in the rural areas and on the roads. People were afraid to leave their homes after dark.

Israel's sovereignty was not respected in any way. The governments from whose midst the terror attacks originated did not pay any price for their complicity. The IDF attempted several reprisal raids, but they all ended in failure. Such was the case in Beit Sira, Beit Awa, Rantis, Idna, Palma, and others. IDF units had trouble reaching their destinations at night, and if they did actually manage that, then they made do with an exchange of fire with the Arab guards and retreated, or perhaps detonated a house on the edge of the village. Neither did complaints to the UN and to the world powers bring results—not so much as a condemnation of the terror or the countries from which it was launched. Once again it was clear that if the Jews did not take care of themselves, no one would do it for them.

FIVE

Paratrooper Commander Sharon (at left), 1957.
(© Avraham Vered, "BaMahane," courtesy of the
IDF and Defense Establishment Archive)

While the terror continued in the early 1950s, my father was just a student. He and Margalit (Gali) lived in a tiny rented apartment in the Beit HaKerem neighborhood in Jerusalem, and he immersed himself in his studies.

Without the Palestinian terror that was growing, his life would have taken a different direction. The fight against terror has been part of every job he's ever held. Perhaps he would otherwise have become a journalist or a writer, a good outlet for his rich imagination and creativity. Or perhaps a lawyer, channeling his overflow-

ing energy and profound empathy into that profession. Or maybe
he would have returned to Kfar Malal, as his parents had hoped,
and continued to work the land. Who knows? But things worked
out differently.

My father remained a student until July 1953, when the murder
of two guards in Village Even Sapir, west of Jerusalem, altered his
academic plans. He was studying for an exam in history when a
messenger arrived with a note from Colonel Mishael Shaham,
the commander of the Jerusalem Brigade, summoning him to his
office. Shaham told my father, who was a battalion commander
in reserves, that the man behind the latest murder, and many of
the earlier ones, was a terrorist named Mustafa Samueli, from the
village of Nebi Samuel, near Jerusalem. (The village is situated in
the former biblical city of Mitzpe, where, according to the Bible,
Samuel the prophet anointed King Saul.)

Shaham, angered and frustrated by the IDF's impotence and
certain that there was no choice but to respond, asked my father
if he thought he could enter the village, which sat on the crest of a
steep ridge, and strike Samueli and his band of murderers.

My father immediately responded in the affirmative, saying that
with seven or eight good men, people he'd served with during or after the war, and with decent equipment, he could get the job done.

Most of the men he took with him on the mission were civilians. The fact that the IDF needed to call a civilian reservist, who
drafted other civilian reservists, speaks volumes about the IDF's
situation at the time.

Eight men headed out after dark and labored up the steep
ridge, taking care not to be caught by the Jordanian legionnaires,
who had nearby posts, from which they had fired artillery rounds
into the Jewish neighborhoods of Jerusalem during the War of
Independence. All of the men remembered the bitter defeat of the
Palmach five years earlier, the sickeningly high casualty rate suffered when they tried to take this very village during the war, but
none of them said a word. They reached the village and found the
house, but the explosive device they placed by the door burned

and sizzled without detonating. The door stood firm. The men threw several grenades into the house. There was no response, and it seemed as if the house was empty. Meanwhile the village woke, and gunfire started to pour out of the nearby houses. Realizing that at this point there was nothing to be gained by remaining in the village, the men used the uproar to sneak out and avoid the Jordanian legion soldiers.

By 1 A.M., they were back on Israeli soil, tired and not happy. My father wrote Shaham a detailed account of what had happened, explaining that in his opinion these types of operations needed to be carried out by professionals, well trained and used to operating at night. More than just a random bunch of guys he rounded up, the IDF needed an elite commando unit.

My father returned to his studies, unaware that Shaham agreed with him entirely and had written as much directly to Prime Minister David Ben-Gurion. What Israel needs is an elite unit, Shaham wrote, an antiterror unit. A force that can strike back against terrorists in the villages they use as bases. My father had no idea that Shaham had not only alerted the prime minister to this need but also recommended my father be appointed as the commander of this unit.

Two weeks later, in late July, my father was summoned to IDF headquarters to meet with Chief of Staff Mordechai Maklef. Maklef explained that, acting on Shaham's suggestion, he was about to found an elite commando unit, and he asked my father if he would be willing to lead it. My father told Maklef that of course he would, but he would also like to finish his studies at some point.

"I can give no such assurance," Maklef said.

On an August afternoon in 1953 there was a knock at the door of my father's apartment. The flat, belonging to the Matot family, had been converted from a chicken coop to a small rental unit in the Beit HaKerem neighborhood.

My father opened the door to the renovated chicken coop and saw two men. "We've heard about your unit, and we've come to join," they said.

After a short talk my father told them, "You're in. When can you be ready?"

"Immediately," they said.

The two men, Yossele Regev and Zevele Slutsky, were both from the Village Nahalal in the Valley of Israel. During their lives they were never divided; only death tore them apart. They were my father's closest friends.

Five months—that's how long Unit 101 existed. At its peak it consisted of forty to fifty warriors, and yet the mark it left on the army, the changes it instilled, continue to reverberate today.

The unit operated in squads of four or five men, the ideal size for the commando warfare they waged. Yossele and Zevele were immediately appointed squad leaders.

Yossele had served in the Jewish Brigade during World War II. The first time he tried to enlist, he was turned down because of his youth. When he returned a second time, he succeeded. After the war, his battalion commander turned to him and said, "We need people like you." Yossele remained in Europe and helped Holocaust survivors make their way to the land of Israel. In 1947 he returned home and was made a squad leader in the Haganah Field Corps.

"Keep your eye on Zevele," they said of the other squad leader, who was two years younger than Yossele. "He's young, but he's a man. He's the only one who wanted to escape from the British prison in Rafah, despite Haganah orders."

Zevele lost his father at age sixteen. He had been driving a mule team home in the dark and gotten into an accident on the road. From then on, Zevele ran the family farm. He was big and strong, and he had a quick and sharp tongue.

At squad leader training in Kfar Yehoshua, Zevele was given special permission to leave the barracks at night to milk the cows and again in the early morning and then to race back to training. He did it with a smile.

In 1947, the two squad leaders from Nahalal would stay up late into the night, talking long after everyone else had gone home.

During the War of Independence they served under another man from Nahalal, Moshe Dayan, in Brigade 8, which was commanded by Yitzhak Sadeh. The war ended with Yossele as a deputy company commander and Zevele as platoon leader.

At a gathering for the brigade's reservists they heard from brigade commander Avraham Yoffe that there was a new unit carrying out reprisal raids to deter terror attacks, a unit that crossed into enemy territory. Right then and there they decided to volunteer.

Yossele was already twenty-eight years old, married to Gaya and the father of a little girl. Zevele was twenty-six, married to Dalya, and also the father of a little girl. My father was the youngest of the crew, a year younger than Zevele.

Perhaps their friendship was rooted in the fact that they were older than the other soldiers, and in that way they helped ease the loneliness of the commander, which my father spoke of often. Perhaps it was because the three of them shared a similarly sharp sense of humor. The three of them shared a similarly sharp sense of humor. Perhaps it was because the two of them were both sons of a village like my father; perhaps because no one had asked them to leave a wife, a baby, and a farm and volunteer—they just came. Whatever the reason, the commander and his two deputies forged a tight friendship that withstood all the tests of events and time.

In 1954, after the merging of Unit 101 and the Paratroopers Brigade, Yossele and Zevele were both made company commanders. Shortly thereafter they decided that the institutional rigidity of military life was not for them. They returned home to Nahalal, where they were made officers in the Gadna, the youth battalions. Yossele was the commander of the northern district, and Zevele, of the Haifa region. Whenever they learned of a terror attack, though, they'd make their way back to the paratroopers' base in

Tel Nof, insisting on taking part in the retaliatory mission that was sure to come. They'd arrive, my father would issue an order to equip them, and they would join the command squad. They were not in the army and had not been summoned—they just came.

Soon enough, the work with the youth battalions grew dull. The two of them, seeking a change, showed up at Moshe Dayan's house. "It's good you are here, help me get this obelisk up," Dayan said, working with some of the antiquities in the yard of his Tel Aviv home.

Dayan listened to their predicament, understood it, and sent the two of them to Isser Harel, the head of the Mossad. During their many years of service they traveled the world and, among other things, assisted Jews who wanted to immigrate to Israel. The particulars cannot be disclosed.

The village was not particularly pleased with Yossele and Zevele's career choices. Life on a collective agricultural settlement was never easy for those who did not toe the company line, certainly not in the 1950s and '60s. They were sanctioned for their absence: the board dried up Zevele's grapefruit orchard and denied his daughters, Ronit and Michal, entry to the village's kindergarten. Yossele received a steady drip of irate letters from the village board, calling on him to return immediately, but the letters were always marked "returned to sender"—the addressee could not be found.

On the eve of the Six-Day War, both of them were abroad. Yossele returned to Israel before the war and showed up at my father's division in Shivta in the Negev, not far from the border with Egypt. As usual, he had not been called to reserves and was not officially a soldier, but that didn't matter at all. He got suited up and rode in my father's half-track. A few days later Zevele arrived, also from abroad. "I've come to replace you," he said. "You have an order. There's a car here, and it's taking you back and I'm staying. The two of us shouldn't be in the same place. So that one of us will remain." Remain alive, he meant. "Why are you staying, and I'm the one who's leaving?" Yossele protested. "We're gearing up for war, and I don't want to miss it," he said at Mossad headquarters, but the head of the Mossad had the final word, and Yossele left for his mission. Zevele spent the entire war by my father's side.

Up until the Yom Kippur War, Yossele served abroad. When Zevele, working for the Mossad in Israel, heard that his friend was in a precarious situation, he told the head of the Mossad, "If anything happens to Yossele—his blood is upon your head."

When war broke out, Yossele was still out of the country. Zevele was in Israel, and when he heard of the general reserves call-up, he phoned my mother and said, "Just tell Arik to wait for me."

He showed up at our house in Beersheba with a uniform from back in the day, but he didn't have army boots, so my mother pulled out a pair of my father's red paratrooper boots from the closet. His arrival calmed my mother; Yossele and Zevele, after all, were my father's good-luck charms.

Zevele and my father rode down to the Sinai together in a civilian truck with the words "Ray of Light Solar Heaters" painted on it. They were joined by Uri Dan.

Zevele was by my father's side in the command armored personnel carrier. Amid the madness, pain, and death that is war, he didn't have to speak; his smile, the face of a friend in the midst of the inferno, was enough.

On the morning of October 17, 1973, a day after crossing the canal, Zevele and Motti Levy, my father's driver and close friend, who was also in the command APC the entire time, were directing rafts to the canal. Suddenly the staging ground erupted in a barrage of artillery fire. Soviet-made MiG jets came screaming in from above and dropped their loads.

Zevele and Motti, taking cover behind an APC, had only one cigarette left. "You take it," Motti said. "No, you take it," Zevele retorted. In the end they decided to split it. They sat there with their backs against the APC, but when Motti turned to hand the cigarette over to Zevele, he was already dead, killed by shrapnel from a shell. Shortly after, Motti was severely wounded. This was a decisive moment in the war: my father had just realized that the Tirtur-Lexicon junction, on the Israeli side of the crossing, was under Egyptian attack. Had the Egyptians taken the junction, it would have put the entire canal crossing in danger. My father rushed over with the small power he had at his disposal—the command APCs—until Amnon Reshef's tanks arrived from the south. Before leaving, he glimpsed a pair of red boots sticking out from under the blanket covering a dead soldier. He did not know that they were his own. In the coming nights, when my father was alone in the desert, the tears would flow, for Zevele and all the others.

Moshe "Kokla" Levin showed up to take Zevele's place in the APC, where he would sit with his long legs dangling off the side. Kokla was not much of a conversationalist; he had come to stand by my father's side. When Yossele returned to Israel after Zevele was killed, he headed straight to the Sinai to join my father. When he showed up, Kokla said, "Now I can go," and returned to his post in the Mossad.

I remember going with my brother and my parents to the temporary graveyard near Kibbutz Be'eri to pay respects at Zevele's grave before he was laid permanently to rest in Nahalal.

I think of Zevele. He was forty-six, not a soldier or a reservist. He had a wife and three children at home. No one had called him, but as always, he had simply showed up to be with his friend, my father. I think of all the pain and hardship, for him, for us, for all of them.

Yossele remained in the Mossad, continuing to work in the field; paperwork had never been Zevele and Yossele's thing. At fifty-three, Yossele was still prowling around Tehran. At the last instant, when Khomeini took power, he finally returned to Nahalal.

Yossele was always a quiet, reserved man. He had two close friends, and that's a lot. Anyone who has known the privilege of having a close friend, the kind you can count on, the kind you can trust with anything, is a happy man. Anyone who has had a friend like that and then lost him feels the absence, the lack, all the more acutely. He knows what he has lost.

After not visiting for a while, I went to see Yossele and his wife, Gaya, in Nahalal. Their house looks just as it did when I was a child. The cozy atmosphere that always permeated it remains unchanged, too. It puts one at ease to know that the people and the places you love have remained just as you remember them. The yard, though, is different. Where there were once calves, there is now a playroom for the grandchildren.

In the 1950s and '60s, when my father asked Yossele how his father was doing, Yossele would say, "He's out playing in the sandbox." More recently, when my father and Yossele spoke, my father would ask him with a smile, "So, Yossef, how are you? Playing in the sandbox?"

When they finished talking, Gaya, Yossele's wife, turned to her husband and said, "How can you speak to him that way? He's the prime minister."

Yossele responded, "I was talking to Arik."

Zevele Amit (in the center), on October 16, 1973,
the day before he was killed. (Uri Dan)

T raining in Unit 101 was hard, and results came quickly.
    The idea was to respond to terror attacks, to exact a price
for the taking of Jewish lives in Israel. Although the IDF had
taken measures to safeguard the border, founding the Border Police unit, there was truly no way to guard the long, winding border with Jordan and with Gaza.

The only way to accomplish that was by establishing new
ground rules—Jewish lives were not just for the taking; the harming of Jews was a crime that would be paid for. The terrorists
would no longer enjoy safe havens, and the armies of Egypt and
Jordan, who aided the terror cells and even actively ran them,
would be forced to pay a price, too.

O n August 31, 1953, a squad from the unit raided the al-Bureij
    region of Gaza and blew up a building used by the *fedayun*.
The following day, the first of September, warriors from the unit
raided Beit Mirsam, in the southwest Hebron Hills, and later they
conducted cross-border recon missions in the Kidron Canyon region, along the Jerusalem-Ramallah road, east of Rosh HaAyin,
Beit Horon, Tzurif, and Tarkumeya.

On October 13, 1953, at two in the morning, terrorists threw
a grenade into the Kanias family home in Yahud. The explosion
killed Sultana (Susanne), a mother of seven, and two of her children,
Shoshana, three, and Benny, one. A third son, Yitzhak, thirteen,
was severely injured and died three years later from his wounds.

That same day my father was called to Central Command
headquarters in Ramla. The deputy battalion commander of the
paratroopers was present, along with staff officers from the Central Command and representatives of the General Staff. During
the briefing my father learned that the General Staff had decided
to carry out a reprisal raid in Qibiya, the village that was a few
miles from the border from which, according to the army's intel-

ligence, the murderers had set out. The plan called for the paratroopers to raid the village while Unit 101 created a diversion,
drawing the Jordanian troops away and setting up roadblocks on
the routes in and out of Qibiya to block the passage of reinforcements.

At the end of the Central Command briefing, the paratrooper
officer reported that his men were not ready to carry out this type
of mission. As the questions came pouring down as to why his
men were not up to the task, my father interrupted: "We'll do it,"
he said, "and I'll be more than happy to command these unready
paratroopers as well as my own men." His words had two immediate consequences. He was given the command of the mission,
which was unusually large and complex for the time and unusually successful, too—a military achievement in which my father
proved his skills and his capabilities, and so the road was open
for him. The second consequence, however, was the resentment
and envy of men like the paratroop officer, which plagued him
throughout his career.

M y father's frequent forays into the field, his intimate knowledge of the territory, and his belief in his ability to execute
missions, along with his calm under fire, made the men under
him feel they could trust him with their lives. They would follow
him anywhere. He instilled in the young army the importance
of extreme perseverance, exacting standards, an unwillingness
to leave wounded behind, and sharp, painfully truthful debrief-
ings. These standards, he signed on for. But it was a package deal;
these traits were not always accompanied by great measures of
patience for officers who could not meet his standards, and there
were plenty such men.

"The mediocre also want to live," he used to say. But back in
those days, with mission following mission, sometimes night after
night, during those stormy years, I'm not too sure he had much in
the way of patience for the mediocre.

His incredible capacity to push the men under his command to the peak of their capabilities, along with the steadfast backing they received from him, his unwillingness to roll responsibility onto others, made some love him for the rest of their lives. But it also brought resentment and hate. Everyone, even the mediocre, wanted to live.

As prime minister, he would tell me and others, whenever he wanted us to be patient, "It took me seventy years to learn that, seventy years." But this was back then, some fifty years before that lesson had been fully learned.

That day, on October 13, 1953, at the Central Command headquarters, they accepted his offer. The paratroop officer was instructed to show up immediately with his paratroopers at Unit 101's base in Sataf, near Jerusalem.

All night long the mission plans were prepared, and the following morning one hundred paratroopers and twenty-five men from the 101 were driven down to the staging area in the Ben Shemen Forest. In the afternoon Moshe Dayan—the IDF's chief operations officer at the time, and slated to replace Maklef as chief of staff—asked to see my father. At first Dayan had been against the founding of the 101. "The army does not need special-ops units," he had said; "every unit should be able to execute these types of missions." With time, he changed his mind.

That afternoon Dayan wanted to see my father. My father told us many times, Dayan said to him in his typical manner, "I understand you're taking this very seriously." My father said this was the case. And he responded, "Look, if it turns out to be too difficult, just blow up some of the houses on the outskirts of the village and get out."

"No," my father said, "we're taking six hundred kilos of explosive, and we will carry out the mission."

The army orders were clear: Take out as many members of the National Guard and the Jordanian Legion as possible; capture the Jordanian fort and detonate houses in the village so that the Jordanians realize that terror has its price.

The operation was a success. The roadblocks worked, as did

the diversion; the Jordanian fort was taken, and ten Jordanian soldiers were killed. Two more were killed when they approached with a jeep. In addition, forty-two homes were blown up. The success, though, was linked to a tragedy that for years has been used against Israel and my father.

Villagers, it turned out, had been hiding in some of the houses that were destroyed. By the following day, it was clear that innocent civilians had been killed.

The village had seemed empty. The demolition team went from house to house. Word had come from one of the roadblocks that a large number of villagers had fled. A radio, warbling in the village's abandoned café, was the only sound they heard. If anyone had the intention of harming civilians, they never would have let so many of them pass untouched through the roadblock.

One boy was found in one of the houses. Later, after the wick had already been lit on the detonation device, crying was heard from another house. Shlomo Hefer (Gruber) ran inside and saved the little girl. Aside from those children, who were unharmed, the men heard and saw nothing. The civilians in the houses had no way of knowing about the new commander or the army's change of strategy right there in their village and that from then on the IDF would no longer content itself with the symbolic detonation of a single house on the far reaches of the village—so they stayed in their homes, ignoring warnings to evacuate the houses before detonation. The loss of those civilian lives disturbed and pained my father.

I remember my father telling me how his men had employed a socially progressive outlook in deciding which houses to detonate, placing explosives in only the biggest and fanciest of the homes, not among the poor. I also remember him saying how Yossele Regev opposed blowing up the school, even though there had been an enemy position on the roof. Zevele Slutsky explained why as he stood next to my father in the middle of the village, "because Gaya's mother [Yossele's mother-in-law] is a teacher." The school was not destroyed.

These were people of the highest moral character, people who would never intentionally harm civilians, certainly not while arguing about whether to take out an enemy position on the roof of a school. What's more, it's unfathomable that such a secret could be kept for so many years by so many people who were there.

After the mission, Dayan sent a note to my father: "There is no one like you."

I n a sharp departure from the common response after Israeli civilians were killed, an international outcry ensued.

Several days after the operation in Qibiya, my father was summoned to meet Ben-Gurion in Jerusalem. It was the first of many meetings. My father remembers it as thrilling. Ben-Gurion asked a few questions about the operation and then asked my father, "Where are you from?" I remember how my father could do a great Ben-Gurion imitation, nailing the intonation and the accent. "Kfar Malal—two townspeople of mine are from there, both sick."

Ben-Gurion asked about the other members of the unit, and whether this was the kind of unit that might get out of control. He was a suspicious and careful man. My father said that these boys were the cream of the crop, and there was no chance they'd get out of control.

"It doesn't really matter what they say about Qibiya in the world, what's important is how it will be looked at here, in this region. That's what will give us the possibility to live here," Ben-Gurion said.

Ben-Gurion knew that the nations of the world could not be counted on to stop the murders and terror attacks in Israel. But he understood that the IDF was no longer helpless, and that he and the government now had a true military option at their disposal. This, however, did not deter him from denying IDF involvement in the operation in Qibiya. On October 19, 1953, Ben-Gurion addressed the nation: "We have performed a careful examination and it turns out that no army unit was missing from its base on the night of the attack on Qibiya."

From left: Avraham Yoffe, my father, and Ben-Gurion,
August 1957. (Photograph by Ephraim Erde)

My father detests improvisation. He is a man of details, of careful and meticulous planning. In Unit 101 and later when he was commander of the paratroopers, this manifested itself in the form of exhaustive training, extensive intelligence gathering, and an acutely acquired familiarity with the field. The paratroopers had dossiers ready with intelligence data, sketches, and diagrams that they had collected and assembled themselves during countless cross-border raids. This enabled my father to receive an order in the morning and carry out a mission that same night, sometimes night after night, in different locales—and there was nothing improvisational about it.

One of my father's major traits is his understanding of the human soul, the warrior's soul. Unit and army spirit, both ours and our enemies', were the tools he used and one of the keys to his great success. People would follow him anywhere, undertake any mission he assigned them. They trusted him, knew that he demanded of them only the things that he demanded of himself. "Arik never sent us on an operation that he himself would not undertake. When he briefed us, I always felt that he was able to envision what was going to happen," said Shimon "Katcha" Kahaner, one of the more famous of the 101's warriors, who has remained a close friend of my father's and the family's to this day. On the one hand, my father was unwavering and somewhat severe with discipline, when necessary; on the other, he created an esprit de corps. The fact is that even today, some sixty years hence, the men who served under him in the 101 and the paratroopers, are still our family's closest of friends. As commander of the paratroopers, he said, he could recognize the different companies by the songs they sang as they marched past his office.

The fighters and companies fought for the right to take part in missions. It was considered a stiff punishment to be excluded from a mission, and the stiffest of all was expulsion from the unit. "Take off your [red] boots [only paratroopers wore red boots] and walk straight out through the gate," my father would order. This was a command he did not hesitate to give if he felt that an officer or a soldier had not shown the necessary perseverance or failed to meet his stringent standards in some other way. As David Ben Uziel, "Tarzan," a soldier in Unit 101 and the paratroopers, said: "Our first law is to execute. You did not return without executing the mission. And if you ran into Jordanians and did an about-face and on the other side had Arik Sharon waiting for you, then you'd be better off facing the Jordanians anyway. And if you turned around for a good reason, then you earned the right to go on the mission the next night. Till you reached your destination. Even if it meant going out three or four times in a row. That is [what it means] to stick to the goal of the mission."

Another element of his success as a commander was his intimate knowledge of the enemy. My father never belittled or made light of Arab enemies: "If you attack them the way they were trained to defend, they will fight to the death." His notion, as I heard him say countless times, was to "throw them off balance." He always attacked from an unforeseen angle, always surprised the enemy. Another element that he told me he relied on was to make sure that his forces always had at least one protected flank.

Such was the case during the retaliation raids, where he always used roadblocks to cut off any reinforcements, and again during the Yom Kippur War, when he picked the area near the Great Bitter Lake as the crossing site, so that at least the southern flank of the bridgehead was protected.

Unit 101 and later the paratroopers shouldered the entire burden of the retaliatory raids. This went on for more than three years, from the unit's inception in August 1953 to the Sinai War in October 1956, after which there was relative calm.

Yitzhak Rabin, in his autobiography, wrote of Unit 101, "In 1954 Chief of Staff Dayan merged the warriors' unit 101 with the Paratroop battalion. The actual work was done by a young officer, Arik Sharon, whose star was on the rise as an excellent commander. The unique spirit of Unit 101 restored the IDF's confidence in itself, which had been shaken by failures. The unit's success was carried on every tongue and it became a lodestone for brave youngsters."

D ayan's decision to merge the paratroopers battalion and Unit 101, and to place the new unit under the command of my father, was not well received by the standing commander of the paratroopers, Yehuda Harari. In truth, the use of the word *merge* was strange: Unit 101 never consisted of more than forty or fifty warriors, roughly one-tenth of the size of the paratroopers battalion, and Harari, who was older than my father and outranked him, had expected to command both my father and the 101. Dayan, though, had other ideas. Tactless as ever, he asked Harari, a brave soldier, "How long have you been commander of the paratroopers?" When Harari told him that it had been three years, Dayan said, "Okay, then, that's enough, I'm turning the command over to Arik."

Zvi Yanai described the change of command that took place on January 14, 1954: "The ceremony of transfer of command only reinforced the sense of disaster among the battalion's veterans. On a Friday morning the soldiers of the battalion stood in formation on the parade grounds facing the flag. When the battalion went to 'present arms,' lieutenant colonel Yehuda Harari entered the parade ground in a pressed uniform, cleanly shaven, stepping in time to the unheard sounds of a military march. In the presence of the silent parade Harari briefly read out the orders of the day and ended with an expression of his certainty that his separation from the battalion would be brief . . . At the end of his remarks Harari read the names of the officers who had chosen, in a demonstration of loyalty, to leave together with him. Everyone whose name was called stepped out of the ranks of the parade and stood beside him . . . With his line of loyalists standing near him, Harari turned to Arik Sharon, who was waiting at the edge of the grounds. Arik ambled toward him in steps that were at half the pace of Harari's silent march . . . Within a few weeks it became clear to everyone that Unit 101 had not melded into [Paratrooper] Battalion 890, but rather Battalion 890 had melded into Unit 101. If in Harari's day, the veteran soldiers had been treated to easy desk jobs prior to the end of their service, with Arik Sharon the worst punishment was the transfer of a soldier from a combat company to command headquarters."

My father could sense mutiny in the air. He did not mind losing the departing officers; what he was concerned with was preserving the unit's airborne abilities. Thus he made sure that the parachuting instructors and packers, as well as those trained in airborne logistics and deployment, all stayed on. Insofar as the warriors were concerned, the first thing he did was send the companies deep into the field in different parts of the country, where they trained harder than they ever had before. My father sent them on long recon patrols, often across the border, testing both their readiness for battle and their physical stamina. Each team consisted of several warriors from the 101 and several paratroopers officers. Before long, around half of the original officers corp had left. He wanted to turn those who remained into a confident and ready unit.

Within a short period of time the battalion changed drastically. The custom of keeping the handsome officers' quarters near the commander's room and the ugly ones far in the distance was abolished. The paratroopers officers were shocked to learn that my father had opened the officers' club to sergeants. "Sergeants in the bar? He's gone too far," was the whispered protest. More important, the new regulations allowed entry to the warriors from the 101. Almost immediately a series of actions and cross-border raids occurred. Operations came in rapid succession. In February, the unit carried out five raids in Gaza. Later that same winter, after the massacre of Israeli civilians on Ma'ale Akrabim, the unit struck in Nahalin, southwest of Bethlehem, and then in Gaza again, taking an Egyptian soldier captive as ransom for an Israeli one being held by Egypt. The mission was known as Operation Cigarette, for the Israeli soldier who had been taken captive while giving or receiving a cigarette from an Egyptian.

On June 29, 1954, several days after a Jewish farmer was murdered in a field near Raanana, seven of my father's men set out on a raid deep into Jordanian territory, striking the legionnaires' base in the village of Azun. The patrol neutralized several threats on the way to the base and managed to penetrate the Jordanian stronghold and kill several soldiers, but Sergeant Yitzhak Gibli, a warrior from the 101, was wounded during the battle. Gibli's insistence, toward the end of the night, that the men carry on and leave him behind earned him a medal of honor; his request that they leave him with a hand grenade, which meant certain death, was denied.

Gibli's captivity threw my father into a whirlwind of activity. He initiated a series of hostage-taking operations—known as GIL, an acronym for Free Yitzhak Gibli in Hebrew—against Jordanian troops for the purpose of an exchange of prisoners. The Jordanians were sure that the man they held was a senior officer; although it seemed strange to them that he greeted the forces that took him prisoner with a thunderous rendition of an Israeli song, "The Roar of the Cannons Has Fallen Silent."

Gibli's red boots also made them think he was an officer, since in those days only Jordanian officers wore red boots. In addition, they could not believe and understand that an entire country could be up in arms over the captivity of a single sergeant. Gibli's father, who was sick and dying, asked my father to bring his son back to him before he passed away, but Gibli's release came too late. He named his son Gil to commemorate the operations that freed him.

Gibli was a dear family friend up until his death some fifty-four years later. At the end of the funeral service the old paratroopers stood around his grave and sang, in deep and gruff voices, "The Roar of the Cannons Has Fallen Silent," the song Gibli had sung to his captors.

On July 10, 1954, Operation Kissufim was launched, the first attempt by the Israeli army to take a fortified Egyptian position. The base was ringed with barbed wire, mines, and trenches. That night, as my father stormed into the trenches along with his men, he was shot in the thigh. "The physical pain was very similar" to that he had felt at Latrun, my father said, as well as the area in which he had been shot, but the feeling was radically different, because this time he was carried home by his own men, victorious.

He received the first report of the results while lying on a stretcher in the Egyptian fort they had captured. Later, while going over the battle during the usual debriefing, he realized that he had been wounded during a particularly vulnerable moment in trench warfare. When considering combat situations, my father was able to zero in on the moments of acute vulnerability, the instances where a warrior might be prone to fatal hesitation, and he would immediately construct a drill that could be pounded into a soldier, made into second nature, so that in battle it would be a reflexive response. The new technique they developed on this occasion, in which the attacking troops would enter the trenches as fast as possible, is still today taught to each and every one of the IDF's infantrymen.

"Maybe it's time things changed," I've heard him say over the years. He always detested analytic rigidity.

On February 28, 1955, the unit launched Operation Gaza, its most difficult operation to date, an assault on the main Egyptian army camp in the Gaza Strip, which was teeming with soldiers and Palestinian terrorists.

On October 27, 1955, in response to an Egyptian attack on one of our bases, the unit launched Operation Kuntilla, targeting a remote outpost some twelve miles inside Egypt. The objective was to take Egyptian prisoners to exchange for our soldiers in their hands. The plan was to penetrate into Egyptian territory via Wadi Faran, a desert valley. Most of the distance would be covered by motorized convoy. Soon enough, though, it became clear that the vehicles could not make it through the soft and sandy valley floor. Rather than turning around, my father decided to continue on foot. There was every reason in the world to cancel the mission: the stuck vehicles; the short time available to get there, execute the mission, and return before daylight; and the unusual light and activity emanating from the fort, which might have been a sign that the paratroopers had been detected.

But my father was determined. The operation was carried out. Twenty-nine Egyptians were taken captive. The reason for the light was a changing of the guard at the base. The fresh soldiers protested meekly, "We just got here, we don't have anything to do with this."

On December 11, 1955, after repeated incidents of Syrian fire on Israeli fishermen in the Sea of Galilee, the unit launched Operation Kinneret [Kinneret is the Hebrew name for the Sea of Galilee], a complex and multifaceted assault on the Syrian positions on the northern and eastern shores of the lake. Some of the forces crossed the lake by boat, some crossed the Jordan River farther to the north, and some set out on foot from Ein Gev.

In 1954–55, at the height of the reprisal raids, my father stopped for a bite to eat at a restaurant in Beersheba called the

Last Chance. He was with his first wife, my aunt Margalit, and on his way to a meeting with Ben-Gurion in Sdeh Boker. On the way out of the restaurant, they were assaulted by a group of thugs. My father fought back, but soon enough found himself pinned to the ground. The leader of the group knew what he was doing, my father said, which was probably why he was the leader. The thugs were known as the Jaffa Gang.

Back at the base, the soldiers were furious when they heard what had happened to their commander. Some, like Shlomo Gruber, had tears in their eyes. The paratroopers assembled quickly and set out to settle the score. They put up roadblocks on the way out of Beersheba, but didn't find the gang. My father, who trusted his men behind enemy lines, worried about them here. He didn't think that soldiers who excelled on the battlefield were necessarily fit for this type of battle.

The gang was not captured by the soldiers and returned to their hideout, but eventually they were caught. In a police lineup my father picked out the head of the gang, who was surprised that someone would dare to pick him out of a lineup. It had never happened before. "You'll never set foot in Jaffa," he hissed at my father on the way out.

"I found myself in an untenable situation," my father said. "I had the ability to cross the border and enter any Arab village or garrison, but Jaffa, smack in the middle of my own country, was off limits."

Their reconciliation took place at a club in Jaffa, after all of the perpetrators had been released from prison (at least for this crime). The party was festive, and every few minutes someone would come up to my father and proudly say, "You remember the guy who pinched you from the back? That was me." Or, "You remember getting kicked from the back? That was me." And so it went all night long. He'd been beaten up, but he had fought like a man. They respected him, and in the end, he conquered Jaffa, too.

With Moshe Dayan after Operation Kinneret, December 1955.
(© Avraham Vered, "BaMahane," courtesy of the
IDF and Defense Establishment Archive)

This is but a brief description of the dozens of operations the
paratroopers carried out during this period. Aside from the
raids, they conducted hundreds of intelligence-gathering cross-
border recon patrols, which allowed them to compile intelligence
dossiers and draft mission plans and also kept the soldiers sharp
and ready for battle.

During those stormy years, from the first raid in Nebi Samuel until the Suez War in 1956, my father was immersed heart
and soul in the passions and rigors of battle. He was young, just
twenty-five when he assumed command of Unit 101 and twenty-
eight when the Sinai was taken in 1956. Only someone with his
sweeping enthusiasm and charisma could have pulled off what he
did during those years: spearheading the army's reprisal raids on
all three fronts, carrying out virtually all of the IDF's offensive
action while the rest of the combat officers looked on in shock,
green with envy.

For his part he did not take any great pains to conceal his attitude toward those officers. During the lead-up to one operation in Gaza, it was decided that an additional force, from the Givati Brigade, was needed to man a roadblock. The brigade commander, Yoske Geva, showed up at command headquarters in full battle dress, helmet on his head, and informed the officers that his men were not up to the task, as they had not yet been through the requisite training. My father immediately said, "We'll do it."

"How, exactly, will you do it?" he was asked. "All of your men have already been assigned to the mission."

"I'll get the platoon of girls assigned to pack the parachutes to do it," he said.

It was a quip, and a funny one at that, but it did not particularly endear him to the brigade commander.

In principle my father was right. How could a brigade of combat soldiers not be up to the task of manning a roadblock, especially when all other responsibilities were being shouldered by another unit? My father enjoyed sniping at other commanders who were incapable of doing what he did regularly, but that satisfaction had a price.

I remember hearing the story of the parachute packers many times, and my father always had a good time telling it, but those kinds of comments cost him dearly in promotion and popularity. This was not the way to climb the chain of command. He remained stuck at the colonel rank for close to ten years. It seemed that the IDF was willing to forgo his services, despite all his accomplishments and capabilities, to get rid of the hugely energetic and enterprising officer who seemed to overshadow his peers.

Not everyone resented him. One not unimportant person who was always in his corner and held him in the highest esteem was David Ben-Gurion. Ben-Gurion wanted to get to know the young officer who, unlike his predecessors, carried out the reprisal raids rather than just returning to Israel empty-handed.

The ties between the prime minister and unit commander became warm, direct, and close. The prime minister severed the chain of command, which was hardly sacrosanct to my father, letting him know that his door was always open. During army conferences the prime minister would ask my father to attend and to sit by his side. Sometimes, when one of the army's high-ranking officers would approach him, Ben-Gurion, who was quite the actor, would look him in the face bewilderedly and inquire, in his sharp voice, "Who are you? What are you doing here?" After that the unfortunate officer had nothing left to say.

Such ties between a young officer and the prime minister and minister of defense were exceedingly rare. My father also enjoyed close ties to Ben-Gurion's military attaché, Nehemia Argov; to his chief of staff, Yitzhak Navon; and to Shimon Peres, the general director of the defense ministry. This, too, did nothing to increase his popularity with the other officers.

With Ben-Gurion at paratrooper training, 1957.
(© Avraham Vered, "BaMahane," courtesy of the
IDF and Defense Establishment Archive)

According to Moshe Dayan, "Ben-Gurion had a special relationship with three officers—Haim Laskov, Asaf Simchoni and Arik. He didn't merely like them, but 'melted' . . . firstly because all three are great soldiers; and secondly—and I think this is the key to Ben-Gurion's special relationship with them—they each represented the new Jew, the antithesis of the Diaspora Jew, the bold, self-confident warrior Jew who knew his profession well—the field, the Arabs, the weapons and the ways of war."

Dalia Goren, Ben-Gurion's secretary in the 1960s, recalled the effects of my father's visits to the Knesset. "Every time Arik arrived in the Knesset in uniform, 'Arik's here, where is he? Here he is,' Ben-Gurion would say, his eyes alight with excitement. He loved him so much, you don't know how much he loved him, he would get as excited as a little child. He saw him as a field marshal."

My father's relationship with Dayan was far more complex. On the one hand, Dayan didn't trust anyone but my father to get the job done, and as the army chief of staff he assigned all of the major missions to him. When things went smoothly, Dayan personally brought the news to Ben-Gurion, conducting his own informal meetings with the newspaper editors. When there were complications—and there always are at some point—he would send my father to deliver the news. Fighting an enemy night after night is not playing a solo on a violin; there are bound to be complications. After Operation Gaza, during which eight of our men were killed and thirteen were wounded, my father was sent to meet with the newspaper editors and to discreetly explain to them what had happened in the field.

A few years later, when Dayan had retired from the army and entered politics, Ben-Gurion asked my father what he thought of the former chief of staff as minister of agriculture.

"He can be an excellent minister, but never prime minister," my father said.

"Why?" Ben-Gurion asked.

"Because he is never willing to take responsibility."

My father used to say of Dayan that he was "brave to the point of insanity" on the battlefield, but, he always added, in public life he "was a coward." Dayan, he would say, "was downright terrified of Ben-Gurion." My father had the greatest respect for Ben-Gurion, but he feared no one.

———————

In December 1954, a squad of five soldiers, made up of three paratroopers and two from the Golani Brigade, set out for the then Syrian Golan Heights in order to change the batteries on an Israeli listening post. The paratroopers had been carrying out similar missions—pioneered by Meir Har-Zion, Shimon "Katcha" Kahaner, and Yoram Nahari—for some time, and the IDF, wanting to train Golani soldiers to accomplish this task, had sent Lieutenant Meir Moses and Uri Ilan to join three experienced paratroopers, Sergeant Meir Yaakobi, Jakie Lind, and Gadi Kastelanetz. The squad was ambushed by a large Syrian force and captured. The soldiers were taken to the Maze prison in Damascus, where they were held in solitary confinement and subjected to brutal torture. Uri Ilan took his own life in prison. When his body was brought back to Israel, a note was found on his person: the words "I did not betray" had been punched into the paper.

More than a year later, once the army had secured enough Syrian prisoners, the other four were returned home. Upon their return, Meir Yaakobi and Meir Moses were forced to stand trial for divulging, under pain of torture, the nature of their mission. Sergeant Yaakobi, the commander of the mission, was charged with shameful and unbecoming surrender and revealing state secrets, and Moses was charged with divulging state secrets as well. My father, despite his stringent demands and his toughness, was outraged by the charges. Yaakobi's lawyer, Amiram Harlaff, wrote the following account:

In the spring of 1956 my childhood friend, Lieutenant-Colonel Ariel Sharon, the renowned commander of the Paratroops who is known as

Arik, asked if I would be willing to represent Sergeant Meir Yaakobi, his soldier, who had fallen captive to the Syrians while on a military operation across the border and who, after his return, was forced to stand trial. I told my friend Arik—"if it's a request from you, then I volunteer." He took me to the borderlands, beneath Tel Azizat, and pointed toward the mountains of the Golan and showed me through the binoculars a line of telephone poles at the top of the mountain and said, "I sent Sergeant Yaakobi to change the batteries that had been fixed to one of those poles."

The trial went on for many days. Since we refused to submit a guilty plea, the prosecution called Colonel David Elazar [later army Chief of Staff], the head of the history department of the IDF at the time, as a witness, and he testified, as an expert on fighting protocol, that it was unbecoming for a group of soldiers to surrender without a battle. Lieutenant-Colonel Ariel Sharon was called as a witness for the defense to dispute that assertion. He spoke of his considerable experience in face-to-face combat and agreed that, in theory, a soldier should fight and not surrender, but under what circumstances? Only when a fight is possible. If you are surprised and surrounded, as the squad was, it's better to surrender than to return in a coffin. According to his testimony, a burst of fire from them would have led to their immediate slaughter because they were surrounded by many soldiers armed with automatic weapons trained on them, and that in this type of extreme circumstance, surrender was the proper course of action.

Arik said, "Meir Yaakobi is made of the same kind of heroic stuff as Meir Har-Zion and if he surrendered—then there must have been no other option. I prefer to have him alive than in a coffin."

Arik was not afraid of testifying against the military establishment, which wanted a conviction at all costs, on both charges, sending out a precautionary signal and ensuring there will be no more surrenders within the ranks of the IDF.

I explained to him that he was putting his position [in the army] at risk, but he said, "I do not leave the wounded behind in the field, and I do not desert my soldiers when they stand trial."

Meir Yaakobi was convicted of unbecoming surrender and stripped of his rank. He was found not guilty of divulging classified information. Both sides appealed, and both appeals were accepted. The verdict was changed to not guilty on the charge of unbecoming surrender and guilty on the charge of divulging state secrets. Several weeks later, the Sinai War broke out. Meir Yaakobi was killed in battle, fighting bravely. Two months after his death, he was pardoned by the president of Israel, as requested by the chief of staff, Moshe Dayan.

In September and October of 1956, terrorists from Jordan entered Israel near the Dead Sea Works and in the Sharon region, several miles north of Kfar Malal, and killed several Israelis, posthumously mutilating their bodies. The Israeli response was an assault on the Jordanian military headquarters in Qalqilya. The number of reprisal raids and the fact that, since Qibiya, they were aimed only at military bases, along with the growing experience of the Arab troops and their readiness for a reaction after terrorist attacks, made the missions more complex and dangerous.

The IDF General Staff decided to limit the scope of the mission. At the last minute, as the paratroopers assembled, the General Staff altered some of the tactical elements of my father's plan, canceling the assaults on the nearby Zuffin Hill and the positions south and east of the police station. That left the forces susceptible to attack. My father did not accept then, or ever, the notion that the higher command could dictate to the officers in the field how to run a mission without being on the spot and seeing and understanding the battlefield situation. This came into sharper relief—leading to an even greater loss of life—seventeen years later, during the Yom Kippur War. "Their job is to say what to do and to interfere as little as possible in how to do it," I've often heard him say. They can cancel an operation, but keeping the objective the same while altering the tactical elements of the plan, without seeing the field and the dangers that lurk there, was, simply, a fatal mistake.

Israeli paratroopers leading the Jordanian army horses to safety
before blowing up the Qalikila, October 1956.
(© Avraham Vered, "BaMahane," courtesy of the
IDF and Defense Establishment Archive)

The mission was accomplished. The Jordanian army headquarters was taken and razed. Some one hundred Jordanian soldiers were killed, and another two hundred were wounded. But
we lost eighteen soldiers of our own, with another sixty wounded.
Among the dead was Yirmi Bardanov, Battalion 890's demolitions officer, who was already on severance leave from the army.
My father had stopped at his parents' farm to say hello before going on to the staging area. Bardanov found him there and asked
to join the action. My father told Yirmi, "You've already been in,

you've been wounded, and now you have your discharge." But Yirmi was adamant, and in the end my father allowed him to join the paratroopers for one more mission. While trying to rescue wounded soldiers under fire on Zuffin Hill, the position that the chief of staff had not allowed the paratroopers to take, Yirmi was shot and killed. When my father served as prime minister, Yirmi's brother sent him his Australian rancher hat, which my father hung in his bedroom.

The debriefing, as always, was forthright and charged. Although a representative from the General Staff was present, my father spoke openly about what he felt were the factors responsible for the mission's failures. The representative passed these sentiments on to Dayan, who, furious, gathered my father and the rest of the officers to reprimand my father. The result was different. All of the commanders supported my father, who spoke his mind again. Dayan did not like it. However, he did send my father to the informal briefing with newspaper editors to explain the number of casualties.

February 24, 1954

My Dear Little Girl!

I must head south today to take part in an operation. In addition, my lovely little girl, I am to parachute tomorrow. For these reasons you'll have to go by yourself to Jerusalem. Go tomorrow morning to the doctor, it's necessary, and we'll meet tomorrow afternoon, healthy, happy and sound. Lovely little girl, don't be angry at the fat and loving Arik. Soon we'll take our vacation and spend days together.

See you soon my dear one,
Yours Arik

My father in his office at the paratrooper base. From right:
Ben-Gurion, Shimon Peres, and an unnamed woman.
(Courtesy of the IDF and Defense Establishment Archive)

SIX

The Sinai War was planned by Ben-Gurion, the prime minister and defense minister, and Dayan, the chief of staff. Ben-Gurion believed that we needed to cooperate with the superpowers France and Great Britain. His mistake was that those were the powers of the past, both in decline. By 1956 the United States of America and the Soviet Union were the countries dictating world events. Israel's war objectives were as follows: to strike back against the terror stemming from Gaza; to attack the Egyptians for their activation and control of that terror; and to force open the Straits of Tiran, which the Egyptians had recently closed to Israeli vessels. France and Great Britain sought to reestablish control of the Suez Canal, which had been nationalized by President Gamal Abdel Nasser of Egypt.

My father (right) sharing a drink during the Sinai War, 1956. (© Avraham Vered, "BaMahane," courtesy of the IDF and Defense Establishment Archive)

A primary mission was assigned to the paratroopers. On October 29, 1956, Battalion 890, under the command of future chief of staff Rafael Eitan, was parachuted into the Sinai, close to 150 miles into Egyptian territory. They landed close to the Parker Memorial, alongside the eastern opening to the Mitla Pass. My father was to lead the rest of the brigade overland, battle through the Kuntilla, Themed, and Nahel Egyptian outposts, and then link up with the 890. Over the radio the men heard the army spokesman report that Israel had launched a reprisal raid against the terror bases in the Sinai—a deceptive attempt to paint what was happening as an ordinary reprisal raid and not an all-out war. Although about 150 miles of desert and enemy forces separated the 890 and the rest of the brigade from the Israeli border, they knew. "If Arik says he's coming, he's coming." And he did, within thirty hours.

The reunited brigade was positioned on the flatlands to the east of the Mitla Pass, exposed to attacking Egyptian warplanes and under the threat of an armored convoy that, according to the Israeli Air Force, had left the Bir Gafgafa region and was heading due south, toward them. This information troubled my father: he had only a few real antitank weapons, and the flatlands were an indefensible position for infantry against armor. He asked for and received permission to send a recon patrol into the pass. His orders were to avoid a battle.

The air force pilots had not seen any enemy presence in the pass. Resistance was deemed unlikely. The plan was for the recon force to enter the pass and secure the western entrance; the rest of the brigade would follow and dig itself into the mountains to defend against the Egyptian armor. At any rate, the seizure of both ends of the pass was the paratroopers' original mission, and that had not changed. However, at the last minute the parachute point was changed because of a report that Egyptian forces were on the west side.

Thus, the recon team came under fire from the dominant positions on the north and south sides of the pass.

Throughout the 31st of October, heroic attempts were made

to enter the killing zone and rescue the wounded while trying to snuff out the Egyptian forces burrowed deep into the cliffs and canyons of the desert pass. Only at night, in fierce face-to-face combat, were the Egyptians defeated. Some 260 of their soldiers were killed, but we lost 44 and had 120 wounded. Twenty-three soldiers were decorated for valor. Morning revealed that at some point during the night, the Egyptian armored force that had been going south had turned around and retreated.

The paratroopers' war did not end there. One battalion was sent to the oil refinery town of Ras Sudar; another was parachuted into A-Tor; and a third jump, into Sharm el-Sheikh, was canceled once the Ninth Brigade, under the command of Avraham Yoffe, made it there. All told, less than a week had passed since the 890 was sent into Egypt, and already the peninsula was in Israeli hands.

After the campaign, an inquiry was launched to determine whether my father had acted properly or disregarded an order. Dayan felt that my father had sent too large a force into the pass and had sought a battle, despite orders to the contrary. When the commission completed its hearings, Ben-Gurion summoned my father and asked him: "If you had to make the decision right now, what would you do?" "You know, sitting here in your warm room drinking a nice cup of tea, and having all the information, maybe I would have done something different, but that day, I was commanding all those people in that open area, with almost no anti-tank guns, with no news about what the French and British were doing, and with the nearest Israeli forces a hundred miles away. In that situation I was by myself, and responsible for all those people. I had to make the decision, and my judgment then was that we would have to move deep into the pass and do whatever was necessary to defend ourselves," my father said.

Ben-Gurion's ruling—"I don't feel that I'm in a position to judge between two commanders on this issue"—was a blow to Dayan. The two commanders were not peers; one was a twenty-eight-year-old brigade commander, the other, the chief of staff.

Soldiers that had been caught in the ambush and others that

had fought among the rocky crags of the Mitla—men like Medal
of Honor winner Dan Ziv, who saved wounded men pinned under
murderous fire; Avsha Adam, who received a citation for bravery
for his action that night; and Moni Meroz of Beit Hashita, who
single-handedly cleaned out a fortified machine gun nest, killing
eight Egyptian soldiers—all have remained close to my father to
this very day.

As prime minister and defense minister, David Ben-Gurion
visited the Paratroopers Brigade on April 25, 1957, and said the
following:

> *Friends, in terms of the unit itself I couldn't possibly add a single
> word to what Arik has already said. Nor could I say the things that
> he told you here, things that were edifying to me. He founded this
> unit, trained it, stood at the head of its operations, and perhaps most
> important of all—served as a personal, living example . . .*
>
> *You know what the American general Marshall said of this op-
> eration, that it was a battle that was unparalleled in the world.
> Perhaps he went a bit too far in his praise of the operation, but he
> was not far from the proper definition.*
>
> *There is no doubt that the chief of staff and his loyal and skilled
> aides have played no small role in [building] the IDF's capabilities,
> but there is no doubt that the operations carried out by this unit over
> the past three years played a major role in terms of the IDF and in
> terms of the nation. And I have no doubt at all that the successful
> operations carried out by this unit over several years entrenched in
> the people a great faith in the IDF. This is an enormous asset whose
> worth can hardly be overstated. But to no less an extent was this
> unit's influence on the IDF itself.*
>
> *One of the main goals of this Sinai campaign was the crossing of the
> Red Sea. We decided to forge a path to Eilat, the Red Sea, the Indian
> Ocean, and indeed during the previous month the port in Eilat has
> received more tonnage than in previous years in Eilat. Yesterday was a
> great day for Beersheba; the oil brought to Eilat flowed into Beersheba.*

And I can let you in on a secret, and this is not a military secret but a civilian one, and you must keep this secret to yourselves, this may not be brought to the press: another oil tanker is on its way to Eilat.

———————

May 30, 1957

Hello Little Gur [Gur was five months old],
  Your fat father will not be back before eleven, eleven thirty tonight, because he will be watching a drill prepared for IDF officers.
  Daddy terribly, terribly misses you and waits for the moment when he can see you again happy and smiling. Send warm regards to Mommy, I miss her very much too and am willing to miss all the drills in the world for her.

Many Kisses,
Yours, Arik

———————

During the coming years, there was quiet. One year later, my father finished his service as paratroop brigade commander and was sent by the army to the British army's staff college in Camberley, England. Throughout his time there, he kept up a correspondence with Ben-Gurion's associates.

My father always said he had no trouble remembering when he was promoted to the rank of colonel: "On the day I joined the party [Israel's ruling party, Mapai]. All I have to do is look at my membership card." The army was heavily politicized then, and although he had been a brigade commander, a post that goes hand in hand with the rank of colonel, he had filled the position as a lieutenant colonel, and was promoted only a year after finishing his term as commander of the Paratroopers Brigade. It took a long time for the paratrooper unit to be declared a brigade and the title of Commander of Brigade took even longer. The General Staff was not eager to see my father rise in rank. But if earning the rank of colonel was difficult, general was far harder. It took close to a decade.

My father's path to promotion was blocked by two events. The first was the November 1957 suicide of Ben-Gurion's military attaché Nehemia Argov, who was close to my father. Second was the continuing fallout for Ben-Gurion from the Lavon Affair, a failed Israeli covert operation that took place in Egypt in the summer of 1954. Ben-Gurion had intended to promote my father all the way to the top, to IDF chief of staff, but he was distracted by the Lavon Affair, and my father was lucky to find himself still wearing a uniform.

November 21, 1957

State of Israel
Ministry of Defense
Lieutenant Colonel Ariel Sharon

Arik my dear, much thanks for your letter, I was happy to hear from you and without a doubt I shall be happy to "keep you in the picture." I've asked that you be sent now and again news or material that might be of interest to you. There is no doubt that the passing of Nehemia, may his memory be blessed, was and is a great blow, acutely painful to us all and to "the old man" in particular. I'll happily send "the old man" your get-well wishes and continue to write. [In his own handwriting he adds] In the meanwhile I gave your [get-well wishes] and he is most interested to know how you are doing.

In Friendship
Shimon Peres

Director General

December 6, 1957

State of Israel
Prime Minister's Office

Hello Arik, and Blessings, only now have I found some time,
[UN Secretary-General] Hammarskjöld has done no small share of
nagging. This time his visit was truly a blessing. The small mini
king who sits in Amman [King Hussein] sank into a swamp, and
he came and dragged him out. We could have done it with the force
of our arms, but it is better this way. The convoy went up to Mount
Scopus and all is in order and at peace, for now.

The Nehemia disaster was crushing. The old man has recuper-
ated some and gone back to work "as usual," most vigorously.

Unfortunately I cannot put down in writing much of what is
happening and what people are thinking. In the meantime, the world
carries on as usual: parties are already preparing for elections, if the
mutual defamations are any indication.

I will of course send regards to the old man. I'll be delighted if you
write details about the entire family.

Yours, in Friendship,
Yitzhak [Yitzhak Navon, prime minister's office chief of staff at
the time]

P.S. In the meantime, sit and learn. I will remember when to call you.

My father spent many years in limbo in the army. Despite his efforts, he wasn't able to advance, and despite the army's desire
to do so, it wasn't able to get rid of him. During those years he was
sent to serve in the armored corps, which turned out, quite clearly,
not to have been a waste of time. During the Six-Day War, his familiarity with commanding armor led the division to remarkable

accomplishments in the Sinai, and in the Yom Kippur War he commanded the complex battle during the crossing of the Suez Canal.

During Haim Laskov's three-year term as chief of staff, beginning in January 1958, my father had no chance at promotion. He served as commander of the Infantry School and of a reserves brigade.

My father's desire to extricate himself from the deep freeze led him to the house of our neighbor on Yoav Street in the Tzahala neighborhood of Tel Aviv—Moshe Dayan, who had since retired from the army.

I remember Dayan's house through the eyes of a child. He kept a lioness in the backyard, and that made a deep impression on me back then. At the time, Dayan was minister of agriculture, and my father came to ask his advice. "I found him in the archaeological garden behind his house, gluing together ancient pottery shards," my father said. "As I talked, he raised his head slightly and regarded me from one narrowed eye. 'Arik,' he said, his voice tired, his eye closing to a slit, 'they will only let you out if a catastrophe occurs.'"

In early 1961 Tzvi "Chera" Tzur was appointed chief of staff. His son, Dani, was friends with Gur. They were in sixth grade when Gur was killed. It is now more than forty years later, and Dani still comes to the memorials. I try to remember the kids who were in my sixth-grade class, and I think about Gur—what a remarkable kid he must have been if his friend, now over fifty years old, still comes to his grave site year after year! I think of the loyal Dani and how his eleven-year-old heart must have been broken. I think of the strength of their friendship. Something of Gur stays alive if his friend still remembers how they used to play and misses him. Our fathers were not such good friends, but my feelings for Dani surely temper what I have to say about his father's reign as chief of staff, during which, it was clear, he wanted to oust my father from the army. In the end, many of the officers who did their utmost to hold up my father's advancement and force him out of the service left no real mark of their own.

In 1962, under pressure from Ben-Gurion, Tzur finally appointed my father commander of an armed brigade

SEVEN

From left: Deputy Chief of Staff Chaim Bar-Lev, my father,
and the commander of the southern front Yeshayahu Gavish,
June 1, 1967. (© David Rubinger,
Israeli Government Press Office)

I n early 1964 Yitzhak Rabin was appointed chief of staff. Ben-
Gurion asked him to be sure to promote my father. "You know
I have a special regard for Arik Sharon," Ben-Gurion said to
Rabin before stepping down as prime minister and defense minis-
ter in 1963. Levi Eshkol was prime minister.

In 1964 Rabin appointed my father to the position of chief

staff officer of the Northern Command, under Major General Avraham Yoffe. At the time there was constant friction with the Syrians, who fired on our farmers as they worked the fields in the demilitarized zone and sought to alter the flow of the Jordan River, drying up one of Israel's primary water sources.

It was an intense period of action, and my father's relationship with Yoffe was excellent. Avraham, a big man, the son of farmers from Yavniel, had confidence to spare and not a jealous bone in his body, and that was why he had no problem with a strong and assertive chief of staff.

In late 1964 Avraham Yoffe retired from the army and founded the Israel Nature and Parks Reserve. There was no better candidate for the position—Yoffe was a devout lover of nature. David "Dado" Elazar, the man who would be IDF chief of staff during the Yom Kippur War, was made the new commander on the northern front.

For my father this meant a return to the needless intrigues and ceaseless attempts to halt his military career. Once again, this campaign was led by a former Palmach officer who had spent most of the 1950s as a staff officer, looking on at the paratroopers' missions from the sidelines. My father was not part of their clique.

My father took a leave and traveled with Yoffe. Their destination was the nature reserves of East Africa. They spent an unforgettable five-week vacation traveling through Uganda, Kenya, Tanzania, and Ethiopia. I remember my father telling the story of how he and Yoffe, two large men, crammed into a VW Beetle, had gotten stuck deep in the Danakil Desert in Ethiopia. Yoffe went to get help, and my father stayed with the car. He had heard of a Danakil custom: "There is not a nicer present that a Danakil man can bring to his fiancée than the testicles of the enemy. She proudly hangs this precious gift on her forehead." He had no intention of leaving any souvenirs behind. After all the wars and operations for the country, it seemed like complete stupidity to end up like that in a godforsaken place in Africa. Eventually, at dusk, a convoy of Land Rovers approached, and an antimalarial team scolded my father and rescued him and Yoffe.

On the way from Jinja to Moroto, Uganda, December 29, 1964, his little orange notebooks also noted:

December 31, 1964, we headed out on a boat up the Victoria Nile, having sent the car to wait for us on the other end of the river. On the banks we saw, in utter tranquillity, hundreds of elephants, hundreds of hippos, dozens of alligators, many with their jaws open wide. (The women in the boat looked sorrowfully at the alligators and then back at their plastic wallets . . .)

January 2, 1965, Jinja, a nearly full company parade + band. Waiting and waiting, at long last minister of interior and commander of the army Brigadier General Apollota. Minister of Interior Anmo is also minister of defense. I sat beside Colonel [Idi] Amin, enormous, deputy commander of the army.

In October 1965 my father completed his term of service in the Northern Command and awaited his next appointment. Rabin delayed his decision for three months. My father spent much of that time on the back of Gur's horse, Ayala, which Gur had received for his ninth birthday. I was not yet born at the time, and my parents, Gur, and Omri lived in Zevele Slutsky's house in Nahalal, near the Northern Command headquarters in Nazareth.

Eventually, Rabin called my father to a meeting. He enumerated all the mistakes that he felt my father had made. Listening to the lecture, my father was sure his military career had come to an end. But after all Rabin's criticism, much to my father's surprise, Rabin informed him that he was being promoted to the rank of major general and that he would serve as the commander of a reserves division and as director of military training. That was it: he had been made a general. Credit should be given to Rabin for extracting my father from the deep freeze. In the end, Rabin and Israel benefited; my father's military prowess was one of the major factors in Rabin's victory as chief of staff during the Six-Day War.

When Ben-Gurion heard the news, he sent my father a handwritten note:

Sde Boker, February 20, 1967:

Dear outstanding Arik—
I was happy to hear you became a general. In my eyes you were already a general several years before . . . There are exploits yet ahead of you!

In esteem and in friendship,
D. Ben-Gurion

Over the years, my father and Rabin maintained a good personal relationship. Rabin was his commander during the battalion commander course in the early 1950s, chief of staff during the Six-Day War, and a two-term prime minister. While he was prime minister in the 1970s, Rabin asked my father to serve as his adviser on the war on terror. "He's a real redhead," my father used to say about Rabin, but he always treated him with respect. That said, he knew Rabin was not a strong man. An example of this was the way he allowed himself to be manipulated during his second term as prime minister by Shimon Peres, who pushed him into positions he did not want to be in.

*Briefing before the Six-Day War, May 15, 1967. From
left to right: Chief of Staff Rabin, Prime Minister Eshkol,
Minister Yigal Allon. In rear, from left: Ariel Sharon,
Chaim Bar-Lev, and Avraham Yoffe. (© Hanania Miller,
courtesy of the IDF and Defense Establishment Archive)*

On May 14, 1967, as my father and other senior officers were
attending the Independence Day parade, word reached
them that Egyptian forces had advanced into the Sinai. The fact
that large numbers of Egyptian troops were flowing into the Sinai, after years of relative quiet and UN supervision and a reduction in the terror coming out of Gaza, was surprising.

Within days my father was in the Negev along with his reserve
division, in charge of the central sector on the Egyptian front.
Thus began the three weeks of waiting. Rabin, who had built up
the IDF and readied it for war, did not project the necessary confidence to prime minister and defense minister Levi Eshkol, who
was not an army man. Nasser boasted openly of wiping the Zionist entity off the map. On May 19 the UN acceded to Nasser's
demands and without protest withdrew its peacekeeping forces
from the Sinai.

On May 22 Nasser once again shut the Straits of Tiran to Israeli vessels, a move that had long been considered a casus belli. Eshkol, along with Minister Yigal Allon, Chief of Staff Rabin, and the heads of the IDF, visited the Southern Command headquarters. My father's close friend Yisrael "Talik" Tal, the commander of the division in Gaza and the northern Sinai, presented the battle plan to Eshkol and company. "Talik is a man who could take apart every single screw in a tank and lay them out on a military blanket and then put them back," is how my father always described him. But this skill did not serve him well in the briefing. Rather than lay out the main elements of the battle plan, Talik went into obscure detail, throwing around all sorts of military acronyms. My father heard Eshkol lean in toward Allon and whisper in Yiddish, "Vas is das?" ["What is this?"] about one of the acronyms. When Talik described how the Israeli shells would penetrate the Egyptian T-34s, belaboring the point about the angle of impact, my father heard Eshkol whisper, "And what if they don't stand at just the right angle?" My father noted, "I saw right then and there how the self-confidence of the nation was eroded."

The plan put forth by the Israeli army called for an invasion in stages: first the capture of Gaza and the demand that Egypt open the Straits, and then, if they refused, a further, incremental offensive. The option of striking hard and fast against Egypt was not presented, or if it was presented, it was done in such a halfhearted manner that it did not allow the leadership to seriously contemplate that option.

When my father was given permission to speak, he did so forcefully. He said that he believed that the Israeli army could deal a decisive blow to the Egyptian forces. He was not putting on a show. He truly believed it. He thought the phased attack was a dreadful mistake, not least because after the first phase, Israel would come under immense American and Soviet pressure to cease hostilities, and would have a hard time reacting with force if Nasser still refused to open the Straits. What's more, the army, then as now, was mostly an army of reserve troops, and when

PRE-1967 BORDERS

LEBANON

SYRIA

Haifa

*Mediterranean
Sea*

Sea of
Galilee

Jordan River

Tel Aviv
Jaffa

Jerusalem

Dead
Sea

GAZA

Beersheba

ISRAEL

TRANSJORDAN

EGYPT

N

0   40 km

Jordanian occupied

Egyptian occupied

there was a general call-up, the nation was brought to a standstill. It would be impossible for Israel to remain in that situation for long. The idea, he said, was a disaster.

After the meeting, Eshkol called my father over and said to him quietly, "Arik, what you are saying is irresponsible. You are irresponsible." (After the war, he explained to my father that everyone had told him that the most the army could hope to accomplish was the capture of Gaza.)

Several days later, Eshkol spoke to the nation in a radio address known in Israel as the Stuttering Speech. The prime minister managed to sound so terribly unsure of himself that by the time the speech was over, the leadership crisis was readily apparent to Israel's citizens and enemies alike.

That evening the division commanders were summoned to IDF headquarters in Tel Aviv. My father, Talik, and Avraham Yoffe, who had been called back to service, all felt that the lack of leadership at the political level placed the responsibility on their shoulders. Eshkol, who looked helpless, defeated, and distraught, updated them on the situation. Major generals Yoffe, Peled, and my father all spoke of the need to act without delay. Each passing day, they said, would allow the Egyptians to further entrench themselves in the desert. My father made the case that the Egyptian army could be defeated; though ordering the nation to war was difficult, the conflict had been forced on us, and our situation was worsening with each passing day. He also spoke of the dangers of losing our national self-confidence and Israel's ability to deter its neighbors: the country's deterrence had been painstakingly built over the course of years, but it could be hollowed out overnight.

In general this is a topic I heard him speak of often—the manner and the time necessary to build deterrence and the possibility that it could all be washed away in an instant. Deterrence is an abstract concept, but nonetheless it is what stops an enemy from attacking you: the fear that you have managed to place in his heart regarding the price of aggression. If an enemy does cross a line, and nothing is done, a country's deterrence has been stripped of all potency. The whole thing reminds me of our cows: they graze in a pasture ringed by a single strip of electric wire, and they are very careful not to touch it, but if they happen on it and are not shocked by the electrical current, then all fear for the thin metal wire is lost.

Toward the end of the discussion at headquarters, Eshkol asked my father if he thought the situation would change if he appointed Moshe Dayan to the post of defense minister. "As far as I'm concerned," my father told the prime minister, "you could appoint Beba Idelson [an elderly leader of the women's labor union] to the position of defense minister. We will fight the same way. It makes no difference." Deep down inside my father felt that the political leadership with Eshkol at its head was not at fault. It was the military officers, he felt, who were responsible for all of the doubt and hesitancy. Israel's military leadership lacked confidence in its own abilities. For years he had encouraged the army higher-ups not to wait for the political leadership to push them to action. "The political cchelon," he would say, "has to have the freedom to make choices, to take either political or military steps. Our job is to give them the freedom to decide."

Another reason for my father's desire to take immediate action was his understanding of the human mind. The Egyptian soldiers were from the lush Nile Delta: they hated the desert, and they feared it. Each passing day allowed them time to become acquainted with it. This was reinforced several days later when five Egyptian soldiers crossed the border inadvertently and were taken captive. The five, two enlisted men and three officers, one of whom was a lieutenant colonel, had lost their way while driving their Russian jeep and had fallen straight into Israeli hands. While questioning them, my father realized that they were uncomfortable in the desert, and he drew this conclusion: the enemy was lacking in fighting spirit, and the IDF had the ability to defeat them. His soldiers drew other conclusions. They were shocked by the divide between the Egyptian officers and enlisted men. "You should see the difference between them," one Israeli officer told a young Yael Dayan, an IDF spokesman-reporter embedded with my father's division and the daughter of Moshe Dayan; "the officers are clean and well kempt . . . they had perfumed handkerchiefs that they gave us in order to blindfold them . . . the enlisted men were unshaven, dirty, dressed in rags." Of my father's men, she wrote, "The soldiers knew that Major General Sharon and his driver Yoram were outfitted with the same boots and uniforms that they wore and that the brigade commanders and the simplest infantrymen shared the same C-rations in the field."

The days of waiting were not whiled away in my father's division. He put his reservists through intensive drills and close inspections, making sure they maintained discipline and presented themselves like proper soldiers.

In the meanwhile, my father worked on his battle plan. The skittishness at the top of the chain of command did not trickle down to him, although it did take the form of constant changes and reductions in the division's manpower. My father never said, "Without such and such force the job will be impossible." He simply fit the plan to the forces he had.

In preparation for the June 2 General Staff meeting with the government ministers, my father wrote down his main points in a little notebook:

> *The objective—the destruction of forces in order to restore the IDF's deterrence—the hesitations have depleted it, this has long-term significance.*
>
> *Not to get involved in peripheral affairs, all matters not focused on the primary objective—the Egyptian army—will make us seem weak.*
>
> *There is a moral justification for war. We are in a situation of no choice.*
>
> *Joining forces with other powers, when it is not militarily necessary, makes us seem helpless and we lose some of our deterrence for the future.*

The longer we wait, the more difficult and more costly the fighting.

If we do not firmly and constantly stand guard over our rights—we will not continue to exist here for long—I do not consider our running around and our calls for "help" from the nations of the world part of standing guard over our rights.

In wars between us and the Arabs the key is not to numerically match tank against tank and plane against plane but the momentum and speed of the strike. In that realm we have overwhelming superiority.

Under no circumstances can we delay the launch of our offensive in order to receive more planes and more tanks. We have sufficient strength to destroy the Egyptian army and to face the others. All we need to do is decide, and fast.

You decide, we'll do the work, and you can count on us.

A few days later these promises would be kept.

On Friday, June 2, during that meeting in Tel Aviv with the entire cabinet, including Dayan, who had since been appointed minister of defense, my father passed Dayan a note. "Moshe, it seems to me that the plan is still to move in phases. In my opinion we should not undertake an operation that will not break the Egyptian main forces. Gaza is not the target!"

Dayan wrote back, "Arik, I've asked Yitzhak [Rabin] to meet this evening to discuss the plans."

The division's primary task was to open the central axis road, which led from Nitzana on the Israeli side of the Sinai to Ismailia on the Suez Canal, on the other side of the Sinai. The key was Abu-Ageila, which loomed directly over the road. Once that had been taken, the other stronghold of Kusseima could be more easily handled. All told, there were five Egyptian divisions in the Sinai to Israel's three.

The Egyptian Second Division held the Abu-Ageila and Kusseima strongholds. My father's fighting force, Division 38, in number of soldiers was similar to the Second Division, but the Egyptians had far more firepower, meaning that my father's

troops were nowhere near the offensive-defensive ratio of three
to one normally considered to be the bare minimum for an attack against fortified positions. The Egyptians had eighty artillery
guns, which had a larger range than our own.

My father cloistered himself and began working on the battle
plan. He always believed that the early stages of planning, when
the basic approach to the battle had not yet been determined, was
when the commander was at his most vulnerable and therefore
should remain alone. He did not yet have a sense of how he would
accomplish his mission. Doubts, at this stage, could loom large.
His men did not need to see his doubts. Only when the basic approach to the battle had taken shape in his mind did he bring in
the rest of the staff to work out the details of the plan.

His plan was bold and original and, while extraordinarily
complicated on the divisional level, simple and clear but difficult
for each of the attacking forces. The battle would be fought at
night and include a deceptive assault, blocking forces that would
isolate the scene of battle (just as he had always done with the 101,
only now on a grand scale), flanking maneuvers, and coordinated
simultaneous attacks from all sides.

*Forces from Sharon's division moving into Sinai at the
beginning of the Six-Day War, June 5, 1967 (© Benny
Hader, courtesy of the IDF and Defense Establishment Archive)*

The battle required that as division commander he control every element of a large and unwieldy force, that he move the different units through the stages of battle with the precision and control of a conductor. War began the morning of June 5, 1967. My father watched as air force planes struck the stronghold, and he saw the streaks of antiaircraft fire rise to meet them. Soon enough an Israeli jet was downed, and my father called off the air force. "I wasn't opposed to close air support, but I felt that here it wasn't absolutely necessary, that we had other answers. There was no point in endangering the precious fighters."

I think of him then, hours before committing his division to a complex night battle, and of his decision to call off the air support. God, what confidence, what icy calm the man possessed.

By the afternoon his forces had taken the outposts on the eastern side of the stronghold. The screening force of tanks, half-tracks, and mortars took up positions to the south, isolating the battlefield and blocking the advance of reinforcements from Kusseima. The false attack force was situated to the south, too, near Kusseima. In the meantime, a battalion of Centurion tanks under the command of Natke Nir flanked in a northerly direction, via the dunes, and attacked the heavily fortified northern outpost. My father could hear Natke facing difficulties over the radio, but the commander did not ask for assistance, and my father did not intervene. By four in the afternoon, Natke's tanks had broken through, taken the position. The tanks, now refueled, rumbled forward, cutting Abu-Ageila off from reinforcements. With the battlefield isolated, the rest of the division's forces moved into their attack positions.

Yekutiel "Kuti" Adam's infantry brigade arrived on civilian buses. That's what they had back then. They came as close as they could along the road and then got off. Looking back, my father saw a long row of buses that had been towed to the side of the road, ensuring that the route would remain open.

In the dark my father and the rest of his forward command rolled across the dunes and parked alongside the artillery commander, Yaakov Aknin. The two could speak face-to-face. Before them were the coils of barbed wire that marked the beginning of the minefield. From that vantage point, my father could watch the coming Israeli tank offensive.

Everyone awaited H-hour. But just then a proposition came over the radio from the Southern Command headquarters: if the division would delay the attack until daybreak, the air force could provide support. My father considered the suggestion for two minutes and then radioed back his response: "We'll attack tonight."

Slightly after ten at night, my father turned to Aknin and said, "Make the whole thing tremble." "Oh, it will tremble all right," Aknin responded.

Over the next twenty minutes, some six thousand artillery shells exploded in and around the infantry trench, softening it for Kuti Adam's infantry brigade. At the same time, Dani Matt's brigade of reserve paratroopers flew over the dunes in helicopters, witnessing the awesome power of the artillery shells. Mat, who had been an officer in the Paratroopers Brigade during my father's day, was airlifted with his men into the rear of the Egyptian position. His force's objective was to attack and silence the Egyptian artillery. My father kept an officer glued to each of the radio frequencies.

A problem arose. The two minesweepers could not be found amid the battalion of Sherman tanks. This was a critical moment in the battle: the paratroopers were attacking from the rear, the infantry were charging straight down the throat of the Egyptian defenses, and the tanks were held up, looking for the minesweepers. The problem was solved the old-fashioned way, clearing a path for the tanks by hand.

The Egyptian force—ninety tanks—was trapped between Natke's Centurions to the east and Mordechai Zippori's brigade to the west. Suddenly Natke reported that he was taking fire from an easterly direction, and he wasn't sure if it was coming from our forces or the Egyptians.

My father used the radio and told Zippori to hold his fire. Then he took the other radio and asked Natke if he was still under attack. When Natke said yes, my father said, "Then nail them. They're Egyptian."

I heard my father tell that story more than once when he wanted to demonstrate the importance of command and control in battle. His control over his forces was so absolute that he could, during a battle, stop the fire of a whole tank brigade, ensure there was no chance of friendly fire, then coordinate a joint attack against the Egyptians between them.

With the fighting still ongoing, Southern Command headquarters asked my father if one of Yoffe's brigades could use the now-open route through the stronghold. My father approved and ordered all forces north of the road to hold their fire once Yoffe's tanks had crossed through the minefield. In the middle of battle, my father recalled, he saw the unbelievable sight of a brigade of armor rumble through the battlefield unscathed. I remember him talking about that sight. In the morning, there was only sporadic fire from the trenches. By noon, there was absolute silence.

In the morning hours after the battle, my father took over the difficult task of calling in the air force helicopters to evacuate the wounded. Amid the sprawling flames and the blowing smoke, it was hard for the pilots to find them.

All told, the division had lost 40 men. Another 140 were wounded. Many more Egyptians lost their lives in that battle. The mission had been accomplished: the central axis route into the Sinai lay open. Kusseima had fallen, too. The last of the Egyptian forces had fled into the desert. The backbone of their defense had been snapped, and their spirit had been broken.

An Egyptian officer, a prisoner of war, described the attack from his perspective in Um-Katef. "It was like watching a snake of fire uncoiling at us," he said. Another prisoner of war, an Egyptian battalion commander, complained that the battle had not been fairly fought: the assault had come from the flanks and from the rear, and not from the front as they had expected.

It's hard to blame him: tanks from the rear, tanks from the front, paratroopers from above, infantry from the flank, and artillery in thunderous bursts, all at the same time and at night. But even after this towering achievement, my father did not let up. He immediately began to apply pressure and demand new missions from the command. Yael Dayan described his actions: "The majority of the Egyptian armored force was still in the central Sinai, almost entirely intact—over five hundred tanks that could still wage war. Arik was raring to go, eager to proceed as fast as possible to the next battle, before the enemy awakened, before some general took the initiative and stirred the fighting spirit in his troops. The enemy cannot be given any time for rest or recuperation. Speaking over the wireless radio with Shaykeh [OC Southern Command Yeshayahu Gavish], he demanded new missions." His thought process was not merely that of a division commander but of someone with a far wider perspective.

It was the first IDF divisional night battle. Since then it has been taught in military academies all over the world. The second divisional night battle—and the last until now—was also commanded by my father, during the Yom Kippur War. Both were made up of reservists.

The day after the battle, June 7, with the entire division moving south, my father noticed that an Israeli soldier was beating an Egyptian captive. Furious, he tried the soldier right there in the field and sent him straight to the stockade. "In battle you fight and have to kill. That's the nature of it. But once a man is your prisoner, you never touch him."

As the troops moved south, they came across a recently abandoned Egyptian position. Suddenly, three Egyptian soldiers came out of hiding. "They left us, they left us," they cried. My father asked who had left them. "Our officers. They left us here. They just left us." My father noted, "The three of them could have stayed safely hidden, but they were frightened and desperate. They looked to me like orphans."

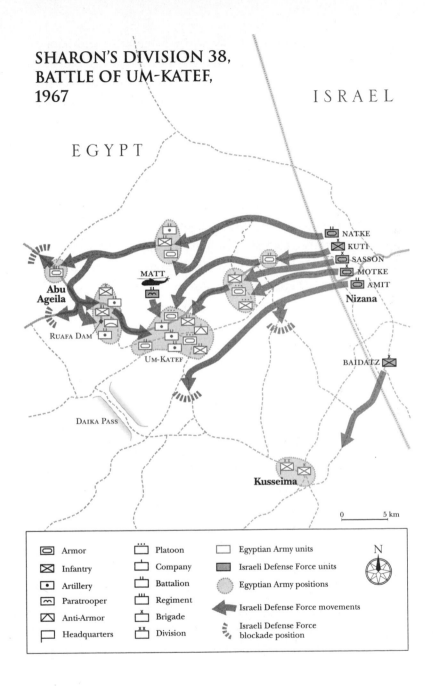

SHARON'S DIVISION 38,
BATTLE OF UM-KATEF,
1967

ISRAEL

EGYPT

NATKE
KUTI
SASSON
MOTKE
AMIT

Nizana

MATT

Abu
Ageila

RUAFA DAM

UM-KATEF

BAIDATZ

DAIKA PASS

Kusseima

0   5 km

Armor
Infantry
Artillery
Paratrooper
Anti-Armor
Headquarters

Platoon
Company
Battalion
Regiment
Brigade
Division

Egyptian Army units
Israeli Defense Force units
Egyptian Army positions
Israeli Defense Force movements
Israeli Defense Force
blockade position

N

The other major action of the war for my father's division was the ambush of the Egyptian Sixth Division in Nakhel. On the morning of June 8 my father received a report that the Egyptian division was in retreat from Kuntilla toward the Mitla Pass, the exact route he had traveled eleven years before when linking up with the paratroopers who had been dropped near the Parker Memorial. The road to Mitla runs through Nakhel, an oasis and an ancient junction that was then dotted with a few houses and an army encampment. While eating his C-rations in the field, my father devised his battle plan. The fleeing Egyptian division was being pursued by a brigade that had come under my father's command. He placed Brigade 14 to the west of the retreating Egyptian division, the battalion of Centurions to the south of the route. The reconnaissance unit, north of the route, attacked the enemy from behind. The result was a killing field. The division, fleeing west, had no chance. By the end of the battle the enemy had lost some 1,000 men, 60–70 tanks, 100 artillery guns, and 300–400 trucks.

The scene of this Egyptian defeat was gruesome, and it left an indelible mark on my father. My mother said that he came back from the war a different man. Yael Dayan described him in her memoir of that war: "Arik, the legendary warrior . . . his face covered in a sort of gray sheen. 'It's awful,' he said, 'very difficult to bear.'"

In four days the entire Sinai Peninsula had been taken. The Egyptian army had been crushed.

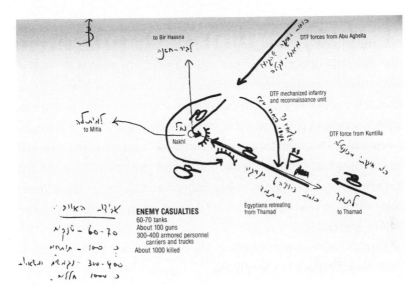

to Bir Hassna

DTF forces from Abu Agheila

DTF mechanized infantry
and reconnaissance unit

to Mitla

DTF force from Kuntilla

Nakhl

Egyptians retreating
from Thamad

to Thamad

ENEMY CASUALTIES
60-70 tanks
About 100 guns
300-400 armored personnel
carriers and trucks
About 1000 killed

*Map drawn by my father of the Battle of Nakhl,
the war of 1967.*

On Saturday, June 10, my father was called to meet with Yeshayahu Gavish, the commander of the southern front. He set off in a two-seater helicopter with no doors for Bir Khassaneh, where Gavish was supposedly waiting for him. "The desert was dotted with figures on foot making their way across the sands to the canal," he recalled. He could see them, the remains of the Egyptian army, from above, heading toward home after the collapse of their units. Occasionally small arms fire rose toward the helicopter. At Bir Khassaneh he learned that Gavish's command center wasn't there. Not long after they were airborne again, going toward Bir Tmade, the pilot informed my father that he was experiencing engine trouble, and they would have to land the aircraft immediately. As they began to descend, the bands of retreating Egyptians started to open fire, and my father and the pilot responded in kind. The helicopter landed on the road and my father considered the irony. "Less than an hour ago I had been secure in the bosom of my division, now I was stuck in the middle of hundreds of desperate armed Egyptians. I felt as I had back in the Danakil Desert in Ethiopia, when the car broke down."

A moment later a large helicopter came into view. Signal flares to mark my father's and the pilot's location were lit, but the helicopter coasted past and then, in a movement that my father remembers as beautiful, banked hard and began to turn toward them. As soon as the helicopter touched the ground, Avraham Yoffe and Talik jumped out. My father hadn't seen either of them since the start of the war, and the three men, most uncharacteristically, hugged.

From Gavish's headquarters they flew to Tel Aviv. Chief of Staff Rabin wanted to see them.

My mother had been told that my father was coming to Tel Aviv, and she waited for him along with Gur and Omri at the airfield. She drove him to army headquarters. I was seven months old, and of course remember nothing. I have heard stories about the sandbags that people filled and stacked all along the Tzahala neighborhood of Tel Aviv, and about our neighbor's parking garage, which the man had turned into a makeshift shelter from artillery rounds. Indeed several Jordanian rounds had been fired into Tel Aviv. My brother Omri, two years older than me, remembers the sandbags vaguely. Our grandmother Vera had come into the city to help out back then, and was, as always, silent and strong.

My father had not been home in a month. The streets of Tel Aviv looked foreign to him. He was jolted by the sudden, sharp swing between the comforts of Tel Aviv and the dreadful things he had seen in the desert.

The meeting with Rabin was full of mutual congratulations and genuine warmth, and by the end my father had been appointed commander of the Sinai Peninsula. The three officers were flown back to the Sinai. Talik got off at Bir Gafgafa. Yoffe got off next. Finally my father deplaned at Nakhel, where he set up his headquarters. He felt as though he had come home.

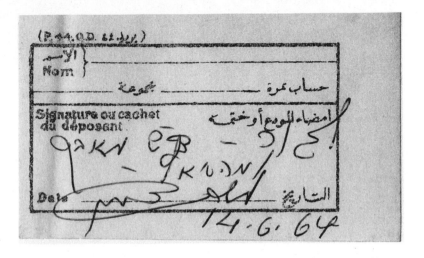

"*To Gur—Regards from your father and from the Suez—Moshe Dayan, June 14, 1967*"

That's how Dayan sent regards to Gur, on a tiny Arabic receipt slip that had been found in one of the Egyptian army bases in the Sinai.

June 15, 1967

Hello, my love,

From the distant Sinai I send you all kisses and love and many many hugs. I hope to finish up here in a few days and be with you again. The address in Arabic on the left says [on the stationery of the Egyptian governor of the Sinai] "Sinai land of victory" . . . I love you and think of you without end. I think we have done a great thing that will influence our fate for many many years to come. I will be home for two days this coming Sunday.

Many kisses to you, beloved Guraleh, wonderful Omri and Gilad who does not yet know what is happening and is delightful and is surely bringing you all a lot of joy.

Kisses and love always, Yours,
Arik

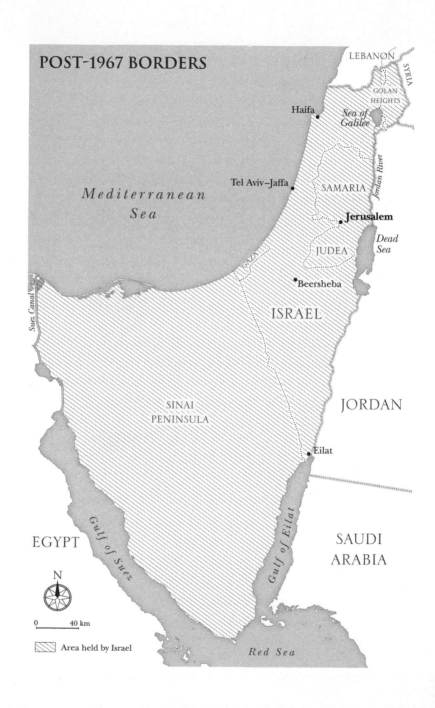

POST-1967 BORDERS

LEBANON

SYRIA

GOLAN
HEIGHTS

Haifa

Sea of
Galilee

*Mediterranean
Sea*

Jordan River

Tel Aviv–Jaffa

SAMARIA

Jerusalem

Dead
Sea

JUDEA

GAZA

Beersheba

ISRAEL

Suez Canal

JORDAN

SINAI
PENINSULA

Eilat

EGYPT

Gulf of Suez

Gulf of Eilat

SAUDI

ARABIA

N

0   40 km

Red Sea

Area held by Israel

EIGHT

With my father visiting Sheikh Ode Abu Momnar, July 1970.

After the war, the division's soldiers—reservists—returned home. They had proved that civilians could execute a complex battle plan and defeat a standing army, but now it was time to go home, back to their lives.

My father returned to his general headquarters job as head of military training. He and the field commanders went over the battles that had been fought on all fronts, trying to draw as many lessons as possible from the experience. As head of all IDF training, he was the commander of all the army's training facilities, and he took immediate and swift action to move these to the now-empty Jordanian bases in Judea and Samaria.

These bases were situated in strategic posts outside the cities and villages: occupying them did not entail taking anyone's land or damaging personal property. What's more, they were important places to hold in terms of controlling the terrain. My father began ordering the troop relocations while still in the Sinai.

In January 1968, Chaim Bar-Lev was made IDF chief of staff. To the best of my recollection, I think it was my father who first got Bar-Lev into horseback riding, and it was Bar-Lev who gave us our first Great Dane, Ogi. And it was my father who supported Bar-Lev over Ezer Weizman for IDF chief of staff. Yet despite all that, the two were not remotely friends. My father's successes during the Six-Day War merely added to the jealousy and hostility that had lingered ever since the reprisal raids of the 1950s.

My father's preference for Bar-Lev was rooted not in his enthusiasm for the man himself but rather in his belief that his friend Ezer Weizman was not the right man for the job. Weizman, a friend for years, had been the commander of the air force and had skillfully enhanced its strength and capabilities. He had served as chief of operations during the war, but my father felt he lacked the judgment and determination necessary for the position of chief of staff. Ezer had gotten the air force ready for the war, my father used to say, "but it's good that Moti Hod commanded it." Moti Hod, our neighbor in Tzahala, a son of Kibbutz Degania, was a calm and collected fellow.

And what of Ogi? He swallowed a pacifier, probably mine, and choked to death. My parents, either on account of oversensitivity or because we had endured too many tragedies already, decided to spare my brother and me this one. They told us that the dog was at the doctor and in the meantime combed the world for another black-and-white Great Dane. The original Ogi had cropped ears, as was customary at the time, but the replacement Ogi had his God-given floppy ears. "What happened to his ears?" we asked. "They grew," we were told.

By 1968 my father and Bar-Lev were once again at odds. Their opinions differed on how best to defend the Sinai from an Egyptian attack. Bar-Lev advocated for a string of forts along the Suez Canal, on the edge of the water. According to Bar-Lev, the fortified positions would keep the soldiers safe and enable them to control the canal. They offered good lookout posts and dominant firing positions, would prevent the Egyptians from ever trying to cross the water barrier, and at the same time would demonstrate Israel's sovereignty over the peninsula. "Our goal is to position ourselves on the water line and to prevent the Egyptians from gaining any territorial access," said Bar-Lev.

My father disagreed. He felt that a stationary line of defense along the canal would force the IDF into a static defense and would offer the Egyptians a bank of targets some two hundred yards from their border. Our forts, and every action taken in and around them, would be under constant surveillance. All of our movements and drills would be known, and the supply convoys would be under constant threat of ambush, mines, and artillery barrages.

The frequent whistle of Egyptian artillery, the sinking of the Israeli navy's destroyer *Eilat* in October 1967, the rebuilding of the Egyptian army, the arrival of thousands of Russian specialists, and the Khartoum summit conference of August 1967, where the Arab nations famously proclaimed their Three Nos—no to negotiation with Israel, no to recognition of Israel, and no to peace with Israel—all these left no doubt that the southern front along the canal would not be quiet or peaceful in the near future. That much was clear to all.

Bar-Lev envisioned only the reality of the current bombardments, and failed to imagine what a massive Egyptian assault might look like. My father felt that in such an event the forts would be flattened or choked with smoke and fire. The Egyptians will cut them off, he argued, and we'll be stuck trying to save them instead of launching the critical element of our battle plan, the counterattack.

"You cannot win a defensive battle on an outer line," my father contended. When the army was forced to defend the outskirts of Tel Aviv or the line of kibbutzim in the Bet Shean Valley, there'd been no choice, but in the Sinai, with a 150-mile buffer zone, it was completely uncalled-for.

If there was to be a defensive battle in the Sinai, my father felt, it should be fought properly, not on a forward line but in depth. Seen in that light, the canal was an important physical barrier in the general defensive plan, but nothing more. It was not a "sacred" site that the army had to chain itself to throughout the duration of a future war.

His plan was that the IDF deploy its first line of defense some five to eight miles east of the canal, along the natural line of dunes and hills that run parallel to the water and dominate the flatlands below. A second line of defense, a mobile force, should be placed fifteen to twenty miles from the canal, at the entrances to the Gidi and Mitla mountain passes. The area between the canal and the first line of defense, he said, should be patrolled at different times and along differing routes so that the forces would not be sitting ducks. In the event that the Egyptians did cross the canal en masse, we could afford to allow them to advance one or two miles into the Sinai, assess their weaknesses, and launch a mobile counterstrike, which was the IDF's forte. Sovereignty all the way to the canal could be established by placing a fort or two along the Great Bitter Lake, for instance, where they would not be susceptible to direct fire.

On December 19, 1968, in preparation for a meeting regarding the strongholds, he wrote out his arguments against the Bar-Lev Line:

The role of the strongholds—

Presence—yes.

Observation posts—During the day and not while under fire, yes. At night and while under fire, <u>no</u>.

Firepower—very limited.

The system is very expensive—the number of weapons trained on the canal is very limited. The effect of a single machine gun is highly overrated. Reject the efficacy of single weapons, their effect will be nil.

What will the picture look like in a time of war, artillery and the neutralization of the positions.

In summation, we are investing a fortune in a product that provides very limited results. Is it justified?

We musn't send too many people forward, rather immediately assume rear echelon formations.

We shouldn't divide the armor.

It is as if he looked five years into the future and saw the breakout of the Yom Kippur War. In his mind's eye he saw the picture of the battlefield. But no one would listen.

Major General Yisrael Tal was the only member of the General Staff who agreed with my father. All the rest fell in line behind Bar-Lev. Five years later, during the Yom Kippur War, Israel paid the full price of that folly. Thousands of soldiers were killed, a national trauma that haunted us for years. As far as Bar-Lev was concerned, this was not merely a professional disagreement, but also a lever upon which he would lean to oust my father from the army. The verbal assaults at the General Staff meetings were personal and had nothing to do with the advantages or disadvantages of what is still known as the Bar-Lev Line.

Bar-Lev's intentions were made all the more clear by a phone call my father received from the Personnel Bureau. The officer on the phone wanted to know how he wanted to receive his retirement compensation, in one lump sum before discharge or in installments after he was discharged.

"I have no intention of leaving the army," my father said.

"Really? But your contract is up in another month."

"Listen," my father said, "I don't have any plans to leave. Just send me the forms so that I can sign up for another ten years."

The forms arrived, and my father signed them. This was a completely ordinary procedure for senior staff officers—it's generally done automatically—but Bar-Lev refused to authorize his reenlistment.

Defense Minister Dayan, loyal to his reputation as a coward in public life and fearless on the battlefield, refused to do a thing to help my father. Golda Meir, who had since replaced Levi Eshkol as prime minister and who presided over Israel's arrogant parade toward the tragedy of the Yom Kippur War, also refused to get involved in keeping my father in the army.

My father considered his options and decided that political life seemed the most obvious course to take. Business, or the accrual of personal wealth, had never interested him. He had been in the service of the state since the age of seventeen, and had been wounded and nearly killed on several occasions, fighting for its existence. He had accomplished great things for the country, and it was natural for him to continue his career in the public sphere.

He was a member of the ruling Mapai (Workers' Party of Israel). There had simply been no other path to promotion for him beyond lieutenant colonel; the army at the time was completely political. This was a known fact, bare and out in the open. But politically my father actually leaned in the opposite direction.

At the King David Hotel in Jerusalem he met with his friend Joseph Sapir of the Liberal Party and with Menachem Begin, the longtime head of the opposition and leader of the right-wing Herut. Listening to Begin speak, my father started to feel like Pinocchio with the cat who wasn't blind and the fox who wasn't lame. He was in a world that was strange and foreign and completely unnatural to him.

When necessary, politicians know how to take swift action: the next day's headlines declared, "Sharon Joins Herut-Liberal Bloc."

Help came from an unexpected source. Pinchas Sapir (who told my father the following story), the finance minister and one of the stronger people in the Labor Party, heard the news during a trip to the United States and promptly called the chief of staff. "What do you think you're doing?" he yelled at Bar-Lev. "What do you think you're doing to the party? Don't you know we're in the middle of an election campaign? Did you not even think to consider that Sharon might bring the Herut-Liberal Bloc a lot of votes? Have you lost your mind?" He told the chief of staff to find my father a position in uniform. That's how things worked back then: the party man (of the dominant Labor Party) told the chief of staff what to do with his generals to deny the adversary party any electoral advantages. Bar-Lev, after retirement, joined the Labor Party, of course, and was elected to Knesset and appointed a minister in due time.

The end result was that my father was given a position that kept him in the army, but out of the country. He was tasked with visiting and lecturing at other friendly armies until a more appropriate position could be found and, more important, until the elections had passed. Israel's national airline, El Al, said that the ticket he held was the fattest one they'd ever printed, with destinations including the United States, Mexico, Japan, Hong Kong, Korea, among many others. The only visit to Israel was on the eve of Rosh Hashanah, for Gur's memorial service.

In December 1969, after Mapai had once again won the election, my father was appointed commander of the southern front.

In March 1970, we left our house in Tzahala for good. At first my parents had thought that after my father's stint at the Southern Command we would return, but we never did. "Too many memories there," my parents explained to me. My father believed that whoever served in the Southern Command headquarters should move along with his family to Beersheba. That way they were closer to the base, and he felt it would help build up the capital of the Negev, which was still such a backwater at the time that the main road to the city had not yet been fully paved.

We moved into a small house on the outskirts of town, near the road to the air force base in Hatzerim. These days, especially after the major expansion in Beersheba during my father's tenure as minister of housing and construction, that neighborhood is pretty much in the center of the city.

I was three and a bit when we moved. It was the middle of the year; Omri was already in nursery school, and I was supposed to go to day care with kids my age, but I refused to leave my brother's side. "It's all right," Naomi Sivroni, the kindergarten teacher said; "leave him here with me today." During storytime, I was completely silent. "You can leave him here with me for a few more days," Naomi told my mother the next day, and that is how I wound up staying for the rest of the year and the year after that with children who were still a year older than me. It's easy to find me in the kindergarten pictures; I'm the one in Naomi's arms.

I remember the first day of first grade at the Massada School in Beersheba. The school was situated between our house and the Southern Command headquarters. My mother brought me to my class. "Look at this kid, he eats with his mouth shut," one of the other kids yelled, pointing at me in wonder.

And, of course, there was the base: the master sergeant in charge of vehicles, Karbo, the female soldiers, the staff officers, and my father's office. I remember him sitting at his desk next to the Great Dane, his feet resting on Ogi's huge back.

Occcasionally a military Jeep or a Wagoneer—and once I remember a half-track—would be parked near our house. We would climb on it, turning it into our personal playground.

When my father took over the Southern Command front, the War of Attrition had been going on for two full years. During the War of Attrition, 721 Israelis were killed, and yet it was only formally declared a war in 2002, when my father was prime minister.

Most of the Israeli dead were killed by Egyptian artillery, and 127 of them were civilians, many of whom were employed by the army to build the costly and completely useless Bar-Lev Line. Israel also lost people to mines and ambushes, and forty-seven seamen were killed aboard the INS *Eilat,* a destroyer that was sunk by Egyptian forces in October 1967. Ten airmen were killed over the course of the war.

The bare facts of the war—the action was far from the center of the country, there were no territorial exchanges, and the death toll was more of a slow trickle of blood—meant that it had no glory and left no marks on the national psyche, but daring commando raids across the canal were conducted and constant patrols were launched, even though it was clear that there was a high risk that it would not end well.

A friend of mine, Tani Geva, from Kibbutz Dorot, which is right next to our farm, served near the canal at the time and told me that my father, as commander of the southern front, would join them on those dangerous patrols. Having the general of the Southern Command by their side emboldened the soldiers, giving them the feeling that they had not been tossed in some godforsaken locale.

The IDF responded. There were retaliatory artillery barrages and air strikes deep in Egyptian territory. Over the course of the war, thousands of Egyptians were killed, including the Egyptian chief of staff, Abed El Munam Riad. "I can't take back the Sinai, but I can break Israel's spirit with attrition," said Egypt's President Nasser on July 23, 1969, explaining his rationale for the war, which he began. In the spring of 1970, after suffering Israeli retaliatory air strikes and realizing that the Egyptian skies were completely open to Israeli fighter jets, Nasser sought Soviet assistance. The Soviets showed up en masse. With the Cold War still at its peak, the Soviet Union sent advisers, antiaircraft specialists, and pilots. The Egyptian army was rebuilt and armed to the teeth with the most advanced Soviet weaponry.

For the first time in the conflict, Soviet soldiers took an active role in battle. In late July 1970, Israeli Air Force planes downed five Russian-piloted MiGs. The Israeli planes returned home safely. On August 7, 1970, the American-backed cease-fire known as the Rogers Plan went into effect. It came as a great relief to the Israeli soldiers stationed along the canal, and to the Egyptians, who had suffered far more casualties.

Soon enough, though, Nasser's rationale for accepting the plan became clear. The Egyptians began moving their antiaircraft batteries closer to the canal. This was a dangerous development and was a violation of the cease-fire agreement. The antiaircraft missiles barred the Israeli Air Force from operating in the area and allowed the Egyptians the option of resuming the War of Attrition. Moreover, if in the future the Egyptians decided to try to cross the canal, as they did indeed three years later, their antiaircraft weapons would effectively take Israel's air force out of the equation.

In August 1970 my mother traveled to London, Paris, and Rome for vacation.

In a letter to her, dated August 15, 1970, my father wrote,

As you know, after the cease-fire went into effect, the Egyptians brought the missiles up to the canal, as well as a vast amount of new artillery. All of this will put us in a very difficult situation if we

do not take measures immediately. (Too bad that our fools did not
properly appraise the situation, even though I contended throughout
that we could not enter the cease-fire under the conditions prevailing
at the time.) Over the last week I was in the Sinai four times. What
a wonderful feeling it is to travel around without helmet and flak vest
and to not have to lie flat each time you hear the whistle of artillery.
On the other hand, I know what could happen at any moment, and
those concerns prevent me from enjoying the existing quiet.

My father recommended firm and immediate action: he suggested we send forces across the canal near Kantara and destroy
the antiaircraft batteries. The General Staff supported the idea
and approved the mission.

This was the first time that my father had considered the practical aspects of crossing the canal. He began checking for possible
crossing sites. Loyal to his own principles of warfare, he considered
only sites that had at least one flank protected by a natural obstacle.

At the southern end of the canal, at Suez, the southern flank
was protected by the gulf, and at the northern end, at Kantara,
the swamps and lagoons to the north and west could protect a
bridgehead. Of the two options my father preferred Kantara,
because a freshwater irrigation canal south of the planned crossing point would make it easier to defend the crossing area. But
Golda Meir's government decided not to authorize the operation, allowing the Egyptians to draw their antiaircraft missiles
close to the canal.

For the Egyptians the move was simply a logistical one. All
they had to do was move trucks eastward and place the missiles
near the canal. For Israel, the situation could only be neutralized
with military action.

Shortly after the cease-fire went into effect, President Gamal
Abdel Nasser died. The personification of Pan-Arabism, the rhet-
orician with the arrogant, hostile attitude toward Israel, died a
defeated man.

*Prime Minister Golda Meir visiting the Sinai
before the Yom Kippur War, September 1970.*

Israel did not launch the canal-crossing operation in 1970, but
the Southern Command began to train seriously for such a scenario. The crossing was a complex military maneuver, and it had
to be practiced under conditions that were as close as possible to
the real thing.

It took some time, but my father finally chose as a training site
the Rueiffa Dam in Sinai, which had been built by the British in
the 1920s. IDF troops deepened the lake and built banks alongside it so that it resembled the banks of the canal. In early 1972
the weather cooperated, and floodwaters filled the lake. The IDF
practiced a crossing under the watching eyes of Defense Minister
Dayan and Prime Minister Meir.

NINE

The sector under my father as the commander of the southern front stretched from the Dead Sea along the Jordanian border all the way through the desert to Eilat and from there to the entire Sinai Peninsula, to the Suez Canal, and all the way back to the Gaza Strip. Three pressing problems awaited him immediately upon his taking over as commander. The War of Attrition with Egypt was at full tilt, Palestinian terror remained rampant throughout the Arava and Negev regions, and last and most complicated, the flow of terror from Gaza needed to be stanched. When the War of Attrition finally ended, my father was able to turn his attention to the terror from the Arava area of Jordan.

Looking across the canal to Egypt, near Firdan
Bridge, August 1970. (© Avi Simchoni, courtesy of
the IDF and Defense Establishment Archive)

The Arava is a remote desert region that spans one hundred miles, from the Dead Sea to the seaside city of Eilat. The road runs alongside the border, sometimes coming within a hundred yards of the border that divides Israel and Jordan. The Israeli villages on the western side of the border are agricultural and quite dispersed. The PLO terrorists took advantage of this, leaving their bases in the Jordanian high desert and sneaking down to the border region at night and then, after spending the day in hiding, crossing the border at night, laying mines or opening fire, and sneaking back to their bases before first light. Life in the shadow of this murderous terror became the norm in this remote region. Once my father had learned the terrorists' modus operandi, he decided that the best way to combat it was by attacking them, throwing them on the defensive. The Jordanian side of the border is populated by nothing beyond a few lonely army outposts, so my father decided to initiate a system that reminded him of his operations with the paratroopers almost twenty years earlier, sending squads as far as twelve miles into Jordan to catch the terrorists before they reached the border, or at least to catch them on the way back. The terror squads were rattled. Suddenly the desert they had owned was crawling with IDF soldiers, making it unsafe for them in any direction. There was a sharp drop in their level of activity, but my father did not stop there.

There was a tracker's path along the border—a smooth road that one would have to cross and thus leave marks that could be read. Bedouin trackers, who volunteered for service in the IDF, proved quite skilled at tracking terrorists. Since it turned out that the best way to track the terrorists was on camelback, the army drafted camels and trained them to sit in the back of a command car and ride to the site where the border had been breached.

A camel standing in a command car, ready for action, circa early 1970s.

The pursuits were also handled in a different manner, with the trackers making educated guesses about which way the terrorists would go, and the army sending troops to head them off.

In March 1970, PLO terrorists fired mortars at the Dead Sea Works factory in Israel from Tzafi, an abandoned village on the Jordanian side of the south end of the Dead Sea. Tzafi was very popular with the terrorists, and with good reason: it had plenty of fresh drinking water, a rarity in this part of the world; the Saudi forces in Jordan sat close by, lending a sense of security; and the Israeli factories were within firing range.

On March 20, 1970, my father sent out a force to capture the abandoned village, forcing the terrorists and the Saudis to flee.

The IDF held the village for three months, during which it killed or captured many of the terrorists operating in the Arava region. Subsequently a deal was reached with King Hussein, whereby the IDF would leave Tzafi, and in return His Majesty would send in Jordanian troops to ensure that no PLO terror attacks would be launched from the region. A Jordanian armored force rolled into the empty village, guns ablaze. Later they an-

nounced that they had pushed back the IDF in battle and forced
them out of Tzafi. It was, as my father said, "entertaining military
theater."

———————————

The situation in Gaza was far more complex. Gaza is a strip
of land some twenty-five miles long, and between three and
seven miles wide. After the Six-Day War the population num-
bered close to 400,000 people, about half of them crowded into
refugee camps. Aside from a brief period after the Sinai War in
1956, Egypt controlled Gaza from 1948 to 1967. During those
nineteen years of Egyptian rule, the Egyptian government never
once extended independence to the residents of the Strip. Instead,
their intelligence services activated the bands of *fedayun* to launch
terror attacks against Israel. It was the remaining *fedayun* from the
1950s and PLO members who were the rising force among the
Palestinians in Gaza, committing acts of terror.

Terrorism in Gaza came, at first, in the form of the brutal
murders of Gazans who wished to work in Israel. The killings
of course terrorized Palestinians. The brutality that Palestinians
inflict on one another, and in general the brutality of Arabs and
Muslims toward one another, is shocking. The Egyptians used gas
against the Yemenites when they intervened in the Yemeni civil
war in the 1960s. The Iraqis gassed the Iranians during their long
war, and famously gassed the Kurds when they fought for independence. World opinion has been muted on these matters, and
these countries' actions have had no effect on their relations with
Europe. Both the Arab world and Europe have different yard-
sticks by which they measure morality, and they choose these tools
on the basis of convenience.

At that time in Gaza awful murders were used to gain control
over the population and to prevent normalization of relations with
Israel, which had brought financial prosperity to many. People
believed to be collaborators with Israel were brutally executed.
Sometimes they were hung on a pole in the middle of the refugee

camps so everyone could see. Nothing has changed in that regard. In 2007, during the skirmishes between Fatah and Hamas
in Gaza, people were thrown off high roofs, and a wide range of
brutal measures was employed by both sides.

Back in 1970, my father had a strong difference of opinion with
then defense minister Moshe Dayan. "Let them kill each other,"
Dayan said with his typical cynicism. But my father wasn't willing
to accept that. If we take it upon ourselves to govern there, then
we are responsible for the lives of the residents, Arabs and Jews, he
said. My father also warned that the matter would not be restricted
to the killing of their own people; it was clear to him that the violence would spread to Jews as well. This is the nature of terror.

Soon enough the terrorists from Gaza began murdering Jews,
laying mines in the fields of nearby kibbutzim. Other Israelis were
killed within the Strip, whether doing business or visiting the area.
Hand grenades were thrown, guns fired, explosives detonated.

On January 2, 1971, the Aroyo family was traveling through
Gaza. The family had moved from England two years earlier. In
the afternoon, on their way home, they got lost. In the vicinity of
Gaza City a terrorist threw a grenade at their car, killing their two
children, Mark, seven, and Avigial, five, and gravely wounding
their mother, Pretty.

Sometime after the murders, my father spoke with Dayan and
said, "Moshe, if we don't act now, we'll lose control, no doubt
about it."

"You can start," was Dayan's response. As usual, he gave no direct orders. That was enough of an order for my father; he didn't
ask for anything more specific. That's another reason why in the
1950s my father and the paratroopers were given the duty of the
reprisal raids. He didn't need to have his orders spelled out. "All
the others ask for written, clear orders. You just do it," Dayan
once confessed to my father. At any rate, my father had the green
light, but he still did not know what he would do with that authorization in the cramped urban quarters of Gaza and among the
20,000 acres of dense orchards.

There were some 700 or 800 terrorists among the close to
400,000 residents. He had no plan, but he did have time: the War
of Attrition had ended, and the Arava frontier had been quieted.
So he went back to his roots, going out on detailed patrols and getting to know the lay of the land intimately.

Even as prime minister, some thirty years after he'd eradicated
terror from Gaza for many years, my father still remembered every orchard in Gaza and every valley and gully. He spent weeks
in Gaza before he decided on a course of action, and some seven
months afterward, until he saw the mission through in February
1972. He spent so much time there that there was a running joke
in the family that he must have had a pretty-eyed Gazan girl, or
maybe a few, and that the local ladies would follow him adoringly
with their eyes.

He decided to address the primary issue—finding the terrorists—without disturbing the lives of innocent citizens.

He took the Strip and parceled it into small, 1½–×–1½-mile
squares, keeping the division clear along natural boundary lines.
Each IDF squad was assigned a square. It was their responsibility,
and they actually lived there. Suddenly the job did not sound so
insurmountable. Each squad knew exactly what they were responsible for—not all of Gaza, just their square. Then they had to be
taught what to do with their square. This was not ordinary army
work. There was no visible enemy. The terrorists had to be sought
out. Each square was also filled with civilians who had nothing
to do with terrorism, and the land itself was densely populated, a
hodgepodge of houses and orchards. There was no drill for this
type of mission, and so my father, yet again, was forced to create
one. To make his guidelines clear, he repeatedly cut the entire
chain of command and explained directly to squad leaders what
he wanted of them: know your square like the back of your hand,
know the people's routines, notice anything out of the ordinary,
notice strangers who come and go.

Soldiers were sent into the refugee camps. They sought out
the terrorists and killed them when they found them. This rep-

resented a shift from the Dayan protocols, which called for the soldiers to stay out of the refugee camps so that the Palestinians could, as Dayan said, "stew in their own juice."

Do not fall into a routine, surprise them all the time, my father counseled the soldiers. Patrols at set times would make the soldiers easy prey; sporadic actions would turn the terrorists from hunters to hunted. "Create a new situation for every terrorist every day."

———

The Gazans are skilled contractors, and they would build double walls in their homes to hide terrorists. In order to deal with that problem, each soldier was given a length of rope to carry in his battle vest. If he had information that a terrorist was hiding out in a certain home, he would measure the walls outside and inside and make sure the dimensions corresponded. The rumor of the ropes traveled fast and forced the terrorists to stay on the run.

Each squad was equipped with a ladder, too. They would set them up against walls and peek into the yards, where most of family life is conducted in Arab homes. Of course peeking in was an invasion of privacy, but intimate affairs are not conducted in the yard, and peeking into the yards was better than knocking on the heavy metal doors and disturbing whatever was going on inside the house. Innocent families were usually merely surprised, but those who housed terrorists were thrown into a full-blown panic.

The troops grew creative, rapidly developing their own procedures. For instance, during patrols in the dense urban areas, a soldier or two would take to the roofs, providing the men on the ground with pivotal information that they could not have gleaned from within the tight alleyways.

My father also made maximal use of the intelligence provided by the security services and the intense activities of the Rimon Unit, whose soldiers could operate undercover, as Arabs, and whose commander, Meir Dagan, would many years later be appointed by my father to head the Mossad. The undercover troops operated in teams of four to five warriors, a hodgepodge of Jews,

Druze, Israeli Bedouin, and occasionally even terrorists who had switched sides. The pressure on the terrorists in the camps was great, and it drove them into the orchards. The Gaza Strip, despite the density of its population, was home to some 20,000 acres of orchard land, representing over a fifth of the territory.

My father took a strong interest in the Gaza Strip's orchards and groves. The manner in which the Arabs tended to their trees was diametrically opposed to what he was familiar with. He was used to intensive cultivation that included pruning.

Gaza Strip's orchards were worked differently; they were cultivated less extensively. As a result, the orchards and groves were denser and more tightly interwoven. They were a problem for the soldiers in that they provided excellent cover for the terrorists. When they opened fire, it was hard for the soldiers to pinpoint the source of the fire and hard to find the attackers once they slipped back into the foliage. The terrorists also proved quite adept at constructing bunkers in the middle of the orchards. Often they were so well concealed that you could walk past or even right over one without noticing. Long barriers of cacti grew along paths and between the orchards, serving as natural barriers and housing the bunkers' ventilation pipes.

The challenges facing the troops in the more cultivated areas were different, but they revolved around similar principles. I recall my father's description of a meeting with several junior-grade officers and squad leaders. They were standing on a small hill facing the vast orchard of Rashid A-Shawa, the mayor of Gaza City at the time and a large landowner.

"What do you see?" my father asked.

"What do you mean? We see an orange grove."

"Look better. What else do you see?"

"Well, there are some palm trees, too."

"Yes, but look more closely at the palm trees."

"Well, two of them have their tops cut off. Maybe they are old trees."

"Okay, when we go down into the grove, go to those two trees first."

"Why specifically there?"

"Here's why. Because we are putting pressure on the terrorists in the towns and in the camps. So we know they are going to hide in the groves. When they leave their old hideouts, they will need to meet someplace. One of them might say, Let's meet at Rashid A-Shawa's groves. Somebody else will say, But it's a big grove. Exactly where should we meet? Let's meet by the palm trees. But there's a bunch of palm trees. Well, we'll meet at the two trees whose tops are cut off.

"You have to know how these people will think. They are moving. They need rendezvous points. They need drops where they can get messages and instructions. They need places where food can be brought to them. They have to use something to mark these places. So what are they going to use? They are going to use something different, something that stands out just a little. That's why if you see two lemon trees in an orange grove, check out the lemon trees. If you see a dead tree among live trees, check that."

The surprises and ever-changing tactics helped unearth many terrorists and eroded the fighting spirit of those that remained. My father threw himself body and soul into this complex war on terror, and he took it home with him, too. I clearly remember how one night my father dreamed that Abdul Jabar, a dangerous terrorist from one of the central refugee camps in the middle of the Strip, a terrorist whom the IDF had been hunting for four months, had been caught in a wadi bed, dressed as a woman. The next day it was reported that Abdul Jabar had been eliminated. Verify it, the operations officers said, there've been plenty such reports in the past. My father hurried to the area and found that Abdul Jabar had been killed along with three other terrorists in the Gaza Wadi. He was dressed in female garb. Motti Zilberboim of Rehovot, a warrior in the Rimon Unit, told me that they saw a woman walking in the valley, leading three men; a woman in Gaza does not lead men. They called to the figure to halt. In response, the person drew a Kalashnikov, and he and the other three tried to open fire. They were all killed.

But Gaza interested my father in ways that stretched beyond the war on terror. I remember him talking for years about razing the refugee camps and building proper and permanent housing for the residents. Shlomi Gruner, a distinguished company commander in the Shaked reconnaissance unit, told me how he once stood next to my father on a hilltop overlooking Jabaliya refugee camp and listened as my father explained how he wanted to build industrial centers and neighborhoods instead of the cramped, run-down refugee camps. But just as my father was unable to convince Prime Minister Levi Eshkol in 1967–68, so too was he unable to convince Golda Meir's administration in 1971 that the upkeep of neighborhoods and proper housing were needed or worthwhile. He recognized that beyond the humanitarian aspect, the upgrade would also represent another way of dealing with terror in the long run.

The civilian population in Gaza realized that the army's intent was not to do them physical or fiscal harm, but merely to combat the terrorists. During the entire seven-month stretch of grueling and continuous warfare, in dense urban areas and in the orchards and groves—as far as I know—only two civilians were killed, one of whom was deaf and did not hear the soldiers' calls to stop, and the other a woman whom a terrorist used as cover while firing on an IDF patrol. She was killed in the exchange of fire. My father insisted that citizens not be harmed, that they be allowed to work and support themselves, and that orchards not be uprooted.

Each commander kept a piece of paper in his shirt pocket on which was listed the names of the wanted terrorists. At the beginning of the operation there were one hundred and ten names on the list. At the end, there were around ten left. All told, one hundred and four terrorists had been killed, and over seven hundred arrested. Quiet reigned through the Strip for ten years.

TEN

Ben-Gurion visiting my father in the Sinai, January 1971.
My father asked him to wear the beret to hide his
well-known white hair from Egyptian snipers.
(© David Rubinger, Yedioth Ahronoth)

In September 1972, after the murder of the Israeli athletes at the Olympic Games in Munich, Zev Schiff, military correspondent for *Haaretz*, wrote:

Insofar as fighting terrorists outside the borders of our region and within the Arab states we have not yet managed to "think outside the box." What we lack is someone obsessed with the matter, someone who will not allow the government to remain idle, who will preempt the terrorists in thought and in planning. We had a similar situation in the Gaza Strip until someone "obsessed" arrived in the form of Major-General Ariel Sharon, who smashed the terrorists despite all of the difficulties and limitations that we imposed on ourselves. At this point in time we need a Sharon-like figure in the more global fight against terrorists.

At the end of his term as chief of staff, Chaim Bar-Lev sent my father a farewell letter. Among other things, he wrote:

As Commander of the Southern Front, it fell to you to bring the Egyptians to a cease-fire, a matter that was albeit the cumulative result of the period that preceded you too, but during your tenure we intensified our operations and in the end attained the cease-fire, which has been ongoing for over a year and a half.

Of late the Gaza Strip has gone quiet and your contribution to that development is vast and I hail you for it. As for the future, I feel that the "Egyptian problem" has yet to be solved and it is possible that even during your tenure the fighting will break out again. I hope that as Commander of the Southern Front you have seen the righteousness of the concept of maintaining the strongholds along the bank of the canal and that in the future it will deny the Egyptians any sort of territorial achievement.

Chaim Bar-Lev, Lieutenant General,
Chief of Staff, December 31, 1971

Less than two years later Bar-Lev had the chance to see for himself the merit of his concept and the bitter fate that was to befall those in his strongholds.

Tel Aviv, January 28, 1971

Dear Arik—

I lack the words to thank you for the visit [to the Sinai] and for what I saw during that visit and for what I learned yesterday. My appreciation and admiration for the IDF has grown and deepened on the basis of the things I saw and learned, and I see yet again that the faith I had in you and in the entire IDF was not excessive. My deep prayer is that by the end of this century or at the least during the decades of the twenty-first century, the prophecy of Isaiah will come into being, "nation shall not lift up sword against nation, neither shall they learn war any more," but so long as peace has not been secured—may the IDF be blessed for its aptitude and loyalty and devotion, and you have no small part in that aptitude.

In Esteem,
D. Ben-Gurion

In January 1972 David "Dado" Elazar was appointed chief of staff, replacing Bar-Lev. Insofar as my father was concerned, it was like switching rooms on the *Titanic*. Dado, Bar-Lev: same difference. And the changing of the guard did not make my father any better liked in the chief of staff's office.

Several months after taking over as chief of staff, Dado told my father that he expected him to retire. He was not yet forty-five. The last thing my father wanted to do was to sit around in cafés in Tel Aviv like some other retired generals and wait for job offers.

Back then, with so much of the business arena dominated by the ruling Labor Party, there may not have been much in the way of offers, either. After all, if they'd wanted him involved, they wouldn't have worked so hard to oust him from the army. During that period my parents looked around and eventually found our farm. In a move that seems to me rather gentlemanly, my father told the nearby kibbutz about his plans to acquire the land so that he would not get in their way if they wanted to buy it themselves.

The kibbutz most certainly had their eyes on that land, but they hoped to be given the territory for free, and so, to our good fortune, my parents were able to acquire the farm. The long-term loan they were given to buy it was paid in full only after my discharge from the army.

Even today the farm is somewhat isolated, but back then it was truly a distant and remote spot. It required unusual courage to take everything they had and mortgage the farm itself to acquire it, but that was their dream, and they did not let anything stand in their way.

One day, while my parents were building our house on the farm, Vera came to visit. She looked at the water-draining holes in the wall surrounding the garden. After examining it for a while, she said, with great satisfaction, "Very good—if necessary to defend." She thought that the holes were there to allow us to use rifles in the defense of the house without needing to raise our heads above the wall.

Meanwhile, my father paid careful attention to the Egyptian army's training drills on the far side of the canal. Again and again the Egyptians practiced offensive crossings of the canal. They paid attention to detail—for example, changing the camouflage patterns of their uniforms to suit the colors of the field in which they trained. His impression was that they were serious and that an Egyptian offensive would surely come. He also was not one of those, in the army and out, that enjoyed belittling the Egyptian army. Aside from the sincerity of their training regimen, he'd also seen their commando forces at work during the War of Attrition, when they would cross the canal at night, lay mines on the Israeli side, and return home undetected. Later, after the cease-fire, the Israeli trackers would find signs of their presence on Israeli soil. This was not the defeated army of June 1967.

In May 1973, shortly before my father was to retire from active duty, the intelligence services reported that the Egyptians were planning an attack and a crossing of the canal into the Sinai. My father went down to the Sinai Peninsula, along with his staff officers. Additional forces were sent down and trained, and the plans were looked over and polished. The army was ready. My father had all of the crossing gear—rafts, bridging equipment, engineering gear, and so on—assembled and organized near Baluza, some twenty miles from the planned crossing point at Kantara. He planned to surprise the Egyptian forces, if they attacked, with a swift countercrossing. He also prepared another crossing spot, farther to the south, just north of the Great Bitter Lake. The trouble was that the fortifications on the Israeli side were so thick, they needed to be breached so that our bulldozers, if needed, could easily push through to the water line. From the outside the embankments appeared unaltered, but my father made sure that red bricks marked the weak spot in the earthen wall, so it could be easily found at night under fire.

The crossing point was protected by the Great Bitter Lake to the south and by a walled-in yard that could shield the troops as they gathered in the staging area. On this site, five months later, the Israeli troops would cross to the Egyptian side of the canal, turning the tide of war and delivering, ultimately, a great victory.

The Egyptians noticed the wartime readiness of Israel's troops and decided to delay the offensive. The Israeli leadership, for its part, refused to believe in the possibility of war. "The mere thought of the Egyptian army crossing the canal," Golda Meir said, "is an insult to intelligence."

My father was concerned. He did not agree, so he asked the chief of staff to extend his term for another year. Dado, as expected, refused. Golda also refused to get involved and sent my father to see Dayan. He too refused. On July 15, 1973, several hours before my father was to turn the Southern Command over to Shmuel "Gorodish" Gonen, my father made one final plea to Dayan: "Moshe, I believe you are making a grave mistake. If we have a war here, and we might have one, Gonen does not have the experience to handle it."

Dayan said, "Arik, we aren't going to have any war this year. Maybe Gonen's not too experienced. But he'll have plenty of time to learn."

A few hours later we were at the change-of-command ceremony. I remember the ceremony and that Gorodish, with his oversize glasses, was very nice to me. Perhaps someone older than my seven years would have been able to detect the seeds of his insanity.

Exchange of command of the southern front, my father saluting, Gonen is second from the right, July 15, 1973.

"I need one good war to advance," Gorodish told my mother, who relayed this to us with a look of disgust. He hurled an ashtray at Zvia the secretary, and she threw it back at him and rushed to ask my father's chief of staff, Itzik Shaked, for a transfer to my father's reserves division.

Many years later, at the Cameri Theatre in Tel Aviv, I saw the play *Gorodish*, which portrayed Gonen as a troubled man and a tragic figure.

It was customary for the army to throw a good-bye party for a retiring general—a dinner, music, songs. The chief of staff makes a speech, and then, as my father put it, "the victim" is asked to say a few words and to receive the customary gift, a clock. "This, I will not let them do to you," my mother said. When they called to set the date, my father said he would pass on the whole affair. "What should we do with the clock?" the officer asked. My father told him, "Put it in the mail."

My father entered civilian life, thoughts about politics not far from his mind. He knew that he could easily be elected to the Knesset on one of the opposition parties' tickets, but the political realities of the time made such an achievement largely insignificant. The possibility that Herut, the Liberal Party, or the other shards of an opposition could win a majority in Knesset was negligible. The Labor Party had ruled the country for decades, since its inception, and before that it had ruled the Jewish Agency. The Labor-dominated Workers' Union controlled the market. Those who were not members of the party had a hard time finding a job in the public sector, if at all, and in many of the country's industries. Even in the army, those who advanced were always "one of ours."

"If I had to do it again today," my father used to say about the founding of the Likud, "even with all the experience I have, I would not be able to." His inexperience, he said, was what allowed him to pull it off. "I was under the impression that politicians were serious people," he would add.

The founding of the Likud, which was composed of Herut and Liberal Party members and other elements of the opposition, was what made the 1977 change in government possible. My father's contribution helped give practical meaning to Israeli democracy: without the Likud, Labor would never have been ousted from office, and Menachem Begin would never have assembled a government. The founding of the Likud was my father's idea. He pulled it off, and in the process received an eye-opening lesson in the workings of Israeli politics.

In July 1973, while working on the farm, he came to the realization that only a unified front could realistically challenge the well-oiled hegemony of Labor, a party that, as in the Communist states, could hardly be discerned from the state itself. My father, in his political innocence, sought to unite Herut and the Liberals with the Free Center, RAFI, and the Land of Israel movement, and it was perhaps just that naïveté that allowed him to succeed.

After thinking the idea through, my father rented a hall in the Bet Sokolov Press Building in Tel Aviv and announced a press conference. The hall was full, though my father hadn't told anyone what he was going to say. The news that he was going to establish the Likud was a complete surprise. The next day's papers were full of speculation, and the wheels of the political system began to turn.

The following day my parents took me and my brother to the Haruba beach in northern Sinai. We played in the sun on the edge of the water. My father lay there in his bathing suit. Another sunbather had a transistor radio, and my father heard a politician ridicule his idea, labeling him "a political baby." My father felt a rush of embarrassment. He wasn't yet used to being openly derided in the press; the army had been a shield from all that. I remember him recalling, "I thought to myself, a 265-pound baby in a bathing suit." He wanted to dig a hole and hide.

But baby or not, the heads of all the relevant parties sought him out. He began seven weeks of marathon meetings, during which he came to understand who he was dealing with. The bickering had nothing to do with complex ideological differences; it was all about the number of spots per party and the ranking of each individual on the combined list. My father belonged to none of these parties, and he owed them nothing. He also understood that he needed them all. Years later, he used to say, "You have to take the political 'game' with humor, but the outcome very seriously." He would not give up on RAFI, for instance, the party Ben-Gurion had founded after being ousted from Labor and before retiring for good. My father believed that by including RAFI, he would signal to middle-of-the-road Labor voters that the new bloc was centrist, not extreme right.

M y father used to say that two principal factors led Begin to accept the strategic plan of founding the Likud. One was the soil-blackened soles of my father's feet. He'd come to meetings in Tel Aviv straight from the farm, still in his sandals, and he'd see Begin look down at his feet in wonder. Begin was a man of words and ceremony, always dressed in a suit and tie, and although he spoke often and with great passion about the land of our forefathers, the only really hands-on connection he had to the land itself was the fact that he lived in a ground-floor apartment in Tel Aviv. Feet darkened by the soil of our homeland were an exotic sight to him.

The other factor that made a deep impression on Begin was Abu Rashid, who used to answer the phone on the farm and take messages for my father when he was out in the fields. Abu-Rashid was an old Circassian stablehand from the village of Kfar Kama in the lower Galilee, where some of Israel's five thousand Circassian citizens live. He had worked at the national park in Ramat Gan, where we kept our horses before we bought the farm. When we moved our four mares—Aziza, Yardena, Noga, and Aya— along with them came Abu-Rashid, a long-standing family friend. He liked my father a great deal and didn't want to part ways with him.

Abu-Rashid worked on the farm. He spoke Hebrew with a thick Arabic accent, and he would write down the messages in Arabic. Begin would call the farm and leave word with him.

"Who is this Abu-Rashid who answers the phone?" Begin asked one day, his curiosity getting the better of him.

"Abu-Rashid? Oh, he's my political adviser and press liaison," my father answered.

Begin wasn't used to dealing closely with Arabic speakers, and the fact that my father, with his soil-stained feet, had one of them answering his phones was an exciting and enticing eccentricity to Begin. The fact that my father was a highly regarded general with many victories was an added attraction. Begin admired military men and officers. They sat nicely alongside his Revisionist Zionist appreciation for honor, pomp, uniforms, insignia, and other such symbols.

In mid-September the mission was accomplished. An agreement was signed, and the Likud was founded. Elections for Knesset were two months away.

My father was chosen as the Likud campaign manager.

PART III

1973—2001

*October 17, 1973—Dayan's first visit to the bridgehead after
crossing the canal.
(Uri Dan)*

ELEVEN

I t is difficult for me to deal with the Yom Kippur War. I was a child of seven, but I remember it. For years I was plagued by the fact that it could have been different. I knew my father was not party to the blinding euphoria after the Six-Day War. I knew that he never belittled the capabilities of the Egyptian army—he knew their strong suits and their vulnerabilities, and he knew how best to confront them.

I know that he had been wronged in the desert during the Yom Kippur War. He grasped quickly what had happened, understood it amid the chaos of battle, in a war that had taken a complacent Israel by surprise. He knew what needed to be done in order for Israel to win, and to win quickly. But the higher-ups didn't understand and didn't want to listen. They did not come down to the field to assess the situation with their own eyes. Not southern front commander Shmuel Gonen, who in a moment of candor during the war, having been asked where he'd gone wrong, said, "I thought it was the seventh day of the Six-Day War." Not Bar-Lev, who effectively replaced Gonen on the fourth day of the war. And not the chief of staff, Dado, the man who had "successfully" ousted my father from the army and who maintained a steady hostility toward him throughout the war.

The General Staff realized that Gonen was in over his head and that he could not control the southern front, but rather than reappointing my father, who had been the commander of the southern front as recently as three months prior, they decided to appoint the very man who had conceived the failed defensive strategy in the Sinai, Chaim Bar-Lev. "What are you going to tell the journalists when they ask you about the Line and why it failed?" Bar-Lev asked my father shortly after he took over as commander of the southern front. This was well before the countercrossing of the canal; the situation in the Sinai was grave, and already Bar-Lev was fighting to save his name.

The defense minister, Moshe Dayan, agreed with my father's strategic plans, but as usual refused to take a stand, instead telling my father, "You convince them."

It's awful, but it's the truth: political considerations swayed battlefield decisions. During the war Pinchas Sapir, the finance minister, the same man who scolded Bar-Lev for letting my father out of the army during the run-up to the 1969 elections, this time made sure that Bar-Lev, the Labor-appointed minister of trade and industry, would command the southern front in the Sinai and not, "heaven forbid," Ariel Sharon, who was running for Knesset on the Likud list.

On Friday, the day before Yom Kippur, my father was at Likud campaign headquarters in Tel Aviv. Elections were about a month away, and he and other Likud Party members were scrambling to tie up loose ends before the Day of Atonement. In the afternoon a call came from the Southern Command. He was to report to his reserves division immediately.

My father gathered up his papers, left instructions for his staff, and headed south for Beersheba, where we still lived, in the small, army-issue apartment near the command headquarters.

As soon as he got home, he ordered all of the division's staff officers to report to base. Then he called the divisional intelligence officer, Yehoshua Saguy, and asked him to come over as soon as possible. Saguy arrived immediately, bearing aerial photos and other intelligence data.

One look at the photos sufficed. "There's no question," my father said, "this time it's war." The Egyptians had amassed all of their crossing equipment along the canal, and the concentration of forces was far beyond anything the IDF had seen during the Egyptians' endless training exercises.

My father called the command and asked whether any changes had been made in the war plans since he had left several months earlier. "No," they said, "it's still the same." That meant that the standing army's tanks holding the line should be concentrated at a distance no more than eighteen miles from the canal, and that the Bar-Lev strongholds should be evacuated as soon as the offensive began. The counterattacks would be carried out in strength, employing an armored-fist tactic, meaning that the attacking forces would be a concentrated force of tanks and not individual tanks, allowing for maximum impact and firepower and playing to the strength of the Israeli tank corps—mobility and firepower.

My father called Gonen and told him that he was ready, and that he thought the evidence was incontrovertible: war would break out soon. Since the chief of staff had not yet drafted the reserves, my parents decided to spend Yom Kippur on the farm, which is not far from divisional headquarters.

I remember being on the farm, experiencing the quiet of Yom Kippur. At around ten in the morning, cars began to line the road outside the farm, an exceedingly rare sight on the holiest day of the year. A few minutes later people from the neighboring kibbutz showed up and said that some of their boys had been called to their reserve units. Then my father got the call. A car came from the division to get him, but my mother packed us in the car and drove him to headquarters herself.

My father's last request of Dayan while in uniform was to have the command of an armored division in the reserves. "I would feel more comfortable with the Egyptian situation," he told Dayan. The truth is, he didn't believe that Dayan would agree: pushing that type of appointment through required a confrontation with Chief of Staff David Elazar, who wanted my father out of the army entirely, and Dayan was happy to avoid that type of confrontation. To my father's surprise, however, Dayan promised, and kept his word.

We left my father at division headquarters. At the base, the reservists began to trickle in and efficiently and calmly assemble their gear. In the afternoon, he drove to the Southern Command headquarters in Beersheba. When he reached the gate of the base, he heard the ominous warble of the air raid siren. He described the sense of war in the air and the worried faces of the officers and soldiers on the base.

The war room was bedlam, with streams of contradictory reports from the front, but all pointed in the same general direction: that very moment, the Egyptians had launched a massive offensive all along the canal, with heavy artillery, air strikes, and land forces streaming across the water. Up north, Syrian tanks were streaking across the Golan Heights. It was a war on two fronts, and one that Israel, under a complacent prime minister and defense minister, had allowed itself to be surprised with.

My father had seen all that he needed to see. He returned to the divisional headquarters, which was now abuzz with citizens turning into soldiers. The storehouses were thrown open, but my father could tell that not everyone had yet internalized the reality of war. Soldiers argued with quartermasters, who, despite the wail of the nearby sirens, refused to hand over the gear before each soldier went through the tiring and time-consuming process of signing for each canteen and shoelace, guarding their gear with their bodies. So many of those soldiers who had to scramble to get the gear they needed never returned from that war.

My father met with the brigade commanders and the staff officers. He updated them with the relevant information and was satisfied with their progress, but time was still of the essence. The Suez Canal is about two hundred miles from the base. Until the reserves arrived, the only armored force in the Sinai was Major General Avraham "Albert" Mandler's, and the less than three hundred tanks under his command could not hold the charging Egyptian army back for long. The Bar-Lev Line proved useless in a real war.

My father made a trip through the base. He stopped and spoke to the soldiers and officers and found them to be in high spirits. Gauging the fighting spirit was always something he could do instantaneously, and it was always clear to him just what hung in the balance—victory or defeat.

When he felt all was in order, he left the base and headed for home to pick up his personal gear and to part with us before leaving for war.

Over the course of that afternoon, October 6, he spoke with Major General Gonen on several occasions, urging him to leave his Beersheba headquarters and head down to his field headquarters in Sinai, so that he could make sense of the flood of information coming his way. Gonen needed, my father said, to see the situation on the ground and size it up for himself.

Gonen's reaction was a strong indication of what was to come. He did not want advice, and he sounded suspicious and hostile. Gonen had been an excellent brigade commander in Major General Yisrael Tal's division during the Six-Day War, but he suffered from a lack of confidence and had issues with authority when dealing with two of his divisional commanders, my father and Major General Avraham "Bren" Adan, both of whom had been his superior officers in the past. All of my father's ideas and suggestions were greeted with hostility and belligerence, underscoring his lack of confidence.

Fourteen months after the war, Uri Ben-Ari, a general who served as Gonen's deputy during the war, published an article in *Yediot Ahronoth*. "What could have been more logical," he wrote,

> *than to have appointed Major General Sharon to the post of Southern Front Commander already in the prewar hours. He had just left that position recently and Major General Gonen could have been appointed commander of the division [replacing Sharon] that he had, up until recently, commanded. The advantages of these appointments are simplicity personified: Major General Sharon*

had lived and breathed the southern front for five years or more, while Major General Gonen had only been acquainted with it, as supreme commander, for several months. Is that not the sensible and natural thing to do? Had this been done, Major General Sharon would have remained Major General Gonen's superior officer, which was how things had been shortly before [the war], and the tension of Gonen serving as Sharon's commander could have been avoided.

But that was more than a year afterward. During the actual war, some in the command headquarters and in the General Staff had toyed with the idea of dismissing my father altogether. Before the war, Chief of Staff Elazar had pushed my father into forced retirement on the grounds that the army had to be made up of younger officers, which was why he had a "young" forty-three-year-old general replace my father, the "old" forty-five-year-old.

During the course of the day, my father sent word to Likud headquarters that there were serious developments, and they would have to make do without him for the foreseeable future. The transition from the civilian world to an army unit readying itself for war is swift and jolting, but my father still felt more comfortable in a uniform. Furthermore, now that Israel was under attack, the importance of politics had dwindled to nothing. He felt that the country needed him on the front, and that was all that mattered. He had come to rout the Egyptians again.

He left for the front in a civilian pickup truck that had been drafted into service along with its owner. His jeep was still being outfitted with the necessary radio equipment, and so, late at night, he set out. His staff officers trailed behind him.

Early in the morning he reached Bir Gafgafa, where the southern front's headquarters were situated. He went down to the underground war room, and as soon as he entered, the officers silently rose to their feet.

During his hours on the road, a journey made without any

radio equipment, a clear picture of the Egyptian offensive had yet to develop. What was clear, though, was that the orders that needed to be issued in the eventuality of war had not been given. Of the three hundred tanks defending the Sinai, two hundred had remained in the rear, far from the canal, utterly useless, rather than in the front where they could attack in hard, fast counterattacks. Moreover, no order had been given to abandon the strongholds along the Bar-Lev Line, and the soldiers remained trapped in their forts as five Egyptian divisions streamed into the Sinai around them.

When the command finally ordered a withdrawal from the surrounded garrisons, it did so in the worst possible way. Single, or at the most, small groups of tanks were sent to the surrounded garrisons, where they were met by murderous Egyptian antitank fire. This flew in the face of all rules of tank warfare. Tanks were to be used as a concentrated force that could move fast and strike hard—but here they were merely moving targets for the Egyptian tank-hunting forces waiting for them. Within twenty-four hours the IDF lost about two hundred tanks. The crews mounted heroic but hopeless efforts to try to reach the trapped soldiers on the canal.

This is what Chief of Staff David Elazar said in April 1973, six months before the war, while presenting material operational plans to Prime Minister Golda Meir: "We do not want to execute any brilliant military maneuvers that might encircle or ensnare the enemy in a trap, and then destroy them . . . we are banking, on both the Syrian and Egyptians fronts, on totally stopping their advances. What we would describe as not merely containing them in order to later destroy them with a flanking maneuver, but barring their entry from the onset. This is so, militarily, because we feel that it is the right solution, and politically, because we want to deny them even partial success."

Dado repeated this like a mantra. He had replaced Bar-Lev at the helm, but the same misguided concepts still prevailed. My father was the most high-profile opponent of this approach.

Elazar, like Bar-Lev, made a dreadful mistake by drawing his line in the sand at the very edge of the border, along the canal. The Bar-Lev and Elazar approach collapsed on the first day of the war.

From left: Bar-Lev, an unknown man, my father, and Dayan, wearing flak jackets for protection at a stronghold near the Suez Canal, March 25, 1970. (© Gershon Gera)

In the spring of 1970, with the War of Attrition still raging and my father serving as southern front commander, Chief of Staff Bar-Lev met with the rest of the General Staff and with Defense Minister Dayan in Bir Gafgafa. The main point of contention back then was about the necessity of the Bar-Lev Line. As usual, my father's point of view was not accepted.

After the discussion, they all went to inspect a fort opposite Port Tawfik called the Mezah (quay). Wanting to avoid kicking up a telltale dust storm and the artillery that was sure to follow, the officers left their vehicles and walked to the base. When they arrived, the Egyptians opened fire with their artillery guns, and everyone ran to the bunkers. Dayan, though, had hurt his leg a while before, and rather than run to the bunkers he lay down. My father, not wanting to leave the defense minister alone, lay down next to him. "Arik, this is a bad mistake. You must convince them to change their concept," he said while the shells were falling around them.

"Moshe, just an hour ago you heard what these discussions are like. You know I can't convince them. Why don't you just give them an order?"

"No, I know you'll eventually do it. Just keep at it. Eventually you will succeed."

That had been three and a half years earlier. On October 7, 1973, my father and his staff arrived at Tasa, their main headquarters. His division was assigned to the central sector. Adan's division, 162, was to the north, and Mandler's to the south.

By the time he reached Tasa, he had been reunited with his now-radio-equipped jeep, and as soon as he stepped out of the vehicle a current of electricity moved through the soldiers. "Arik's here," they said as they clustered around him. He shook hands and strode toward the underground war room.

During the course of the day he was able to grasp what was happening on the southern front. He had seen warriors retreating, and he always remembered the look he had seen on the face of one young tank officer whom he met that day and exchanged a few words with: "They are advancing, and we can't stop them," said the young officer. My father described him: "It wasn't fear on his face but rather astonishment." Speaking earlier in the day with the deputy chief of staff, Yisrael Tal, my father said, "Talik, the IDF is a victorious army. There's a difference between an army that has been victorious and an army that's used to taking a beating. We're only used to winning. Today, we've been beaten, and our problem is shaking off the shock of the beatings and winning, and we can do it, so long as the Jews don't get in our way."

Once he had established a clear understanding of what was happening on the front, my father realized what was necessary for a swift and unequivocal victory. On the afternoon of the seventh of October, he spoke with the chief of staff. "Look, Dado, I am currently moving with my forces toward Tasa. The first tanks are finally here. By evening I'll have more. This is the immediate armored tank fist in the vicinity. I suggest using it, in a nighttime attack, in either of these two options, either to the north or the south. If you can, I suggest you also come to me."

The chief of staff said he would.

"We have a large force in hand. True, only half of it has arrived, but they are fresh, and we have not yet executed a concentrated attack. I could launch an attack at nine this evening. I'd like to add another word. From what I can see, they are moving in their mechanized divisions. The danger is that they may move armored divisions in during the night. Don't forget that the Egyptians are also not in an easy position. They've already been fighting since yesterday. In my opinion we can deal them a debilitating blow in one of the sectors tonight. We will not be able to crush their infantry. We will only be able to crush them entirely once we are able to gain a foothold on the other side of the canal."

After a single day of war my father was already calling for a countercrossing. He understood that the army had suffered a loss of morale and that the Egyptians had seized the initiative and could already smell victory. They had to have the legs cut out from under them, their spirit broken, with a surprise counterattack.

*In the bunker at his headquarters in Tasa, during
the Yom Kippur War. (© Brigade 14 website)*

At approximately four in the afternoon he entered the war
room in Tasa. The radios were alive with desperate pleas from
the strongholds along the canal. It was agonizing. At a stronghold
called Hizayon, near the Firdan Bridge, the officers had been killed
or wounded and a radio operator named Max Maman called out
to my father. "Forty, Forty," he said, using my father's code name
on the radio frequency, "we recognize your voice. We know you . . .
we know you will get us out of here. Please come to us. Please send
us help." His pleadings persisted for three days, until he was killed.

By five that afternoon, an hour after arriving at Tasa, my father had a rescue plan that could be launched that night. "The
plan was based on what we had learned from the previous night.
We would break through on a narrow front, creating a box of
fire of tanks and artillery, which would allow us to approach the
strongholds, send a small force in, get our people out, and then
disengage."

At approximately six in the evening my father spoke with
Gonen, who refused to authorize the mission. "I tried to convince

him on this matter, [explaining] that we would lose the night and that there was sure to be an Egyptian invasion, and that the nights were short. I tried several times. I will not note here in what manner I was answered, but it was not in the affirmative."

It was Avi Yaffe, radio operator at the Purkan stronghold, who said: "The only one who told us the truth was Ariel Sharon. He didn't bluff, and he didn't lie to us."

The abandoning of our men on the front line plagued my father, and he told Gonen again that he felt that he could save the ninety or so soldiers in the strongholds in the central sector. It was not only our military duty, my father said, but a moral one as well.

Gonen told my father that this matter would be raised at that evening's meeting. They were to meet at 7 p.m. at the command center in Dveila. My father called Dayan and asked for his assistance, but was unable to gain immediate authorization. Dayan told him that the rescue options would be discussed that evening at the meeting and decided upon there.

A helicopter was to come and bring my father to the meeting. He set out with a small squad, prepared the trapeze directional markers for the helicopter pilot, and sat down to wait. The appointed hour came and went, and the helicopter did not arrive. Despite the danger of moving around at night because of Egyptian commando forces in the area, they traveled back to the Tasa landing strip, but it too was empty. My father returned to the original site. Command reported that the chopper was on the way. Nearly two hours had elapsed. My father could not escape the feeling that this was not a mistake. Gonen did not want him at that meeting, preferring to spare himself the argument about rescuing the soldiers that were trapped in the strongholds along the canal.

At the entrance to the bunker he met Chief of Staff Elazar and former chief of staff Yitzhak Rabin, who was accompanying him. The meeting had just ended, Dado said, and the best thing to do at this point was to go down to the bunker and get the orders straight from Gonen. My father told Dado that, based on what he had seen during the day, they needed a concentrated attack of at least two divisions.

Conferring with (from left) Chief of Staff Elazar and former commander of the air force Ezer Weizman. (© David Rubinger, Yedioth Ahronoth)

"We just can't do that," Dado said. "The only force we have right now between this spot and Tel Aviv is your division."

My father replied, "The Egyptians are not heading for Tel Aviv. That's beyond them. Their target is the canal and the ridgeline—five to eight miles. They cannot afford to get beyond their [surface-to-air] missile cover." He tried to explain that a concentrated attack could destroy their Second Army bridgehead in the north and that then the IDF could turn south and destroy the Third Army to the south. Rabin, in an uncharacteristic gesture, laid a hand on my father's shoulder and said, "Arik, we're counting on you to change the situation," and with that they departed.

Having failed to convince Dado, my father went down to the

bunker to see Gonen. He repeated what he had told Dado and recommended a countercrossing of the canal to deflate the Egyptian offensive. He suggested Kantara as the crossing spot, on the northern part of the canal. It would be wise, he said, to decide now to begin the preparations and leave troops in the area, as close as possible to the waterline.

He was told rudely that the plans had already been made and would not be changed. Adan's division, in the northern sector, would attack from north to south, parallel to the canal, some two miles away. This would keep them safe from the antitank crews positioned on the ramps on the Egyptian side of the canal. In the meanwhile, he was told, his division would be in position at dawn, northwest of Tasa, and ready to strike in a northwesterly direction, ready to aid Adan as necessary. My father tried to explain to Gonen that the striking force had to be two divisions strong at least, but he saw it was of no use; the atmosphere was too defensive. Instead he took Gonen aside and said, "Look, Shmulik, I've left the army already. My life is going in an entirely different direction. I am not coming back to take your place here. The only intention I have is to defeat the Egyptians. Once we've finished with that, I'm gone. Shmulik, you can win this war. You can come out of it a winner. All you have to do is concentrate your forces against them. You don't have an enemy in me. You don't have to deal with me at all. Just deal with the Egyptians."

Gonen nodded in agreement.

———————————

October 8 was without doubt the most grievous lost opportunity of the war. On that day the IDF could have dealt the Egyptians a debilitating blow and spared ourselves two more weeks of hard battles and many casualties. The failures were at the divisional and command levels. Adan's division failed to fulfill its mission, never mounting a full divisional offensive, as planned.

The commander of the southern front was absent from the field, unaware of the reality of the situation. On the basis of contradictory reports he received and his own intuition, Gonen passed reports on to his superiors that simply had no relation to reality. Gonen issued orders to the officers in the field that can only be described as hallucinatory.

My father flew back to Tasa and at two in the morning on October 8, standing in the light of the half-tracks, gave his division the command to roll out to where Gonen had instructed him to be.

He slept for twenty minutes and then at four in the morning set out. By dawn he was positioned on the high ground along with his staff officers, looking down at the canal plain. Suddenly an Egyptian shell fell one hundred yards in front of the five APCs. A moment later another one fell, this time behind them. The division staff moved elsewhere. Within minutes, though, the shells began falling again. It was clear that the Egyptians had officers on the ground, somewhere nearby, and that they were radioing coordinates back to the artillery batteries. Amatzia "Patzi" Chen, together with Dani Wolf and Meir Dagan, eliminated the eight Egyptian forward artillery observers. Patzi, who used to be the commander of the Shaked reconnaissance unit, joined my father when the war broke out. He formed a volunteer special commando force to aid my father.

Patzi showed up waving his booty of Soviet binoculars. A quick look at the maps that the Egyptian soldiers had in their possession showed that they had only marked lookout points within six miles of the canal. The Egyptians had not planned to advance any farther.

Shortly thereafter word arrived from command: Do not attack. Wait to see how Adan's offensive develops, and then wait for further orders. The armored division under my father's command hunkered down on the high ground, waiting for the order to strike.

The tanks traded fire from afar, but my father waited to be let loose to attack.

Between eight and nine in the morning, my father saw Adan's forces. They were moving in a southerly direction, but much too far to the east, behind my father's forces, and only after they had flanked them did they turn west and engage the Egyptians. Division 162 did not attack from north to south, and they did not attack as a division. At best there were two battalions charging into the Egyptian fire.

My father could not believe his eyes, but while watching the divisional strike-that-wasn't, he received a very strange order from command. He was to capitalize on Adan's success and move south with his division. This entailed retreating east for close to nine miles and then driving south for about another sixty miles in order to attack the southern Egyptian bridgehead near the city of Suez and seize control of the Egyptian bridges there for an IDF crossing.

This order was a complete fantasy. Gonen, inexplicably, was under the impression that Adan had crushed the Second Army, and that all that was left to do was for Sharon to run over the Third Army.

This was insane: Adan had not launched a divisional attack, and he most certainly had not crushed the Second Army; my father's forces held the high ground, which was critical for the rest of the war. Abandoning that ridge meant allowing the Egyptians to occupy it, and once they took it, the ridge would have to be taken back at a terrible price. If it were not taken back, then there would be no counterstrike against the Egyptians in that sector. In addition, it was wishful thinking at best to assume the IDF would find usable bridges in Suez; even if they did, the Egyptian crossing equipment was suitable only for their own Soviet tanks, which were lighter than ours.

My father called Gonen on the radio and said, "Nothing has been accomplished here. There is nothing to exploit. This is a terrible mistake." He told him what hills they were holding and asked him to come down to the field and see for himself. The response came in a squawk over the radio frequency. If you don't obey the order, you will be dismissed immediately. Immediately! "Then come down here and look for yourself," my father repeated. "No!" Gonen yelled. "You will be dismissed. I will dismiss you right now!"

It is no simple thing to set an entire division in motion, especially while certain elements are engaged in a firefight, but the order had been given.

My father set out at the head of a giant column, with two hundred tanks and a long line of APCs, half-tracks, and trucks all rumbling along with him. He did not, however, disengage completely. He left his divisional recon unit to hold what he considered to be two absolutely critical ridges. Code-named Hamadiya and Kishuf, the two ridges, on either side of the Akavish route, loomed over the road that led to the canal in the Deversoir region, where the countercrossing was eventually launched. They also guarded the entrance to Tasa. Without control of those ridges, the path to Tasa was wide open to the Egyptians. He had no intention of handing them over.

Three and a half hours later, the division reached the region of the Gidi Pass, some thirty miles south of where they had been. At two thirty in the afternoon a helicopter landed at the head of the column, and a staff officer from the Southern Command told my father what he had already witnessed: that Adan's attack had failed; that there had been no countercrossing, as had been reported earlier in the day; and that Adan's division had taken losses. The Egyptians were steadily advancing to the high ground that my father had until very recently held. Thus the new orders they received were to turn around and to race north to help Adan and take back the high ground they had relinquished earlier in the day.

My father remained outwardly quiet and calm. You have to

know him to recognize when he is furious. He turned the column around and handed down orders to the commanders as they rolled. It drove him mad to think that he was at the head of a force of civilian soldiers who had been called out of their houses and synagogues two days earlier (Yom Kippur), that they had been ready for battle the following morning, and that since then all they had done was drive up and down uselessly through the desert.

Toward evening, they returned to the region they had left that morning. The recon unit, the Eighty-seventh Reconnaisance Battalion, had managed to hold the same ridges, although the commander of the unit, Bentzi Carmeli, had been killed in the battle. But the ridgelines to the north were now in Egyptian hands. During the course of that night the division suffered grave losses as they clawed their way back up to the positions they had left that morning. Fierce tank battles raged into the night.

The one good thing that happened that day in the central sector was that my father's division had managed to rescue the soldiers from the Lakakan stronghold alongside the Great Bitter Lake. The following day, in a coordinated effort, the soldiers in Purkan, in the Islamiya district, poured out and clung to one of the division's tanks, which carried thirty-three of them to safety.

Several months after the war, Gonen admitted, "Had the offensives of October 8th been conducted under the command of Arik, the results would have been different."

There are those who explain October 6 and 7, the day war broke out and the day after, as either a surprise attack or as Golda and Dayan's inability to see beyond the prevailing "concept," which held that the Egyptians, still smarting from 1967, would not strike first and start another war. But October 8 can be blamed on no one but the IDF. The reserve divisions were in place, the nature of the Egyptian deployment was clear, and the Egyptians had not yet dug themselves in or finished transferring their forces across the canal. They could still have been hit. Victory was within reach. But the IDF did not deliver.

As Dayan wrote in his own memoir,

Arik's division, which had assembled during the day in the Havraga-Nozel region, deployed during the night of October 8 to a new line, Ziona-Hamadiya-Kishuf. The counterattack not only failed to push back the Egyptians but, on the contrary, wound up allowing them to take positions that had previously been in our hands . . . Facing those forces on the southern front were three divisions: Bren's (Adan), Albert's, and Arik's. It is possible that with this balance of forces, we could have had good results with a multidivisional attack. But this was not done. In essence, small forces threw themselves into battle. Bren was badly bruised. He had not lost his fighting spirit, but the blows he had taken had left their mark. Arik was furious. He studied, analyzed, and understood what was happening in the field. His solution was correct— to cross the canal, to destroy the missiles, and to attack the Second and Third armies from the rear. But he emphasized that we could not rely on miracles. We could not plan on the basis of the hope that we would be able to capture a working Egyptian bridge and that we would be able to use it to cross with our own troops. We needed our own bridges and rafts—and those had not yet reached the canal area . . .

This insane situation pushed the military system onto very uncertain footing, where it doesn't know how to function. My father had been an army man for decades. Military discipline is imprinted on his psyche. He obeyed the orders of the commander of the southern front, even though it was clear to him that the best interests of the campaign lay in disobeying the orders.

"More than a few of the orders seem illogical to you and on more than one occasion you begin to wonder about the orders you've received, but that day was our first real day in the field.

"After that I scrutinized every single order I received from

Southern Command two or three times. I am sure that Avraham Adan and Albert Mandler did the same."

That evening, October 8, Defense Minister Dayan suggested to Chief of Staff Elazar that my father and Gonen switch roles. Elazar was the one who had appointed Gonen. Despite the failure of his appointee, he was not willing to authorize the swap or to relieve Gonen of his command entirely. Instead, the former chief of staff, Bar-Lev, was approved to oversee the Southern Command, in effect dismissing Gonen without actually dismissing him. Bar-Lev, the founding father of the failed defensive strategy along the canal, was now called in, four days into the war, to try to right the catastrophe that his flawed military conception had wrought. Bar-Lev was the last thing that was needed on the southern front. He did not suffer from personality disorders like Gonen, but he too avoided coming down to the front. From afar, he assessed the situation wrongly and made many poor decisions. He was, however, a member of the right party, and he had been a member of the same Palmach group, along with Dado, Gandhi, and Bren.

The week following the eighth of October was wasted in what is known as a strategy of holding and containing. In truth, neither the Southern Command nor the General Staff had any idea what to do, so they ordered the troops to remain in a holding position.

This infuriated my father. Why wait and allow the Egyptians to send more armored divisions across the canal? Why give them time to further entrench themselves, fortifying their bridgehead? He urged the higher-ups to attack. The October 8 failure did not alter the fact that the following day, too, required an offensive, this time carried out properly, with two divisions moving in unison and in concentrated efforts to crush the Second Army and eradicate its presence in the Sinai and then proceed south and deal the Third Army a similar blow. Nothing had changed from October 8.

In the early evening of October 9, recon forces from my father's division discovered an open seam between the Egyptian Second and Third armies. My father knew that this was what they had been waiting for. Tanks from the Eighty-seventh Recon Battalion had almost reached the water near the Great Bitter Lake and the shore at Deversoir, where, in May, my father had prepared the yard and ramparts for a possible crossing. Every instinct told him that this situation had to be exploited immediately, before the Egyptians noticed the opening and sealed it shut.

He called Gonen ("Gorodish") and said, "Shmulik, we are near the canal. Shmulik, we can touch the water of the lake." Gorodish began to yell. My father, knowing his mental state, finished the call, waited twenty more minutes, and then called back. He explained that if they tried to cross in the northern sector, near Kantara, they'd have to fight their way to the water, while here it was wide open and theirs for the taking. All they'd need to do is organize the crossing equipment, secure the area along with Adan's division, and then start to send the troops across. Why wait for the Egyptians to discover the seam and shut it down? The response from Gorodish was chilly. We'll think about it, he said.

That was the first time my father had raised the matter of crossing at Deversoir. His stance was not accepted.

Less than an hour after my father's two calls to Gorodish, Gorodish's deputy was on the line: there will be no offensive; the troops are to be pulled back; no pressure should be brought to bear on the Egyptians; they should merely be contained.

My father tried to reach Major General Yisrael Tal, the deputy chief of staff. He was not available. Instead he reached Dov Siyon, an old friend whom we used to call Dubon ("Teddy Bear"). My father said he did not understand the logic. After all, if this development had happened on October 8, everyone would have been overjoyed, so why was it any different now? Why should they pull back? The Egyptians would find the seam, close it up, and dig in even deeper, and then the army would have to do the same thing, only under far worse conditions.

His logic did not sway the command or the General Staff, who were probably still in shock from the failures of the previous day. Sorrowfully, he ordered his forces to withdraw.

Four days passed, from October 10 to October 13. It felt like the waiting before the Six-Day War: the division was bursting with confidence and fighting spirit, and the higher-ups projected nothing but weakness and a faltering resolve.

During this waiting period the Egyptians launched small-scale attacks and used helicopters to land commando troops behind Israeli lines. The Egyptian attacks were all rebuffed, with some of the helicopters shot down in the air, and the commandos that did land behind our lines were taken care of by the Patzi Force. On the night of October 13, Patzi and his men took out more than one hundred Egyptian commandos.

Most of the other commanders apparently liked the situation they were in, of staying put and from time to time sniping at Egyptian armor. It drove my father crazy. He did not feel that the IDF had any reason to be on the defensive. He wanted to put the initiative back in the IDF's hands and attack. By October 10 the Syrians had been pushed out of the Golan Heights, aside from Mount Hermon. Their war effort was a failure. Israeli forces advanced through the Syrian territory.

The Egyptians, on the other hand, were riding high. They had managed to amass and entrench forces on the eastern side of the canal in two areas, and were a cease-fire to be imposed, they would have achieved their primary goal.

On October 14, at 6:20 a.m., the Egyptian army, acting against its best interests, sent one thousand tanks forward, charging east. By afternoon it was over. Amnon Reshef and Haim Erez's brigades in my father's division destroyed 100–120 tanks while losing a sum total of 5 of their own. The results to the north and south were similar; Adan and Kalman Magen, who had replaced Mandler after he was killed by a direct hit, together took part in a massive victory that claimed some 200–250 Egyptian tanks. But the bridgeheads remained unscathed, and the Egyptians still had more than 700 tanks in that sector closely guarded by infantry, and antitank missiles.

The blow to the Egyptians, however, restored the General Staff's confidence. Two days prior, the chief of staff had recommended to the government that Israel request a cease-fire. Dayan accepted Elazar's opinion. The Israeli request was forwarded to Henry Kissinger, the American secretary of state, who thought it was a mistake, because the status quo on the front lines would put Israel and the United States in a poor position for postcombat negotiations. The Egyptians, in their arrogance, rejected the offer.

At this stage, after crushing the Egyptian offensive, the General Staff decided at long last to move to an offense strategy. That night my father presented his plan to Bar-Lev, and it was promptly authorized—Sharon's division would punch through the Egyptian lines and secure a corridor to the canal. It would occupy and secure a crossing point at Deversoir, north of the Great Bitter Lake, precisely where the recon regiment had stood five days prior. The staging ground was in the same yard that he had prepared in May. Dani Matt's brigade of paratroops would cross the canal on rubber boats and secure the western side of the bridgehead. As soon as they had their defenses in place, tanks from Haim Erez's brigade would cross on rafts, and a bridge would be built to allow additional forces to cross.

All this was to take place during the night of October 15–16.

Each of the division's forces had an extremely difficult but clear objective. The division commander had the complex task of controlling the forces.

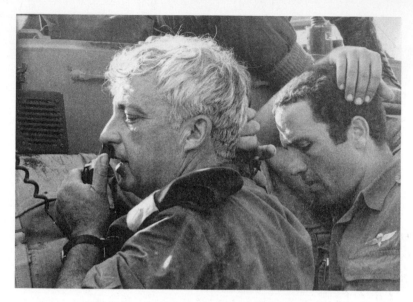

*Giving instructions to his troops, with Motti Levy
at his side, Yom Kippur War. (Uri Dan)*

At six in the morning on October 15, my father convened the
brigade commanders—Tuvia Raviv, Haim Erez, Amnon Reshef,
Dani Matt, and others—but as soon as they came together, Egyptian jets flew over the area, strafing and bombing. My father was
concerned that a shot could take out all of the commanders at
once, and so he scattered them immediately. An hour later they
reassembled, and my father laid out the battle plan.

Given the complexity and danger of the mission, there was barely
any time for prebattle preparations. The IDF had never before
crossed this type of obstacle in battle, and it never has since. When
he was through detailing the plan, my father added, "I'll be there,"
so that the soldiers knew that he was not sending them alone into the
line of fire while he waited somewhere safe for news from the front.

The main problem, though, was prosaic in nature: the crossing equipment was stuck in a traffic jam. The road from Baluza
to Tasa was jam-packed, and the Southern Front command had
not asserted its authority over the route. Adan's division needed to

be moved aside so that the crossing gear could get through. The same was true of the paratroops and their rubber boats. In the afternoon Bar-Lev called and asked my father if he wanted to delay the operation by a day, so that they could be sure that the crossing equipment would in fact arrive.

My father faced a crucial decision. Without the gear there'd be no crossing, and there was no doubt that by the following day the gear would arrive. On the other hand, a day of waiting meant that the Egyptians might well realize what was happening. Moreover, he had very little faith by this point in his superior officers. For the time being, he had authorization to cross. Who knew what the next day would bring? Maybe the authorization would be revoked, and the war would end with the Egyptians on top.

After a few minutes of contemplation, he decided not to delay the mission. He asked, though, that the command take control of the route and prioritize the traffic, and he applied heavy pressure on the staff officers to ensure the arrival of the rafts. As for the paratroopers, he knew they'd be there. Dani Matt was the same commander of a paratroop brigade that had been airlifted into the rear of the Egyptian forts in 1967, and there were men he knew from his days as their brigade commander.

Late in the day on October 15, my father's mobile command of APCs set out. It was Sukkoth, the Feast of the Tabernacles, and the road was dotted with sukkahs, built by the soldiers mostly out of ammunition boxes.

Just before dark he was alongside the Great Bitter Lake. He could see Amnon Reshef's column of tanks. This was the brigade that had suffered such awful losses on the first and second days of the war. They were led by the recon battalion. Tuvia Raviv's armored brigade would strike first, moving east to west. That brigade would draw the attention of the Egyptians, serving as a diversion. At the same time, Amnon Reshef's armored brigade would sweep south and move through the seam between the Second and Third armies, eventually reaching the northern edge of the Great Bitter Lake. There they would split into three separate forces, one occupying and guarding the yard; one attacking to the north, parallel to the canal, driving the Second Army out of the crossing site; and one attacking from west to east, contrary to the Egyptians' expectations, opening up the Akavish and Tirtur roads, which would serve as the corridor to the canal. Off to the east, my father could hear Tuvia's brigade open fire.

Dani Matt's paratroopers would then reach the canal with the rubber boats and cross. As soon as they had established a position on the western bank of the canal, the corps of engineers would set up the crossing equipment, allowing Erez's tanks to rumble across the water.

Company commander Major Yossi Regev, who was on the first tank that crossed the canal, later said, "I served as a tank company commander in Colonel Haim Erez's 421 armored brigade. I was tasked with leading the scout force in Ariel Sharon's division to the chosen area of the crossing and to lead the first group of tanks westward across the Suez Canal. I was told to rush to the intersection of routes Akavish and Tirtur and to make contact with the divisional headquarters. We kept on moving west with the generals' divisional headquarters with his APCs behind my third tank. I recall that as we arrived [next to the canal] I planned to get out of the tank, approach the division commander's APC, and complain about his decision to take part in such a dangerous mission, riding with the lead force. The battalion commander, Giora Lev, told me to relax, saying, 'Arik isn't someone you argue with, you just fulfill the missions to the letter.' Nonetheless, I approached division commander Ariel Sharon's APC." In response to Regev's question, my father replied that it was important that he be there. "That left me speechless."

The paratroopers' half-tracks convoy came through with the rubber boats, the floating bridges, and the bulldozers. The yard was completely silent, but behind them, a few hundred yards away, a brutal battle unfolded near the Tirtur-Lexicon junction. Farther north, Reshef's brigade hit the surprised Egyptian Sixteenth Armored Division from the rear.

Amid the chaos of battle, the Egyptians did not notice the crossing force in the yard. The potential bridgehead was hammered with seventy tons of artillery to ensure that no unwelcome surprises would be waiting for the paratroopers once they crossed. It was an unprecedented artillery barrage. At around 1 a.m. on October 16, the paratroopers were first across the canal in their rubber boats. The other side was quiet and lush, lined with trees and greenery that were irrigated by the freshwater canal. The paratroopers sent word over the radio: Acapulco, meaning success.

In the yard, my father personally directed the bulldozer driver toward the red bricks. He was the only one who knew exactly where the ramparts had been weakened to enable the bulldozers to go through.

A tank crossing toward "Africa" from the yard,
October 16, 1973. (Uri Dan)

My father described the scene he beheld the morning after the crossing.

On October 16th, with first light, the mesmerizing sight of the African bank was revealed to us, the western side of the Suez Canal, visible through the breached embankments.

At that time an armored brigade, under the command of Haim Erez, began crossing on the rafts, with Dani Matt continuing to widen the bridgehead to the north. By 9:00 Haim Erez had managed to destroy five missile batteries and to advance 25 kilometers away from the canal in the direction of Cairo. "The road to Cairo is open," Haim radioed back to me. Egyptian reinforcements had been destroyed.

East of the canal the battle had died down a bit. The sight before my eyes was awful. The "bonfires" of the previous night had turned into blackened, charred hunks of steel. Some 150 enemy tanks and some 50 of ours along with hundreds of other vehicles lay there, one beside the other, cannon to cannon, sometimes only a few yards apart. Alongside the burnt-out tanks, lay, at times commingled, the charred crews who had leapt out of their burning tanks to their deaths. Israelis and Egyptians together, one beside the other. I looked at the awful sight and I said to myself in a whisper, the Vale of Tears, the Vale of Tears [in Jewish tradition, a place of sorrow].

The division had three hundred casualties during the crossing battle, and many more were wounded. The Egyptian losses were far heavier.

The Tirtur-Lexicon junction was only taken at 9:45 a.m. This critical route led straight to the yard, and it had taken all night and some of the morning to secure it. On the other side of the yard things looked entirely different. The embankment had been entirely breached, and Egypt, or Africa, as the soldiers called it, lay right before them. The Egyptians were taken completely by surprise. Haim Erez's twenty-eight tanks crossed the canal. Israeli armor had opened the sky to the air force.

The price was grave, but the mission was accomplished. A corridor to the canal had been opened. A bridgehead had been formed, and Israeli paratroops and armor were operating on the western side of the canal. Rubber dinghies and tank-bearing rafts floated back and forth. Tirtur Road itself had not been opened. A division of Egyptian infantry and armor had dug in too deep, and my father's division had been unable to uproot them.

This was the situation on the morning of the sixteenth. But in the rear, at command headquarters and among the General Staff, the feeling was different. "You are cut off! You are encircled! You are surrounded!" were the calls my father got from the hysterical command headquarters. Fearing that the bridgehead had been cut off, they ordered my father not to send a single additional soldier across the canal.

"But I was right there. I did not feel cut off. I was getting supplies, fuel, men, and tanks in. I was getting my wounded out. Akavish was not exactly an open expressway. But it was certainly open."

Command's reading of the situation was both stupid and unbearable. The Egyptians still didn't understand what had happened. The rafts and the rubber boats floated freely across the canal. The bridgehead was not under assault. The Egyptians had not yet mounted opposition to Haim Erez's force, which was routing the Egyptian forces on the west side of the canal. And yet those critical hours, during which the surprise could be exploited, were spent, yet again, in a holding pattern.

If Bar-Lev or Chief of Staff David Elazar had come down to the field, to the site of the major action on the entire southern front, they would have seen that the bridgehead was not cut off, that there was a corridor open—not two and a half miles wide, as they had wanted, but open nonetheless—and that the idea of widening the corridor, pushing it farther north, while bolstering the force on the west side of the canal, was not practical for a single division. Several miles up the road an entire and entirely fresh division—Adan's—waited for orders. The right thing to do was to send one of the divisions, either Sharon's or Adan's, forward to encircle and defeat the Egyptians, and leave the other one in the Sinai to guard and widen the corridor.

They should have done one of two things: come down to the field, or leave it up to the ranking commander in the field. Bar-Lev didn't make it down to the bridgehead until October 19, four days after it had been established. The chief of staff came even later. But on the sixteenth they did not come, and they did not give my father the freedom to act.

Thus another critical day was wasted. October 16 should have been the day that Israel moved large forces to the Egyptian side of the canal, winning the war. It should have been a great day of triumph.

Instead, command continued yelling over the radio frequency that my father was surrounded, in peril. They wanted him to head back to command headquarters to discuss how best to proceed. But my father maintained that he could do no such thing, both in terms of the soldiers' morale and for the obvious reason that the situation required the commander to remain in the field. "You come here!" he told them. "Look at it for yourselves."

On Tuesday, October 16, the division was still in a holding pattern, barred by the General Staff from capitalizing on surprise. During the course of the day, the Egyptians realized what had happened. My father, frustrated by the waste of their efforts, fell asleep on the warm hood of a tank engine.

On October 17 the yard turned into the yard of death, as it was known. The Egyptians, finally realizing that the IDF had crossed the canal, pounded the yard with everything they had. The bridgehead held fast, but it was hell. Early in the morning, the yard was full and the engineers were working feverishly to assemble the bridge when a hail of Egyptian artillery began to fall, followed almost immediately by a formation of MiGs. The yard turned into an inferno.

As more jets swooped into the yard and the artillery fire intensified, Zevele Slutsky stopped directing the self-propelled rafts into the water and yelled to my father, over the roar of the jets, "Get into the APC." The MiGs were doing a dry run. My father jumped into the APC, grabbed one of the three mounted machine guns, and fired wildly at the returning jets, along with antiaircraft guns on the edge of the yard.

After the war he remarked that he fired more rounds as a general in the Yom Kippur War than in all of his military service up until then. Many men were wounded that day, including some of the people who had shared my father's APC since the beginning of the war.

Zevele Slutsky, my father's dear friend, was killed in action that morning. When my mother used to say, "Come on, Zevele, feed the boy," he would gently guide the spoon to my mouth while opening his own mouth and making airplane noises. And Motti Levy was badly wounded. My father's radio operator, Shalom Galuvei, was wounded, too.

During the Egyptian assault, my father decided to roll the mobile command out of the compound. The antennae were being destroyed by the shrapnel, and he was afraid he'd lose his radio and his ability to control the forces. While moving out, his APC shuddered hard from a nearby shell. My father felt a sudden pain in his forehead. He heard someone say, "Our friend just bought it." He opened his eyes. There was blood pouring from his head. "Does anyone have a bandage?" he asked. Other than that, he seemed fine.

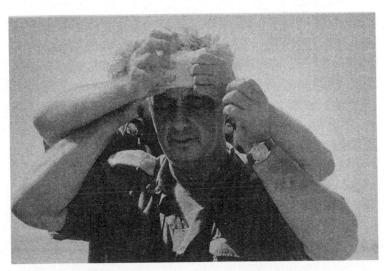

Receiving first aid after being injured during the Yom Kippur War, October 17, 1973. (Uri Dan)

At the entrance to the yard they saw tanks towing more rafts. One of them was on fire. He looked up and realized that while artillery was pounding the inside of the yard, the outside of the yard was under direct tank fire. He couldn't see how Egyptians tanks could be so close. He looked through his binoculars and was shocked to see an Egyptian tank and infantry assault on the Tirtur-Lexicon road at the rear flank of the yard, a few hundred yards away. This was a critical moment. If the bridgehead was cut off, the entire counter crossing could fail. The only force at his disposal at the time was his own mobile command, five APCs strong. They rumbled straight toward the junction, with my father standing in the APC. As they advanced, an APC exploded in flames to their left. To their right was a minefield that the IDF had laid years earlier. My father called Amnon Reshef over the radio and asked him to immediately send tanks to the junction. While speaking on the divisional frequency to Reshef, my father, who was standing and therefore had the best view, shouted directions at his driver. "Driver, to the right," he yelled when they approached the burning APC. The driver veered hard, toward the minefield. "To the left, drive to the left," he yelled. He didn't know that the entire front could hear those directives, and that it rattled them.

Everyone was used to hearing his calm, confident voice over the radio. The yelling at the driver caused ripples of worry. Throughout the war soldiers turned their radios to his frequency so that they could hear his voice and let it soothe them. I still meet people who tell me that amid the carnage they would turn to his frequency to hear the rock-solid calm of his voice.

The five M113 APCs charged toward the junction, machine guns blazing. In a few minutes Amnon and his tanks arrived. My father tried to think how it was possible that the junction had been left unprotected. "What had happened to me?" he thought. "How had I left that place unguarded?" It took a moment, but then he recalled that the previous night he had gotten an order from command that an attack was to be launched to open Tirtur Road, that it would come from east to west, and that he should remove his tanks from the junction and the line of fire. The attack was carried out by the storied 890th Paratroopers Battalion, part of Adan's division. They were sent without tank support and without any intelligence into a heavily entrenched divisional complex of trenches and tanks. It was a suicide mission. No one from command had sent word that the mission had failed and that the junction had to be reoccupied.

My father and Patzi near the yard on October 17, 1973.
(Uri Dan)

Taking their first breather since the beginning of the Egyptian artillery barrage, my father wondered where Zevele and Motti were. Some of the men in the APC knew already then that Zevele had been killed, but they didn't want to tell him. They told him he was wounded. The two had been friends for twenty years.

Lying in the APC and waiting to be evacuated along with Motti, Shalom saw a radio above him and turned the dial to my father's frequency. Motti wanted to let my father know that he was okay. He knew that my father worried about him all the time and that he treated him like a son.

Shalom had a message as well: "Nail them," he said, "nail them hard." My father laughed and said, "It'll be all right, now you're going to have pretty nurses taking care of you." He knew what he was talking about.

A short while later, he was told to come to a meeting with the upper command at Kishuf, some eight miles backward from the yard. The APCs drove on the supposedly shut-down Akavish Road, and the soldiers that saw my father on the way, his head bandaged, yelled out, "Arik, Arik, be well."

The APC reached Kishuf and found Moshe Dayan, David Elazar, Chaim Bar-Lev, Shmuel Gonen, Avraham Adan, and other staff officers sitting on the warm sands of a dune, surrounded by cold bottles of soft drinks. My father approached and was greeted by a stony silence. Only Dayan said, "Hello, Arik." My father had not seen any of them in two or three days. In the interim his division had fought its way to the canal and crossed it. They had done this alone. Been through hell. And still not a single hand was extended, not a word spoken.

Speaking in his slow way, Bar-Lev said, "The distance between what you promised to do and what you have done is very great." My father managed to control himself, but barely. He was wounded, sweaty, dusty. Some of the bodies of the dead had not yet been evacuated, and here, on this dune, surrounded by cold soft drinks, sat a clean, closely shaven, and hostile group of officers.

My father clamped his mouth shut. A short discussion took place, at the end of which it was decided that they would do what should have been done a day and a half earlier. More forces would be sent across the canal. Sharon's division would hold the yard, secure the corridor, and move north on the west side of the canal toward Islamiya and some fifteen miles west toward Cairo. Adan and Magen's divisions would cross the canal, continue south along the shore of the Great Bitter Lake, and surround the Third Army.

Throughout the discussion Dayan was silent. Occasionally he got to his feet or fidgeted, but he held his silence. When it was over, he surprised the officers by announcing that he would be joining my father. Elazar, Bar-Lev, and Gonen boarded a helicopter, and Adan returned to his division. The two of them were left alone. Dayan asked about the wound on my father's forehead, and my father told him that Zevele, who like Dayan was from Nahalal, had been injured. At long last, a human interaction. They got into the same APC, and Dayan began talking about the need to stream troops over to the Egyptian side of the canal. He knew that other than Dani Matt's paratroops and Erez's force, no other troops had crossed. Bar-Lev and Elazar still did not approve the crossing. A day and a half had been wasted.

"You have to urge them to start crossing," Dayan said. "We are wasting time!"

"Moshe," my father responded, "give them an order!"

The scene reminded my father of the scene in the yard during the War of Attrition. And again, throughout the duration of this war, Dayan, the minister of defense, made do with what he termed "ministerial advice." Nonetheless, his arrival at the bridgehead sent a clear signal to the heads of IDF, who still didn't come to the bridgehead themselves. Dayan showed up, and he continued to do so day after day. He also asked to cross the canal, and together the two inspected the other side of the bridgehead. Dayan's visits were significant.

Dayan was a pessimist by nature, a realist who envisioned the worst. His spirits were lifted only after he crossed the canal along with my father, inspecting the troops that had crossed about forty hours earlier.

Meeting in Kishuf. From left: Avraham Adan (Bren), Bar-Lev
(with cigar), and Moshe Dayan, October 17, 1973. (Uri Dan)

Dayan understood that the crossing of the canal was what was turning the tide of the war, and that it would eventually lead to victory.

On October 17, at four thirty in the afternoon, the bridge was completed. Rafts moved easily through the water, the bridge was operational, and still seven more hours passed before the first of Adan's troops crossed the canal. Two full days had been wasted, and the Egyptians had not been idle: what could have been a quicker and easier mission had been turned into a difficult and demanding undertaking.

Militarily, it made more sense to attack to the north on the west side of the canal, encircling the Second Army rather than the Third. The bridgehead was protected to the south by the Great Bitter Lake, and in any event had to be widened to the north to protect it. The Second Army was positioned closer to Israel's interior. Most of Israel's forces in the Sinai were positioned opposite the Second Army, so it could be more easily pressured. Additionally, the Egyptian reserves, guarding Cairo, could threaten the flank of a southern advance (and they did), but not a fanning out to the north and west. Nonetheless, it was decided to advance in a southerly direction. My father could never shake the feeling that the decision was not based on military considerations. That feeling only hardened with the passing years. His division was charged with the northern sweep, and the years of Elazar and Bar-Lev's enmity toward him, coupled with their political rivalry, dictated the course of events. He was a candidate with the Likud Party, and they wanted to make sure he did not emerge from the war with more victories.

"Give credit to military ignorance, too," a military historian told me when discussing Bar-Lev and Elazar's wartime decisions.

But I do not think it was just ignorance. Labor's well-oiled political machine was already at work on October 12, when the "Palmach rhymer" penned an article in *Yediot Ahronoth*. This was only six days after the outbreak of war, but my father was already taking flak on the home front. At the end of the article, titled "The Commanders," the chief of staff and the other generals are all mentioned by name. Adan is described in the article as a lion; Mandler, who was not in the Palmach, as a leopard; and only the general who fought between the leopard and the lion is not mentioned. There is only one name missing—Sharon—as though he and his division had not even taken part in the war.

———

I remember my father calling home one time from the western side of the canal. "I'm talking to you from Africa," he said.

"What, you're a prisoner of war?" I asked. "No," he said, laughing.

My father wanted to cut off the Second Army entirely, but he could not get authorization for the move. He was afraid that a cease-fire would be forced on Israel before the Egyptian army had been thoroughly defeated.

On October 19, Yigal Allon, deputy to Prime Minister Golda Meir at the time, came to visit the division. My father told him, "Time is working against us. We're not going to be able to finish the job in time if we don't hurry."

My father was afraid that a cease-fire would be imposed by the Americans and the IDF would not have sufficient time to fulfill its objectives. "This I can say to you with full authority: there is no problem of time," Allon said.

Three days later a cease-fire was, indeed, forced on Israel.

Roller bridge across the Suez Canal looking toward Africa. (Courtesy of Zvika Nahshon)

During the cease-fire, brigade commander Haim Erez's men found a book put out by Egyptian intelligence in which there were photos and descriptions of the main Israeli officers. Erez gave the book to my father with a dedication:

> *To Arik, notice how the Egyptians also held you in high esteem, maybe more than the Israelis. With great appreciation, from Brigade 421.*
> *October 26, 1973, Haim Erez*

Aside from a description of his family and several biographical details, including the operational successes of Unit 101 and the paratroopers, with special mention given to certain missions, the Egyptian intelligence officers also included the following under the heading "Characteristics":

> *Full bodied, silvery hair, very calm, confident, enamored with the act of self-sacrifice, a talented officer, stubborn, dynamic, brave, in love with the Paratroops and the Special Forces, a believer in the element of surprise, prone toward violence, loves to learn and to know.*

I would sign off on this description of my father except for "prone toward violence." However, from the Egyptian perspective, they would see this as a trait of my father's. After all, one isn't victorious in battle by sending flowers.

*Dossier of Ariel Sharon prepared by Egyptian intelligence
and captured during the Yom Kippur War.
(Courtesy of the Egyptian Army)*

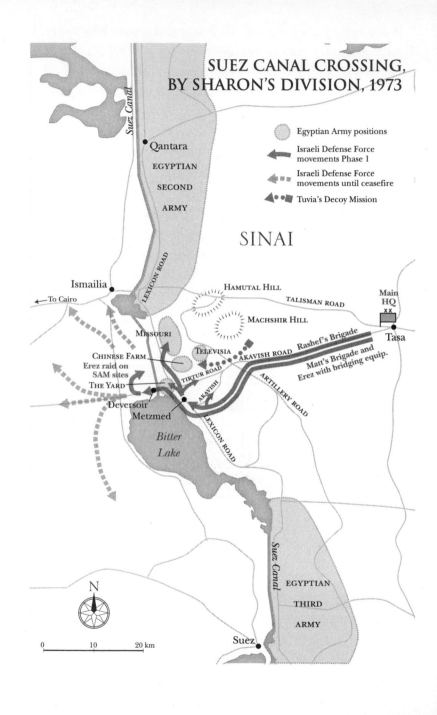

SUEZ CANAL CROSSING,
BY SHARON'S DIVISION, 1973

Egyptian Army positions

Israeli Defense Force
movements Phase 1

Israeli Defense Force
movements until ceasefire

Tuvia's Decoy Mission

Suez Canal

Qantara

EGYPTIAN

SECOND

ARMY

SINAI

Ismailia

To Cairo

LEXICON ROAD

HAMUTAL HILL

TALISMAN ROAD

Main
HQ
XX

MACHSHIR HILL

MISSOURI

TELEVISIA

AKAVISH ROAD

Rashef's Brigade

Tasa

CHINESE FARM

TIRTUR ROAD

Matt's Brigade and
Erez with bridging equip.

Erez raid on
SAM sites

THE YARD

AKAVISH

ARTILLERY ROAD

Deversoir

Metzmed

LEXICON ROAD

Bitter
Lake

N

Suez Canal

EGYPTIAN

THIRD

ARMY

0   10   20 km

Suez

The battle that left the deepest scars of regret on my father was the unnecessary battle of October 21. One day before the cease-fire began, he received an order to attack and push back the Sixteenth and Twenty-first armored divisions. They had entrenched themselves in code-name Missouri, north of the corridor that led to the bridgehead on the eastern side of the canal. The divisions were dug in extremely well. The Egyptian divisions posed no risk to the bridgehead, or the three bridges already in place over the water. In addition, Tirtur Road was open as well as Hakavish. Tuvia Raviv's brigade had been slicing away at the division's forces, but there was no way for their force to destroy the stronghold. Nor was it necessary. The IDF forces were operating in an area on the west side of the canal, moving toward Islamiya, and each time they took another Egyptian position, my father had them raise an enormous Israeli flag, illustrating to the Egyptian forces on the east side of the canal how they were being cut off. This frightened them and broke their spirit. The IDF did not need to attack "Missouri." My father did everything in his power to have the order rescinded. He explained that it was an unnecessary waste of human life, but he was unable to change it.

In the afternoon of October 21, my father stood up on the ramparts of the western side of the canal and watched the offensive on the eastern side. He saw Raviv's tanks penetrate deep into the Egyptian positions, and he saw them pummeled by Sager missiles, rocket-propelled grenades (RPGs), and tank fire. Raviv's tanks were stopped in their tracks. They burst into flames. My father and all the people who were there with him had seen so much death during the two weeks of war, but all of the other battles in which men had died had been crucial to victory, so the awful pain had been tempered by the knowledge that this was the price that had to be paid. But the attack on Missouri was a pointless suicide mission. It sickened my father, and the feeling has only gotten worse with the years. In the evening the command ordered him to launch another attack. He told them that he was on the ground and had a clear reading of the situation. They should come and see for themselves.

Dayan agreed:

> *After midnight I received a telephone call from Arik, who requested that I intervene in the matter of Missouri. The command demanded that he press on with the attack. As far as he sees it, it is a pointless thinning of his ranks. If he mounts another attack, he will continue to lose men without capturing Missouri. I've called Deputy Chief of Staff Tal and asked that either he or the chief of staff get fully involved, clear up the situation, and issue the necessary orders. A request like this from Arik cannot be ignored. If necessary, I too shall get involved. In the morning Major General Tal called my office and left word that he had spoken to Arik and to Gonen and that it had been decided that the command would leave the decision in Arik's hands whether or not to resume the attack on Missouri. It turns out that in yesterday's attack Tuvia lost many tanks.*

Gonen's troubled psyche goes a long way toward explaining his dreadful errors during the war, but that defense does not hold when examining Bar-Lev and Elazar's actions. Those two men were of very firm and settled minds. One can't shake the conviction that the order to attack Missouri was a consequence not of operational ignorance alone but also of their deep-seated hostility toward my father, which was only exacerbated by political considerations. That is what drove them to try to prevent my father from attaining any further accomplishments on the west side of the canal, which is where the war would be won. They wanted the curtain to come down on the war with him mired on the east side of the canal. The years have done nothing to soften this conviction.

The order to attack Missouri was not the high command's only attempt to chain my father to the eastern side of the canal. A few days earlier they had been consumed by another matter—how to ensure that Sharon's division not be first across the water. Arie Baron, Dayan's aide-de-camp, describes what he witnessed from

Southern Command headquarters. "From my observation post at Command headquarters I couldn't help but get the impression [regarding orders about who should cross first]. Earlier on I'd heard most of the top officers there, including the chief of staff, grappling, first and foremost, over the issue of how to stop Sharon from crossing before the others, lest he take all the glory for himself." This was on October 16, in the morning, while bitter battles raged especially around the Tirtur-Lexicon junction, but the high command had immersed itself in the struggle for a glory that had not yet been won.

This struggle is evidenced over the radio frequencies as well. Gonen spoke to Adan (Bren) during the afternoon of October 17:

> GONEN: *Cross with your 1 [. . .] If you don't cross they'll send 2 [Sharon] to cross. You cross right now. Do you understand what I'm saying?*
> BREN: *I understand. I'm completely empty [no ammunition, no fuel] [. . .]*
> GONEN: *Take half of 3 and cross with him. Take 1 as is, not full, and cross with him, too. Am I being understood?*

Later in the evening they spoke again:

> GONEN: *What's going on at the bridge?*
> BREN: *I'm not there yet. [. . .]*
> GONEN: *That's not good, Bren.*
> BREN: *That's how it's going here.*
> GONEN: *That's not good. Move them unfull and unready and get them up.*
> BREN: *Listen, it's not going to work like that.*
> GONEN: *No, Bren, I'm not kidding. I can't talk to you because this is radio. There's a reason. [. . .]*

My father's conscience was racked with guilt and regret over his decision to send Tuvia's men into a battle that should never have taken place. He always felt that legally and morally he should have refused to obey that order.

The war ended with a great but sad Israeli victory. The counter-crossing, which my father had pushed for from the onset and eventually executed, led to the encircling of the Third Army (in a further act after the cease-fire), and the Second Army could have been encircled as well. The IDF was positioned about sixty miles from Cairo with no Egyptian forces in its way. But the price that was paid, the price—thousands of dead and wounded, souls mangled, bodies broken, families torn apart, a national trauma, a deep pain.

The antiheroes, some of whom testified against each other in the Agranat Commission of Inquiry, fared badly—Golda Meir was ousted from power by a disgusted public. Dayan was never himself again after the war, and he died young, at the age of sixty-six. Elazar, a coolheaded officer, was pushed from his position by the commission, and he died even younger than Dayan. Bar-Lev always carried with him the infamous line that collapsed so early in the war and bore his name. And Gorodish was barred from command, and his life unraveled in the jungles of Africa like the crazy Colonel Kurtz in *Apocalypse Now.*

Crossing the bridge back from "Africa." (Uri Dan)

TWELVE

Clockwise from left front: President Anwar Sadat, Prime Minister Menachem Begin, then minister of foreign affairs Yitzhak Shamir, and my father, in Sharm El-Sheikh, June 4, 1981. This summit was held three days before the destruction of the Iraqi nuclear reactor. (© Moshe Milner, Israeli Government Press Office)

On December 31, 1973, two months behind schedule, the country held elections. For the first time, my father was voted into the Knesset. When the war ended, the heads of the Likud Party asked my father to resume his duties as campaign manager, but he felt uncomfortable leaving the front and decided to stay in the field along with his division.

The Likud did well in the elections, earning 39 seats in the

120-seat house, more than Herut or the Herut-Liberal bloc ever received. The electoral change that began with the formation of the Likud was finalized three and a half years later, in 1977, when the Likud won the election and formed a government.

On January 20, 1974, my father parted with his soldiers. The next day he was sworn in to the Knesset.

Not long afterward the Likud organized a demonstration in Tel Aviv's Kings of Israel Square (later to be renamed Rabin Square) against the terms of the interim agreement with Egypt. The agreement, brokered by Secretary of State Henry Kissinger, allowed the Egyptians to occupy a strip of land east of the canal, from the Mediterranean Sea to a point between the city of Suez and Suez Bay. And so, despite Israel's military victory and its ability to destroy the Egyptian army, the Egyptians secured a major victory at the negotiating table. My father and the Likud opposed the terms of the agreement.

On a rainy day people packed the square. After the speeches, my father tried to get back to his car, but the loud and loving crowd wouldn't let him go. He was surrounded by people who wanted to congratulate him, shake his hand, hug him, kiss him. The crowd pushed him back all the way to the storefront windows on Ibn Gvirol Street. After an hour, the police freed him from the hands of the adoring crowd. Throughout his political career my father was always significantly more popular with the public than with the members of his own party. The forming of the Likud Party and the political success that came in its wake was only the first step, as far as he was concerned. The next stage was the creation of a single institution, making the Likud not just an alliance of political parties but a unified and stable entity that would not have to live under the constant threat of extortion and possible dissolution by its parties when the day came and they were asked by the president to form a government.

Other Likud members didn't see why he continued to insist on uniting the various parties into one. The idea had worked; they'd gotten more seats than they would have individually, and he, too, had been elected. What more did he want?

Golda Meir formed a government in March and resigned shortly thereafter, in April. Yitzhak Rabin succeeded her as prime minister and made sure that my father was appointed commander of a reserves armored corps. This was both a promotion and a show of confidence, and the Labor Party, of which Rabin was head, liked neither. The Labor Party politicians devised their own solution, a special decision that forbade any member of Knesset from holding a high-level field command. This was tailor-made; there were no other members of Knesset (MKs) with a high-level field command. All the rest of the military men were staff officers.

With Yitzhak Rabin in the Knesset.
(© David Rubinger, Yedioth Ahronoth)

My father did not feel terribly conflicted. On December 23, 1974, less than a year after his election, he resigned from the Knesset, feeling the reserve command was more important.

In June 1975, Prime Minister Rabin asked my father to serve as his security adviser. Personally, the two men had always gotten along well. Rabin had always been a Laborite, but he was not a fierce Labor ideologue like Golda, whose first question was always "Is he one of ours?" Rabin was not the classic party man, like Shimon Peres, and my father and Rabin shared the same views on certain issues. My father also felt that accepting the post would allow him to have some influence on matters of national security.

Rabin recalled in his memoir, "It was my adviser, Arik Sharon, who recommended, against the prevailing opinion, that the Egyptian warning station be positioned within the passes, as close as possible to our troops as deployed along the new lines. In general, I was greatly encouraged by Sharon's attitude. He said: 'I disagree with your understanding and firmly oppose the interim agreement, but so long as I am your adviser, I will give you the best advice that I am able within the context of your policies.' In this way Sharon proved his fairness and his loyalty."

My father continued to serve as Rabin's adviser until February 1976, when they amicably split ways. After all, my father is not a great candidate for an advisory position; he is a decision maker, not an adviser. But he found that period to be useful, an opportunity to see from up close how a prime minister decides on matters of national security.

Before the elections of 1977, my father made a political error—he founded a political party of his own, Shlomtzion. He decided to run with his own party rather than with the Likud.

Despite his popularity, it was impossible to compete against the big parties and their machinery. Moreover, the desire to oust Labor from government, after nearly thirty years of power, made people hesitant to cast a vote in favor of a small party. Voting for Shlomtzion and not for the Likud would contribute to yet another Labor victory.

In 1969 Ben-Gurion, the former prime minister and the man who had proclaimed Israel's independence, had only received four seats at the head of a small party. "A big party is a big pit of snakes.

A small party is a small pit of snakes," I've heard my father say. If both are snake-ridden, then one might as well be in the big one.

But much was gained by the experience of founding Shlomtzion. It was an important political lesson, and it also allowed us to meet Reuven Adler, who was then a young graphic artist. He came and never left. He designed the party's logo, which looked like a Star of David. He quickly became a close friend and has remained so for more than thirty years.

Likud won the election, ousting Labor from power. Shlomtzion received two seats in Knesset and immediately merged with Begin's party. On June 20, 1977, my father was sworn in as minister of agriculture in Begin's government. My mother, my brother, and I sat along with Vera and watched the ceremony. Vera was nearly eighty, and she had been working the land for nearly sixty years, with another decade still ahead of her. Samuil, her husband, had been an agronomist and an innovator in agriculture in Israel, and so both she and my father felt that the appointment to this post was especially moving and symbolic. My father felt comfortable taking over at the Ministry of Agriculture.

During the session of the Knesset, MK Meir Vilner, a member of the Communist Party, called out a comment. "Go back to Russia," Vera shouted at him from the spectators' seats. She wasn't crazy about the Communists in Russia, and even less so in Israel.

Begin appointed Dayan foreign minister, a decision that seems like the act of a man who had thirsted for power for years and then feared it when it was finally given to him by the people. After all, he disagreed with Dayan on all matters and particularly those pertaining to the interim agreement with the Egyptians, in which Dayan had played a role. Perhaps what Begin sought was a wholly unnecessary stamp of approval from a former Labor minister. On the ground, the appointment made sense—Dayan was an intelligent and experienced man—but the invitation to join Begin's government and Dayan's agreement were a surprise. Dayan crossed the lines, earning himself a cover shot on the weekly magazine

HaOlam HaZeh (This World) beneath a headline that proclaimed: "The Political Prostitute."

Ezer Weizman was named minister of defense.

My father, from the moment he was appointed minister of agriculture, has always done his utmost to aid the kibbutzim and farming villages, especially the ones far from the center of the country. The number of politicians who understand the importance of settlement and its unique needs is dwindling. Civilian settlements are what determined the contours of our borders, and today it is civilian settlements that protect our open spaces. They are far more important than their numbers would indicate. Kibbutz Nir Am, established in January 1943, for instance, situated close to the north of the Gaza Strip, does more for the security of this country than a neighborhood in a large city, even though the total population of the kibbutz could fit into two or three city buildings.

My father understood this and helped whenever he could. There are agricultural communities, he used to say, "that I cradled in the palm of my hand." This never stopped our kibbutz neighbors, all of whom belonged to the Labor Party, from coming out to protest outside the gate of our farm, armed with angry placards.

He used to remind our friends from the nearby kibbutzim, the ones who came to our house, "During the day you stand outside the gate and protest, and at night you sneak inside and ask for help." He would say that with a forgiving fatherly smile. But then come to their aid, always, and even when he was in the opposition and their people, Labor, were in power, they still came to him. The difficulties of agricultural communities such as kibbutzim or farming villages, quite frankly, don't interest the members of either party.

My father's other role in Begin's government was as chairman of the Ministerial Committee on Settlements. In this role he put Likud policy and his own beliefs into practice. He founded many dozens of settlements in Judea, Samaria, the Gaza Strip, Galilee, the Golan Heights, the Negev, and in the Arava. If somebody was needed to speak about our rights to the land of Israel and the security need for settling different areas in Judea and Samaria, there was no better man than Begin. The history of his movement is filled with flaming speeches and ideological directives, but those stand in stark contrast to their record of actual accomplishments.

My father was born into a different culture, pragmatic Zionism, which believed in simply getting things done: establishing another village, laying another water pipe, planting another orchard, tilling another furrow of earth. Political Zionism, which Begin and his people believed in, attached great power to words, to each comma in their ideological constitution, and far less importance to the actual execution of those ideologies. It was only natural, then, that my father would be the one to translate Likud ideas into action.

Settlers and my father and I, summer 1979.
(© Micha Bar Am, Magnum Photos, Inc.)

My father began to consolidate his thoughts on the matter of settlement in Judea and Samaria during his service as Rabin's adviser.
He believed that Israel could not under any circumstances afford to
return to the June 4, 1967, lines. Living within those borders, Israel
was attacked by Jordan and suffered for years from Palestinian terror. Pre-1967, Israel's width along the coastal plain at the country's
center, where the majority of the population lives and where the national infrastructure such as power plants and the airport is housed,
is only a few miles across. That is not a defensible border. The plan
that my father drafted and brought before the government for approval offered solutions to several problems—Israel's lack of depth
along the coastal plain, its vulnerable eastern front, and the safeguarding of Jerusalem. Holding a large map, he presented his vision to the ministerial committee in September 1977, three months
after being appointed minister of agriculture. What he showed
them was a line of settlements along the high ground that looms
over the coastal plain. In that way Israel was given depth at its most
vulnerable point and it secured control over the dominant terrain,
which could no longer be occupied by hostile forces.

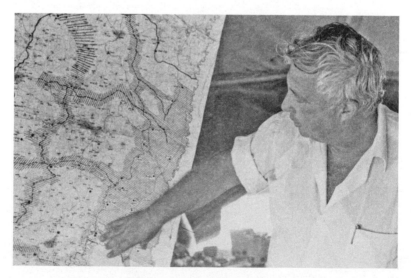

With map of planned new settlements, circa late 1970s.
(© Micha Bar Am, Magnum Photos, Inc.)

Jordan, Syria, Iraq, and Saudi Arabia all waged war against Israel in 1948, 1967, and 1973. They constitute what is known as the eastern front. Even Labor governments have recognized the need to create a line in the Jordan Valley, which is nearly entirely empty of Palestinian villages. A Labor government had already erected a thin line of settlements along the Jordan River. My father's plan called for fortifying the hills to the west of the Jordan Valley with additional settlements, building a cross-Samaria road that would be protected by settlements and serve in a time of need as emergency routes for troops heading to the eastern front.

The third element of his plan was Jerusalem. The question was how to secure Jerusalem as the eternal capital of the Jewish people, especially in light of the post-1967 wave of Palestinians flocking to the city. In the decade following the war, the Arab population increased by more than 50 percent.

The solution my father presented was a ring of Jewish settlements around the city. This would preserve the demographic character of the city and would prevent the threat of making Jerusalem a part of an urban Arab bloc stretching from Bethlehem in the south to Ramallah in the north.

On October 2, 1977, the cabinet authorized the plan, putting it into motion. My father and his aide Uri Bar-On, a brigadier general in the reserves who was also a close friend, began surveying the terrain, mountain by mountain, hill by hill.

The points chosen were state-owned lands that were untilled and uncultivated. These lands had been the property of the Ottomans during their rule, then the British, followed by the Jordanians and Israel. He worked with the Ministry of Justice, accompanied by Plia Albeck, the head of the civil department of the state attorney's office. As Albeck explained, "My job in regards to the settlements was to make sure that the land upon which they want to build a settlement is state land and that no individual rights are infringed upon."

My father would laugh when recalling his trips with her on

helicopters and on rocky hillsides, her hair covered according to Orthodox tradition in a kerchief and her feet in boots. Her rulings regarding state land all stood up under appeal to the Supreme Court.

During the following four years my father spearheaded the effort to found sixty-four new settlements in Judea and Samaria. But the rise of the Likud to power and the fact of his service in government were not enough to get the project off the ground. They needed people willing to settle the land, too. These were found in the form of the Gush Emunim loyalists. These God-fearing religious nationalists felt that settling in the biblical land of Israel was a commandment of supreme importance. Years later, my father would remark with a smile that they viewed him as "the Messiah's donkey," the man who would help them realize their ideals and faith.

This relationship became far more fraught years later when my father orchestrated the withdrawal from Gaza, underscoring the ideological divide between them. The dispute about the proper path for the state of Israel strayed far from the parameters of a mere disagreement, and many of the settlers in Judea and Samaria grew very hostile toward him personally.

Alongside his efforts in Judea and Samaria, he also built fifty-six new settlements in the Galilee. The story of the massive growth spurt in the Galilee began in our kitchen one Friday night in 1978 after dinner, as we watched the news. A story came on about the Galilee, and a young Arab man from one of the villages in the region spoke straight into the camera, saying that the only Jews that came into the village were truck drivers for the dairy company and mail trucks, and that soon even they wouldn't come. The young man described the reality of the Galilee at the time, noting that Arabs were building houses, both legally and illegally, that Jewish settlements had not been built for years, that there was hardly a Jewish presence in the region, and that talk was circulating among the Arab population about Arab autonomy in the Galilee.

The people of Israel owe their thanks to that young man, who triggered a quick solution to the problem. Two days later, on Sunday morning, my father began working on a plan. The settlements established in the Galilee were small at first, and they drew their share of ridicule, but they are today in very high demand and they have changed the reality of life in that region. Without them I am not sure to what extent the Galilee would have remained Israel's.

In November 1977 Anwar Sadat, president of Egypt, made his historic visit to Israel. He spoke before the Knesset in Jerusalem. This visit was a prelude to the Camp David Accords in September 1978 and the final peace agreement in March 1979.

My father first met Sadat on that visit. Before Sadat deplaned on his arrival at Ben Gurion Airport, he asked, "Is General Sharon here?" He was, and when the Egyptian president greeted my father, he smiled appreciatively and said, "I tried to catch you at the canal." To which my father replied, "Now you have a chance to catch me as a friend."

In May 1980 Ezer Weizman resigned from his position as defense minister, leaving the post in Begin's hands. My father soon found himself serving as the primary channel of communications between Begin and Sadat. During the negotiations at Camp David, at a critical juncture, with Begin forced to decide if he could agree to cede all of the Sinai in return for peace with Egypt, he called my father in Israel. My father's support helped him make the decision to sign the peace treaty.

*With Kamil Hassan Ali, the Egyptian minister of defense,
in Samaria, 1979. (Courtesy of Issac Ismach)*

One of the thorniest issues during the negotiations was the matter of the Palestinians. It was important to the Egyptians to show the rest of the Arab world that they had not forsaken their Arab brethren. The ever formal man of words, Begin, advocated in favor of autonomy for the Palestinian population in the West Bank and Gaza. It was important to him that the word *state* not be used. My father disagreed. "Better to have a Palestinian state on part of the territory than autonomy across all of it," I heard him say countless times. Autonomy is, by its nature, likely to lead to statehood, and therefore the formalities and the semantics were less important than the likely reality. My father contended that autonomy could lead to Palestinian statehood across all of the territory, and it could be recognized internationally, but a Palestinian state, which would have seemed like a greater achievement, would allow us to keep the places most crucial to the security of

Israel, and this would make the conflict less severe, for then it would be a territorial dispute between neighboring states and not a question of occupation and forced rule.

Begin remained true to himself, valuing the written word over reality on the ground. In any event, neither occurred.

In early 1981 the Egyptian minister of agriculture, Dr. Mahmoud Mohamed Daoud, called my father with an unusual request—could he send an Israeli team to set up an irrigation system on an Egyptian farm "in a very important place" within ten days?

My father was under the impression that he was talking about Sadat's family farm in Mit el-Kom, a small village along the Nile Delta where he was born. It was not clear to my father why the Egyptian minister was in such a rush, but he felt that the Egyptian minister was uncomfortable explaining over the phone, so he accepted the very tight schedule and merely told Dr. Daoud yes.

An Israeli team set out the very next day. They were taken to Sadat's farm. Within ten days they managed to install an advanced irrigation system for the orchards and for a new planted vineyard. The filtration system, pumps, pipelines, and drip irrigation system were all very unusual in the villages of the Nile Delta, where they still used the ancient agricultural methods, irrigating with the help of floods and yoked oxen that pulled buckets of water into irrigation ditches. On the tenth day, after round-the-clock work, the president's wife, Jihan Sadat, arrived and was very pleased with what she saw. The entire matter was kept secret until April 1981, when Sadat invited all of the major newspaper editors in the country to his farm and said, "Look at what the Israelis are capable of doing," pointing out that the entire project had been done in ten days flat. He explained that Israeli-Egyptian cooperation in agriculture was of crucial importance. This fell in line with Sadat's vision of making the desert arable, increasing the country's ability to feed its own rapidly growing population.

Minister of Agriculture Sharon with President Anwar Sadat at
his palace in Cairo, May 1981.
(Uri Dan)

A month later, in May, my father was in Egypt and visited Sadat's farm in Mit el-Kom, which now looked like an Israeli farm. When he met Sadat at the presidential palace in Cairo, the two discussed agricultural cooperation and innovation. When one of Sadat's ministers suggested that once they were shown how, they would be able to do it themselves, Sadat cut him off: "We have the land, we have the water, and now we have Sharon. He will help us do it." Sadat exhibited a true greatness of spirit. When he turned toward peace he did so in earnest, as when he came to Jerusalem and showed the Israeli people in his actions that he meant what he said. He went the extra mile and it endeared him in the eyes of the world—unlike his successor.

When Sadat was done speaking, he clapped his hands once and then two more times. He did this very softly, and my father never knew how the people outside heard him, but they did. An aide appeared from behind a cream-colored velvet curtain and was asked to bring a map. Within a moment it was spread on the floor. My father and Sadat bent down and looked at the map together. Sadat pointed to the places where subsurface aquifers had recently been found, mostly near the border of Sudan and the western desert. "These are the areas where we would like to develop modern farming," Sadat said, and he asked my father if he would be willing to go and see the places and form his own opinion. My father agreed, and Sadat made sure that his private plane was readied for the following morning. Toward the end of the meeting, my father asked Sadat if they could meet in private. Once everyone had left, my father sent greetings from Begin and asked whether he might agree to a summit meeting between the two leaders. Sadat agreed, and it was decided that the details would be ironed out at a later date.

The next morning my mother and father and a small Israeli delegation set out for the nearby Egyptian air force base, where Sadat's sleek and shiny Antonov jet waited for them.

My father sat up front, in the rear middle seat of the cockpit, between the two pilots. Talking to them, he learned that they were both fighter pilots who had taken part in the Egyptian assaults on the bridgehead during the Yom Kippur War. They flew through the desert, with my father telling them when to drop down, when to circle, when to leave, as he looked over the territory, and there, between the two fighter pilots, he had a realization about peace. It all came down to this: two fighter pilots, who only eight years earlier had been doing their best to kill him, were now following his requests to be flown around the country, searching for arable lands so as to feed the Egyptian population.

My father and mother crossed back to Israel through the new tunnel the Egyptians had built under the Suez Canal, which Sadat wanted my father to see before leaving. There they crossed the Sinai toward el-Arish, where there was an Israeli checkpoint. (The Israeli withdrawal from the Sinai was only completed one year later.) The familiar roads, the bloodstained places, the bombings and burned-out vehicles—all of those were still fresh in his mind.

Several days later, the Egyptians sent word that they would like the summit to be held in Sharm el-Sheikh. But then, for reasons unclear, they began to drag their feet.

For Israel, the timing of the meeting was crucial. Israel planned to strike the Iraqi nuclear reactor outside Baghdad. Attacking before the summit was problematic because certainly the meeting would be canceled. Or if they did attack after the meeting, it would incriminate the Egyptians in the eyes of their Arab detractors, making it seem as though they had condoned the attack. They needed some space between the two events.

But the strike could not be delayed. It had to be executed before the reactor went "hot," and that, according to the information Israel had, might have been days away. After that, any strike against the reactor would cause unspeakable environmental damage.

My father's stance regarding the Arab states' attempts to acquire nuclear weapons has remained constant over the years: Israel cannot abide such a development. Action must be taken to prevent it from becoming reality. According to Shlomo Nakdimon in his book *Tammuz in Flames*, "Sharon advised his colleagues to set a policy whose main point was that nuclear weapons in the hands of the Arabs is a red line in the sand for Israel. According to the theory he developed, Israel will not be able to accept a balance of power as was between the superpowers: a balance of power between Israel and the Arabs will only make it difficult for Israel."

In my father's opinion, anti-Israel terrorism would increase under such a balance of power if it occurred. Israel's freedom to respond to attacks would be greatly constrained, because Israel, a rational actor, would be hemmed in by its concern that too forceful a response could lead to a disaster. Moreover, the acquisition of nuclear arms by countries like Iraq, Syria, and Libya, allowing weapons of mass destruction into the hands of men like Gaddafi, Assad, and Saddam Hussein, who never veiled his intention to wipe away the Zionist entity, repeating it again and again, was a completely unacceptable scenario. My father saw that kind of development as something that Israel simply could not afford.

The French, in the mid- to late 1970s, were the ones who built the Iraqi nuclear reactor. Early in 1980, with the plans progressing smoothly, it was clear that Saddam Hussein sought to attain nuclear weapons. My father applied constant pressure during the meetings of the ministers' security committee to prevent this, and by 1980 the entire forum felt the immediacy of the threat. France signed an agreement obligating it to sell Iraq a nuclear reactor, as well as supplying enriched uranium and training Iraqi physicists, engineers, and nuclear technicians. Italy also signed an agreement with Iraq. As Nakdimon noted, "The important part of the agreement was the Italian commitment to sell to Iraq four facilities including a radio-chemical laboratory and hot cells. From now on, there was no need for sophisticated secret service investigations to know where Iraq was heading. Indeed, already in mid-December 1975, Hudson Institute researchers estimated that the French reactor built for Iraq could produce enough plutonium to produce two small atomic bombs a year." On March 20, 1980, London's *Daily Mail* wrote, "Next year Iraq will be capable of manufacturing a nuclear bomb with the help of France and Italy." All this put Saddam Hussein on a fast track toward the manufacture of a bomb, whose components had been supplied by Western nations, thereby neutralizing any real hopes of disarming him with the tools of diplomacy.

Against this background, starting from April 1980, my father began a concerted effort to convince the cabinet to destroy the reactor entirely with a military strike, rather than merely damaging it by sabotage. Sitting in the Knesset cabinet room, under Reuven Rubin's painting of olive trees in Safed, a light fog hovering over the trees, he told the other cabinet members, "By any feasible means it is vital that we execute a military operation that will ensure the reactor's destruction." He turned to Begin time and again about this issue, expressing himself in writing and orally. But the truth is, Begin did not need much convincing. He had come from Poland. He'd seen what had happened to the Jewish people on European soil when a dictator threatened annihilation and did his utmost to carry it out. He was acutely aware of the danger posed by Saddam Hussein. For thirteen months the cabinet met and discussed the issue in secret. But when preparations reached their final stage, the information was leaked.

On May 9, Peres wrote to Begin: "I add my voice, and it is not mine alone, to those who tell you not to act, certainly not under the present timing and circumstances."

I srael still had the problem of the Begin-Sadat summit, which had been set for June 4, 1981.

The summit took place at an Israeli hotel in Sharm el-Sheikh. Sadat was in an expansive mood, and he suggested that perhaps the Israeli staffers at the hotel could stay in the future. At dinner my father sat next to Kamal Hassan Ali, Egypt's former defense minister and current foreign minister. During the meal Ali leaned in toward my father and whispered proudly, "You see, Arik, I was the one who did it. I arranged it so we could meet as close as possible to your elections." He said this and began to laugh with pleasure.

This explained the delays. The Egyptians wanted to help the Likud in the coming elections. My father nodded and tried to show his appreciation, but he knew that in a few days he would likely have to explain to Cairo the necessity of the operation.

On Sunday, June 7, 1981, three days after the summit, Israeli jets took to the air. Members of the cabinet convened at the prime minister's residence. An hour and a half later, they received a report that the mission had been a success, and that the planes were on their way back to Israel.

After the congratulations my father remained alongside Begin. As he got ready to go, Begin put a hand on his shoulder, a rare gesture of affection, and said, "Arik, if it weren't for your persistence, I don't know if we would have done it." But my father knew that, at the end of the day, the decision had been Begin's, and a courageous one.

Iraq was in shock, as were the other Arab countries. The UN, the United States, and the Europeans—indeed, the whole world—condemned the strike. The *New York Times* on June 9, 1981, wrote, "Israel's sneak attack on a French-built nuclear reactor near Baghdad was an act of inexcusable and short-sighted aggression." The *Washington Post* wrote, "The Israelis committed a severe action." The *Boston Globe* called it "a stupid act by a desperate politician." The *Wall Street Journal* was one of the only supporters: "We should all join together and send thanks to the Israelis." Ten years later, during the Gulf War, some retracted their hasty condemnations. In 2005, former Secretary of State James Baker said, "We were wrong and Israel was right when it attacked the nuclear reactor in Iraq." But the worst of it was not the international response. The truly alarming condemnations were the ones issued by the Peres-led Labor Party. This was the first time that the consensus surrounding matters of national security had been broken. This was not some vague war in a distant land in Southeast Asia. This was a purely defensive action, certainly from an historical perspective. I don't think that Peres or any of the old Labor MKs are overly proud of their behavior back in June 1981. They condemned the operation and claimed that it was both an error of judgment and an election ploy staged in advance of the coming elections. In a pre-election ad in the newspaper *Yediot Ahronoth*, the Labor Party accused the Begin government: "The political decision to bomb the Iraqi nuclear reactor [raises] questions that cannot be publicly aired, and necessarily points to the gloomy conclusion that the Likud's electoral considerations are what decided the matter."

Begin was stunned. For twenty-nine years he'd been in the opposition and had backed the government during all of its wars and military operations—in 1948 and again during the reprisal raids in the 1950s and Operation Kadesh in 1956. In 1967 he accepted Eshkol's invitation to join a national unity government as a minister without portfolio, and during the Yom Kippur War he remained loyal to the government, careful not to criticize it while under fire. And then, with the tables turned, he found himself being accused of sending Israeli pilots into harm's way to better his results in the coming elections. "Would I send Jewish boys to risk death, or captivity, which is worse than death? Would I send our boys into such danger for elections?"

This was the first time that an IDF operation had been alleged to be driven by political purposes. Not only was it not backed by the opposition, but it was used by Labor to bludgeon the standing government. This should have signaled to Begin and to my father that the rules of the game were not as they once were, and they had not been informed of the changes. The opposition's reaction to the strike against the nuclear reactor was merely a precursor to what was to come, a year later, during Operation Peace for Galilee.

———————

E lections for the Knesset took place on June 30, 1981. The Li-kud won again. Begin appointed my father to the position of minister of defense. He left the Ministry of Agriculture with a feeling of satisfaction. Not only had he founded dozens of new settlements during his term as minister, but the agricultural production was greatly increased by dozens of percentages. As had happened many times before in his life, he went from agriculture and settlement to security. These were two sides of the same coin, as far as he was concerned.

The Zionist understanding of "settlements dictating borders" was one I learned at a young age from my father. I remember in seventh grade writing a paper with that exact title. We worked on it together. History provided plenty of examples: Hanita, Metulla, the settlements of the Negev. The places that were held only by army forces were less secure.

Taking up the post of defense minister did not feel foreign. That world, like agriculture, was intimately familiar to him.

He continued to be involved in relations with the Egyptians. "Why are you so difficult," they would ask, "on any possibility of our moving some outpost two or three hundred yards from where it is supposed to be? Why does it make such a difference to you?"

His answer was constant—Israel would always be in an inferior position to Egypt, and that could have destructive consequences for Israel. "According to the agreement, you are entitled to keep 240 tanks east of the Suez Canal. Now let's assume that one day we wake up and find three hundred tanks there. For you it's a simple logistical operation, moving sixty tanks from the west side of the canal to the east. But for us, in order to restore the situation we have to be ready to go to war. And that is a difficult decision indeed. Our own people would very rightly say, 'What, you are dragging the country to war over fifty or sixty tanks, as strong and well armed as Israel is?' So we would have a problem. Now let's say that a month later there are another fifty or sixty tanks there. For you this was another simple matter of logistics. The result is that you will be sitting in Sinai with your tanks and we will still be sitting on the horns of a dilemma, either to give up our deterrence ability or go to war. We do not want to face that dilemma, and that is why we will not tolerate the slightest deviation from our agreement."

I heard him explain this often. In the background was his recollection of how the Eshkol government had continued to waver before the Six-Day War even when Nasser had eliminated the UN presence and inserted troops into the Sinai, and the way that Golda Meir's government had resigned itself to the new situation once the Egyptians had moved their surface-to-air missiles closer to the canal after the War of Attrition. He would not allow that to happen again.

My father inspecting troops in Zaire. (Uri Dan)

THIRTEEN

For years Israel had kept up ties with nations in Africa and Asia. They were known in Israel as the "countries of the periphery." The intent was to improve an isolated and enemy-surrounded Israel's foreign relationships. Such was the case with Iran until the revolution, with the Kurds who fought for independence against the Iraqis, and with the Yemenis who fought the Egyptian incursion in the 1960s. Israel provided much agricultural, military, humanitarian, and intelligence cooperation to these countries as well as to nations of Africa. As defense minister, my father continued this tradition, feeling that it was in Israel's best interests and that it was helpful to the United States in its global campaign against the spread of Soviet influence. This was a common interest for Israel and the United States. It was clear that countries in need of assistance would not sit by idly. If help did not come from the West, then they would seek it from either the Soviet Union or one of its proxies, perhaps Cuba or Libya.

In 1981 my father traveled to Africa, accompanied by my mother. Their first stop was Gabon, where my father met with President Omar Bongo Ondimba, hoping to persuade Gabon to reinstate its diplomatic relations with Israel, which had been cut off during the Yom Kippur War. After these talks, the leaders signed a memorandum of understanding that my father hoped would lead to normalization of relations between the two nations. From there my parents flew to the Central African Republic, where he met with General Andre Kolingba. There he was haunted by the thought of what might be left in the palace refrigerators. The previous ruler, Emperor Bokassa, was rumored to have had a taste for cannibalism. The night of their visit, they were served monkey heads, eyes and all, and neither my parents nor anyone else in the Israeli delegation required warnings from the delegation physician, Professor Boleslav "Bolek" Goldman, to know that it would be best to stay away from the food.

In this poverty-stricken place my father met with Shmuel "Goro-dish" Gonen, the former southern front commander, who was hunting for diamonds and living in miserable conditions. Their meeting, despite all their bad blood, left my father saddened and shocked. This was not the way that Gonen, a hero during the Six-Day War, should live out the rest of his days, my father thought.

Fishing with Mobutu Sese Seko, president of Zaire, 1981.
(Courtesy of IDF and Defense Establishment Archive)

Next on the itinerary was Zaire. Mobutu Sese Seko, who proudly wore IDF paratrooper wings on his chest, was completely taken by my father and invited him to come and see his ancestral village, a rare honor. Mobutu piloted the presidential plane, flying low, skimming the treetops, and then raising the nose of the plane and breaking into a booming laughter. He did this repeatedly. The experience was on the one hand frightening and on the other, insane.

As a token of his friendship, Mobutu invited my parents to go fishing with him. Each of them sat under the shade of an umbrella alongside a fishing pond. "The president has caught a fish," my father said with great ceremony whenever Mobutu caught one. "The president has caught another fish," he said again and again. The trick behind the president's skill with the rod soon became apparent. It had been placed in a barrel of fish so as to ensure that no one ever took more fish than he did. Nonetheless, a diplomatic incident almost occurred when my mother, who had never gone fishing before—or after—caught more fish than the president. Mobutu's servants, however, solved the problem, moving fish from my mother's basket to the president's.

There was something weird about the chickens, my mother said after dinner: "Their heads were on backwards." They turned out to be bats.

A while later I accompanied my father on a different trip to Africa. We flew on a heavily loaded military plane to Kenya, where he had a meeting, and then continued on to Zaire. I remember the red earth on the road between the trees, the pineapple fields, and of course Mobutu himself, who was immense. In the end, Zaire was the first African nation to restore full diplomatic relations with Israel.

Tarzan (me) in Zaire, 1982.
(Courtesy of Professor Boleslav Goldman)

My father returned to Africa in May 1982, this time to Kenya to meet with President Daniel Arap Moi. The small Israeli delegation spent the night at the home of the Kenyan head of secret service. In the morning, Oded Shamir, my father's military adjutant, discovered that his pants had been stolen. The following day they set out for a meeting with Gaafar al-Nimeiry, the president of Sudan. They drove deep into the jungle to a fabulous villa owned by the Saudi millionaire Aadnan Khashugi, where they met with the soft-spoken and exceedingly polite Nimeiry, whose face, furrowed with deep tribal tattoos, was at odds with his demeanor. Completing the scene were young blond women who giggled up on the second floor, perhaps guests of the owner.

The conversation with Nimeiry revolved around two central issues. The first was Soviet expansionism, through Libya, where the oil-rich Qaddafi, with help from the Russians, funded and promoted acts of terror against Israel and, with his dreams of increased influence in Africa, threatened Nimeiry's regime. The other matter was the Jews of Ethiopia.

Ever since he was a child, my father had been fascinated by the story of Jews in Ethiopia. Since 1977, Israel had been involved in dangerous and covert missions to bring Ethiopian Jews to Israel despite the strongly anti-Israel sentiment in Sudan and Ethiopia. Nimeiry was amenable to my father's request for help on the matter of the beleaguered Ethiopian Jews. He kept his word.

*My father with Caspar Weinberger, signing a
memorandum of strategic cooperation, November 1981.
(Courtesy of the Pentagon)*

FOURTEEN

In November 1981 my father flew to the United States to sign a memorandum of strategic cooperation, which had been in the works for three months. This was the first time such an agreement had been signed between the United States and Israel. Ronald Reagan was president at the time.

Aside from the formal purpose of the visit, the signing of the agreement, my father also took the opportunity to speak with Secretary of State Alexander Haig, Secretary of Defense Caspar Weinberger, and CIA director William Casey. They spoke of how best to combat Soviet expansionism in Africa. They also spoke of the Iran-Iraq War, which, as opposed to the prevailing notion in Israel, my father was not happy about. True, both of Israel's enemies were spilling each other's blood, but in the meantime their armies were growing, as was their operational experience, and one day that war would end and the armies would remain. He warned against Western support for Iraq; although he had no love for the ayatollah-controlled regime in Iran, it was the Iraqis who had fought against Israel in all past wars. He also warned of the possibility of the Russians gaining a foothold in the Persian Gulf if the balance of power shifted. They spoke of Lebanon, too. Civil war had erupted in 1975. Christian Lebanese battled Muslims. This was not new. This time, though, the background was the PLO's control of South Lebanon, the coast, and parts of Beirut. They had created a state within a state, disrupting the balance of power and pushing the Christians to the brink. In 1976 Syria invaded Lebanon, a move that elicited no response from Israel, the United States, and the nations of Europe.

My father had been opposed to allowing the Syrians to gain a foothold in Lebanon in 1976. Ever since the War of Independence in 1948, Lebanon had not been a battle frontier during any of the wars against Israel. He thought that Syria's entry into Lebanon was a worsening of Israel's strategic standing in the region. Rabin was hesitant. Several times Rabin marked out a red line, as he called it, a line that Israel would not permit Syria to cross in a southerly direction. The Syrians crossed all of the lines, and nothing happened. It would have been better to say nothing. Conditions are best not laid down if there is no rock-solid intention of following through on them. All that does is weaken your country's credibility and deterrence. The Syrians finally halted just south of Sidon and eastward.

At first the Syrians fought the PLO forces, but soon enough it was apparent to all, even those who had felt they had come as peacemakers, that their true goal was to take over Lebanon. The Syrians have never recognized Lebanon as a nation; they see it as part of an entity called Greater Syria. (The Syrians only appointed an ambassador to Lebanon for the first time in 2009.) Shortly after their invasion, they turned their guns on the Christians and joined hands with the Palestinian terror organizations, which thrived under Syrian control.

In 1964 the Palestine Liberation Organization (PLO) was founded. In its original charter, written in June of that year in Jordan-occupied East Jerusalem, the organization stated, in Article 2, that "Palestine with its boundaries at the time of the British Mandate is a regional indivisible unit": in other words, there is no room for an additional Jewish state. Article 4: "The people of Palestine determine its destiny when it completes the liberation of its homeland."

This is from 1964, before Judea, Samaria, and the Gaza Strip became part of Israel. The complete liberation spoken of here translates into the destruction of the state of Israel.

Article 15: "The liberation of Palestine, from a spiritual viewpoint, prepares for the Holy Land, an atmosphere of tranquillity and peace, in which all the Holy Places will be safeguarded, and the free worship and visit to all will be guaranteed, without any discrimination of race, colour, tongue, or religion." Oh yes, freedom of religion is truly close to their hearts. For hundreds of years Jews were not allowed beyond the seventh step at the entrance to the Tomb of the Patriarchs in Hebron, and in the name of the same respect for freedom they were barred from the tomb itself, where the forefathers of the Jewish religion are buried, a tomb that Abraham bought in silver coin. There were also times when Jews were barred from Jerusalem entirely and forced to pray from the Mount of Olives. The Temple Mount was off-limits to all Jews until after the Six-Day War, and even prayer at the Western Wall was limited; in fact the request to place prayer benches near the Wall was one of the excuses for the riots and murders of 1929. The truth of the matter is that only since 1967 have Jews, Muslims, and Christians enjoyed the freedom to worship as they see fit in Jerusalem.

Article 16: "The liberation of Palestine from an international viewpoint is a defensive act necessitated by the demands of self-defense."

In other words, all of their terror attacks are but self-defense.

Article 17: "The Partitioning of Palestine in 1947 and the establishment of Israel are illegal and false regardless of the loss of time."

Article 18: "The Balfour Declaration, the Mandate system, and all that has been based upon them are considered fraud. The claims of historic and spiritual ties between Jews and Palestine are not in agreement with the facts of history or with the true basis of sound statehood. Judaism, because it is a divine religion, is not a nationality with independent existence. Furthermore the Jews are not one people with an independent personality because they are citizens of the countries to which they belong."

These are strange claims to be leveled against the Jewish people, who had already lived as a free nation in their own land more than three thousand years ago, and it is all the more strange when it comes from a group of people who during the course of history have never had any type of national independent government of any kind until they were given such by Israel in 1993 during the Oslo Accords. Neither Jordan nor Egypt thought to grant the Palestinians independent status between 1948 and 1967, the years in which they ruled the West Bank and Gaza, respectively.

In general you could say that this last article is a good example of the level of arguments they use. What could they possibly mean, there is no link between the Jewish people and the land of Israel? And what could they possibly mean when they say that they want to deem null and void events that happened more than forty-seven years ago? What kind of behavior is it to say, Stop, rewind history, we're not happy with what has happened?

Article 19:"Zionism is a colonialist movement in its inception, aggressive and expansionist in its goals, racist and segregationist in its configurations and fascist in its means and aims. Israel in its capacity as the spearhead of this destructive movement and the pillar for colonialism is a permanent source of tension and turmoil in the Middle East in particular and to the international community in general."

One gets the impression that the Palestinians might not really like us at all.

Article 24: "This Organization [the PLO] does not exercise any regional sovereignty over the West Bank in the Hashemite Kingdom of Jordan, on the Gaza Strip or the Himmah Area. Its activities will be on the national popular level in the liberational, organizational, political and financial fields."

This is an interesting article that needs clarification.

Why would the Palestinians write in their own charter in 1964, in the very document in which they define themselves and their aspirations, that they harbor no claim to sovereignty over the West Bank and the Gaza Strip, controlled by Jordan and Egypt? Those lands are precisely what they now demand of Israel.

This charter was published three years before the Six-Day War in 1967, before Israel acquired these territories. Now they maintain that the "occupation" of the West Bank is the reason for the conflict. But in 1964 those very same territories were in Arab hands, and they were still not satisfied and tried to eliminate Israel. In addition, when in November 1947 they rejected the UN partition resolution, there was not a single Palestinian refugee, and of course up to 1967 the Arabs controlled East Jerusalem. This proves that the territories, the refugees, and the status of Jerusalem are the excuse for the conflict and not the real reason behind it.

The PLO's first terror attack was an attempt to blow up Israel's major water pipe, which stretches from the Sea of Galilee to the Negev. The attack was carried out by Arafat's Fatah wing, and the terrorists left from and returned to Jordan.

During the next two years, Palestinian terrorists murdered Israelis who lived near the border and laid explosives in Jewish neighborhoods in Jerusalem.

And then came the Six-Day War. The armies of Egypt, Syria, and Jordan, aided by the Saudis and Iraqis, sought to wipe Israel off the map, but as we all know, that's not how things played out, and Israel managed to pull off an incredible military victory. From that moment on, all the thousands of murdered Jews and the tens of thousands of terror attacks were forgotten; the Palestinians now had an excuse, their acts of terror deemed legitimate. Suddenly the terrorists were freedom fighters, and the fact that there was nothing new about their modus operandi was forgotten. Israel's victory was the reason for the terror, and if Israel would only withdraw to its pre-1967 boundaries, then violence would abate, there'd be peace with the Palestinians, peace with all our Arab neighbors, peace throughout the world.

The thousands of Jews murdered before 1967 will testify in silence that this is not the case.

My father would say: "The primary problem is that the Arabs have yet to recognize the Jewish people's inalienable right to have a Jewish state in their homeland." That explains why Israel is not featured on the maps in geography classes all across the Arab world, including among Palestinians, who further exacerbate the problem by indoctrinating their students with hate for Jews and Israel, ensuring that there will be no true acceptance of us as a national entity in the region.

From 1968 on, all throughout the 1970s, the Palestinians engaged in, aside from their usual terrorism, the hijacking of planes and attacks against Jewish targets all over the world. In July 1968 an El Al flight en route from Rome to Israel was hijacked and taken to Algeria. The passengers and crew were held hostage for thirty-nine days, until a deal was reached. That was the beginning of the Palestinian air offensive. In December 1968 terrorists opened fire on and threw grenades at an El Al plane in Athens. Leon Shirdan, a passenger on board, was killed, and a stewardess was injured.

In February 1969 five terrorists attacked an El Al plane in Zurich, hitting it with gunfire and grenades and wounding two crew members, one of whom, pilot Yoram Peres, died of his wounds several weeks later. An Israeli security guard on board, Mordechai Rachamim, killed one of the terrorists and was forced to stand trial in Switzerland, sitting on the same bench as the terrorists had during their own trial, until he was acquitted.

The attack on El Al planes was greeted with equanimity by the world. No steps were taken to deter the countries hosting terror organizations in their midst, and no condemnations were heard. In general one could say that the world, and especially Europe, saw the hijackings and attacks as a strictly Israeli-Palestinian affair, an Arab war on Jews taking place in a different region, away from the Middle East. It was seen as unpleasant but not their business.

In August 1969 a TWA flight from Rome to Athens was hijacked and put down in Damascus, Syria. All passengers were freed aside from two Israelis, who were held in Syrian jails and only released months later as part of a prisoner exchange. The plane was blown up, and as usual Syria was not sanctioned or sternly reprimanded for linking hands with terror. The next stage was inevitable.

On February 21, 1970, the Popular Front for the Liberation of Palestine (PFLP) seized control of a Swissair flight en route from

Zurich to Tel Aviv and blew it up, killing all forty-seven people on board, including fifteen Israelis.

September 6 was known as the day of the Dawson Field hijackings.

That day a TWA plane en route from Frankfurt to New York was hijacked by another squad from the same organization. My dear editor, Claire Wachtel, was aboard. The plane was forced down in Dawson Field, an isolated airport in the Jordanian desert. A Swissair flight from Zurich to New York met a similar fate and was also taken to Dawson Field. Three days later, on September 9, a BOAC flight from Bahrain to London was hijacked.

That plane, too, was forced down in Dawson Field so that there were now three Western planes and hundreds of hostages, with Jewish men separated from the rest.

In the end the planes were destroyed, and the hostages were set free as part of a prisoner exchange. The freed terrorists included the attackers of the El Al plane in Zurich.

On May 8, 1972, a Belgian Sabena Airlines plane was hijacked on the way to Israel. Four terrorists, two men and two women from the Black September wing of the PLO, took control of the plane and forced it down in Israel. The next day commandos seized control of the plane, killing the two male hijackers and capturing the female ones. Ehud Barak commanded the mission, and Benjamin Netanyahu took part in it.

On June 27, 1976, Air France flight 139 from Tel Aviv to Paris Athens was hijacked.

After takeoff, four terrorists—two Palestinians and two Germans—hijacked the plane and steered it first to Libya, where they refueled, and from there to Entebbe, Uganda.

After several days in Entebbe, the terrorists released all hostages who were not Jews; the pilots and the crew chose to stay with the remaining hostages, a decision that later earned them medals of honor from the prime minister of Israel. On July 4, 1976, Israel rescued the hostages in a daring raid that left the world amazed.

During the raid, the IDF freed the hostages and killed the terror-ists. Lieutenant Colonel Yoni Netanyahu was killed during the course of the operation, and a soldier named Surin Hershko was severely wounded and paralyzed.

Western countries praised Israel for its actions, found them re-markable, but the Soviet bloc and the Arab countries condemned the raid, and the UN, under the direction of Secretary General Kurt Waldheim, condemned the rescue operation, labeling it a "flagrant aggression." This from a man who had a Nazi past.

International Palestinian terror did not restrict itself solely to hijackings and the assault of planes. In 1968 El Al offices were attacked in Zurich. In September 1969 grenades were thrown at Israel's embassies in Bonn and The Hague and at El Al offices in Brussels and later in Athens. Israel's consul in Istanbul, Efraim Elrom, was assassinated. Terrorists took over the Israeli Embassy in Bangkok. Israel's consul for agricultural affairs in London, Ami Schori, was killed with a letter bomb. On May 30, 1972, three weeks after the Sabena Airlines hijacking, three Japanese tourists arrived in Israel on an Air France plane. They waited for their suitcases, took them off the conveyor belt, extracted automatic weapons and grenades, and began firing. The attack, known as the Slaughter at Lod Airport, was perpetrated by a Japanese group known as the Red Army, acting on behalf of the PLO's Popular Front, which trained them in Lebanon and provided them with their weapons. They killed twenty-four people and wounded over eighty, including seventeen pilgrims from Puerto Rico.

On September 3, 1972, in the midst of the Olympic Games in Munich, eight terrorists from the PLO's Black September wing snuck into the Olympic Village and seized control of the Israeli delegation's quarters. Two members of the delegation were killed in the early stages of the assault; the other nine were murdered during the failed and amateurish German rescue attempt. Five of the terrorists were killed during the rescue operation. The remaining three were arrested in Germany and released seven weeks later as part of a deal for a hijacked Lufthansa flight.

The rest of the Israeli delegation departed with their dead, but the IOC decided, after a one-day suspension, to let the games go on.

The international terror, of course, did not come at the expense of the usual terror here in Israel. Most of it came from Jordan, where the PLO had set up its own kingdom of terror. That ended in September 1970. After a few attempts on King Hussein's life and many exchanges of fire between the Jordanian army and the PLO gunmen, the Jordanians launched what has become known as Black September. During that month the Jordanians killed thousands of Palestinians (there are no reliable or definitive numbers) in refugee camps in Jordan. Of course there was no international outcry, and the UN did not launch a fact-finding mission of any sort.

After fleeing Jordan in 1970, the PLO established its headquarters in Lebanon, where it relied on support from the refugee camps and conspired, as always, against the local government. No one embraced this modus operandi more fervently than Yasser Arafat, who sprang into action as soon as he felt he had sufficient power. Lebanon, on account of the permanent religious tension and the weak central government, was easy to unsettle. On May 22, 1970, a squad of terrorists knowingly ambushed a bus of schoolchildren on the way from the village of Avivim on the northern border to the school in Village Dovev. They fired antitank weapons and automatic weapons at the bus, killing twelve children.

These events were harbingers of what was to be visited upon Israel from the north throughout the 1970s and the beginning of the 1980s: a hail of rocket fire and a long series of cruel terror attacks.

On May 15, 1974, a squad belonging to the Democratic Front for the Liberation of Palestine perpetrated one of the worst terror attacks Israel has known. The squad snuck into Israel from Lebanon and made its way to the city of Maalot, several miles from the Lebanese border. The terrorists reached Maalot and broke into the Cohen family's home, where they killed a man, a woman, and their four-year-old son. From there the death squad continued on to the Netiv Meir school, which had more than one hundred children and ten teachers, seized control of the school, and held the children hostage. Before the terrorists were eliminated, they managed to kill three adults and twenty-two children. The Israeli army team that charged into the building lost one man, bringing the death toll to thirty, twenty-three of whom were children. Some seventy people were wounded.

On Saturday, March 11, 1978, eleven terrorists from the Fatah organization from Lebanon docked their two rubber boats at the Maagan Michael beach south of Haifa. They promptly killed nature photographer Gail Rubin, who was taking pictures along the coastline. From there they proceeded to the coastal highway, the main road between Haifa and Tel Aviv. Two of them seized a taxi, and the remainder commandeered a bus loaded with families of bus company drivers on their way back from a trip. Near the city of Hadera the other terrorists joined them on the bus, and as the bus traveled south to Tel Aviv, the terrorists sprayed gunfire at passing cars, killing four people.

During the journey, the terrorists managed to stop an additional bus and load the occupants of that bus onto the one they had already seized. Their journey was brought to a halt near Gelilot junction, just north of Tel Aviv, when police officers shot the bus tires.

During those chaotic moments the terrorists opened fire on the passengers and blew up the bus, turning it into an inferno: thirty-five babies, children, women, and men were killed, and approximately seventy were wounded.

On April 22, 1979, Samir Kuntar led a squad of terrorists to Nahariya's shores. On that night policeman Eliyahu Shahar and Danny Haran and his two girls, Einat and Yael, lost their lives. Danny was killed in front of his daughter, Einat, who was bludgeoned to death by Kuntar. Smadar Haran, wife and mother, was the only one who survived.

Nearly thirty years later, in the summer of 2008, Kuntar was released from Israeli prison as part of an exchange for the bodies of Eldad Regev and Ehud Goldwasser, two reserve soldiers who had been abducted from Israeli territory in the summer of 2006.

Kuntar is considered a hero to Palestinians and members of

the Shiite terror organization Hezbollah. After his release, President Bashar al-Assad of Syria and President Mahmoud Ahmadinejad of Iran saw fit, on November 24, 2008, and January 30, 2009, respectively, to award him a medal of honor. In their eyes the murder of a four-year-old girl is worthy of commendation.

My father stayed in touch with Smadar Haran, who remarried and visited my father along with her second husband during his term as prime minister. In 2004 he personally made sure that she would be told that a prisoner exchange at the time would not include the murderer of her family.

March 15, 2004

> *To: The Prime Minister*
>   *Smadar Haran called and said Abu Abbas died, and the Palestinians want to bury him in Ramallah. [Smadar Haran] is reminding you that he was the one who sent Samir Kuntar. The military advisers said they have not as yet received an official request.*
>
>   *Marit*

In his handwriting, at the bottom of this note he wrote, "To the military adviser. We have to make sure that this murderer will not be buried here. Arik." (On January 4, 2006, on his last day in the prime minister's office, Smadar Haran called him, and they talked.)

About five years after my father's illness, at the farewell event for Meir Dagan, head of the Mossad, I met Nadya Cohen and Sophie, the wife and daughter of Eli Cohen, the Israeli agent who was caught and hanged in Damascus in 1965. For decades the Syrians have refused to return his body for burial in Israel. Sophie told me about my father "he was the only one who cared I said harsh things to him . . . He said he understands and respects what I've said. I saw in his eyes that he cares. I was deeply moved."

A year after the shocking murders in Nahariya, another squad of terrorists infiltrated the Galilee from Lebanon, this time to Kibbutz Misgav Am, situated on the Ramim Ridge near the border with Lebanon. This was on April 7, 1980. At the time, my parents, my brother, and I were staying with friends in Kibbutz Lehavot Habashan, in the valley below. That morning the rumor of what was happening in the nearby kibbutz began to make the rounds, and I remember the kibbutz members from Lehavot Habashan seeing my father and saying, "Arik's here," taking comfort in his presence: he's here, everything will be all right, he'll work it out.

The terrorists burst into the nursery and holed up among the sleeping babies. They murdered Sami Shani, the kibbutz secretary, who arrived on the scene, and they shot Eli Gluska, who was two years old. For crying out loud, the child was two years old, how could they? How can one come along after that and say that these terrorists are idealistic freedom fighters?

Throughout 1981, Katyusha rockets continued to claim civilian lives in Kiryat Shmona, Misgav Am, and Nahariya.

On Thursday, June 3, 1982, Israel's ambassador to Britain, Shlomo Argov, was shot and gravely wounded by a terrorist from a Palestinian organization.

At the time my father, serving as minister of defense, was in the midst of a trip to Romania. The government there had asked to keep the trip a secret, so my father used me, my mother, and my brother as his cover, saying that we all had come to see Braşov, the city where my mother was born. We did visit the city, but that was in addition to his meetings with Nicolae Ceauşescu, Romania's president at the time, with the ministers of defense and industry, and with other senior officials. The Romanians wanted to cooperate with Israel on several technological projects and to escape, if only slightly, the smothering Russian embrace.

We landed in Bucharest and set out for Braşov in a convoy. The road was entirely free of traffic, and all along the route we saw long, orderly rows of cars waiting silently at the intersections. They waited like this for a long time. The sight was disturbing. "What do you say to these people?" my father asked his Romanian escort. "Simply that a delegation has come," he responded. Seeing close up, but also from a safe distance, a communist dictatorship, the obviously oppressive nature of the regime, left a mark on all of us. At the hotel the mood was different. We were aware that the Romanians were listening and perhaps watching, so Omri and I went around the room, telling the mirrors and the lampshades how happy we were with the accommodations. It amused us, and maybe some of them, too, since they were forced to listen to our jokes.

News of the assassination attempt on Israel's ambassador to the UK reached my father in Romania. It stands to reason that Operation Peace for Galilee would have been waged without the assassination attempt: the PLO's ascendant power in Lebanon, its growing martial might, its stores of artillery and rockets, and the continuing export of terror against Jews from Lebanon were no longer tolerable. On May 16, 1982, some two weeks before the assassination attempt, the IDF presented to the government its plans on the Lebanese front. "The objective of the mission," my father said at that cabinet meeting, "is to cleanse the area within thirty miles to the north of Israel's border. Will this bring about the clear and complete end to the campaign? I am not sure. Within twelve hours, in my opinion, the cabinet will have to reconvene. At any rate, it will have to meet continuously throughout the war and to take the cardinal decisions—what are aims . . . Insofar as the Syrians are concerned, our aim is not to get entangled with them. But I am telling the cabinet that I believe the Syrians will get involved."

———————

The long years of terror from Lebanon, the infiltrations, the rocket fire, the attacks that were planned there and directed at other parts of the country and against Jewish and Israeli targets internationally—all those, along with the exploitation of the inherent Lebanese sectarian vulnerabilities, made the campaign against the terror state to the north inevitable. Over the years, Israel carried out several antiterror operations in Lebanon, including the IDF operation in the Beirut airport in 1968.

On April 10, 1973, just hours after an attack on the Israeli ambassador to Cyprus's house and an attempt to hijack an Israeli Arkia plane, Operation "Spring of Youth" was launched. The daring raid was aimed at terrorists, terror headquarters, and terror weapon workshops in the heart of Beirut. The operation was carried out by commandos, with the assistance of Mossad agents who met the seaborne troops on the beach.

The daring raid stunned the PLO terrorists. Dozens of them were killed, some senior and some who were responsible for the murder of the Olympians in Munich. They were taken by surprise in their apartments, in buildings under armed guard.

During the 1970s the IDF carried out many raids in South Lebanon. On March 14, 1978, one of the largest of them, Operation Litani, was launched in response to the awful terror attack on the coastal road that claimed thirty-five lives. Three days later, two infantry brigades set out along with an armored force, engineers, and artillery, with the support of the navy and the air force. In a matter of days the IDF managed to take all of the territory south of the Litani River, excluding Tyre. During the course of the fighting some three hundred terrorists were killed, and the IDF lost eighteen soldiers.

The United Nations Interim Force in Lebanon (UNIFIL) was founded in response to the operation and was deployed once the troops withdrew in June of that year.

The UN force, of course, did not prevent further terror attacks launched from the area. In fact, their mere presence made it more difficult for Israel to respond to those attacks, as care had to be taken that they not be harmed. Just as it had after Israel's full withdrawal from Lebanon in 2000, here too UNIFIL policed only the law-abiding side. In other words, it imposed restraints on Israel, while the terrorists, the reason the force was deployed in the first place, continued to operate freely.

The problem Israel faced in the early 1980s was very similar to the one that presented itself in the summer of 2006: a terror organization that established a state within a state—then the PLO, and in 2006, Hezbollah—whose members, recognizing Israel's reluctance to harm civilians, hide among them and intentionally direct their fire on the civilians of Israel, although the north of Israel has military targets as well. But in the 1980s the situation was far more complex.

The difference was twofold. Hezbollah, for one, can be pressured: their constituency is the Shiite sector, spread among the villages of South Lebanon, and any harm inflicted on those villages during a war can potentially weaken their local support, as could a strike against Lebanese national infrastructure, since Hezbollah plays a role in the government. On the other hand, the PLO took over all the villages in the south, although the people there were not the base of their support. Thus, the PLO could fire on Israelis and not worry about the retaliation against these villages. They couldn't care less about these people. The second profound difference was that the PLO terrorists stationed themselves in areas under Syrian control and enjoyed their aid and protection; an attack on them in these regions made a confrontation with Syria a real possibility. In 2006 Hezbollah was aided by Iran, but it did not live under the physical protection of its army, and there was no threat of war with Iranian national forces as a result of Israel's engagement with Hezbollah.

In May and June 1981 the PLO rained rockets and artillery down on Israeli villages all across the northern Galilee. During one tough week in July, thirty-three Israeli villages were under fire. This reality forced citizens to live in bomb shelters and in fear, with many businesses closing down and many residents leaving for the center of the country. The terrorists' goal was to destroy the normal routine of life in the north, and they were successful.

In July 1981 the Americans were able to arrange a cease-fire between Israel and the PLO, brokered by Philip Habib, a U.S. State Department official who mediated between Israel, Syria, and the PLO. However, the agreement was explicit only regarding preventing terror from Lebanon, which is why my father encouraged the cabinet not to accept the offer as presented by the Americans.

He knew that the Palestinian terrorists would interpret it a green light to go ahead with terror from other locales—Jordan, Judea, Samaria, and Gaza—toward Israel and Jewish targets the world over. He wanted an absolute cease-fire, not the kind that would hinder Israel's ability to respond while allowing the terrorists to operate against us in different sectors, with the planning, the orders, and sometimes the actual squads being sent from Lebanon. In late July 1981 a cease-fire agreement was reached, but Israel had no illusions: across the border there were 15,000 to 20,000 terrorists, armed with some ninety 122mm and 130mm cannons, one hundred trucks bearing some thirty to forty Katyusha launchers per truck, an array of short-range artillery options, and more. Why did they need to amass such vast stockpiles of weapons? They had the equivalent of four to five divisional artillery batteries. The majority of the stockpiling occurred during the eleven months of the cease-fire, and it was clear that all of their arms were intended for Israel.

The cease-fire, as both the PLO and the Americans saw it, did not include terror attacks stemming from Lebanon and carried out against Jews in Europe and other locales. In a meeting my father had with Alexander Haig and Philip Habib on May 25, 1982, Habib repeated what he had already said many times before: "Terrorist attacks against Israelis and Jews in Europe are not included in the cease-fire agreement."

During the eleven-month cease-fire, many terror attacks in Israel and abroad either began in Lebanon or had their orders sent from there, resulting in the murder of 15 people and the injuring of 250. The attacks and the attempted attacks included an assault on a synagogue in Vienna; an attack on a Jewish restaurant in Berlin; an explosive-packed car in Antwerp; the murder of 3 tourists in Jerusalem; the attempt to blow up a kindergarten in Holon; and the attempt to fire rockets on Eilat from Jordan. The attempted assassination of Israel's ambassador to London, Shlomo Argov, may have set the date, but it most certainly did not single-handedly bring about the war.

FIFTEEN

Surrounded by Israeli soldiers in Lebanon, June 1982.
(Uri Dan, courtesy of the IDF and Defense Establishment Archive)

On Friday, June 4, the day after the assassination attempt on the ambassador, while we were still in Romania with my father, the cabinet decided on an air force strike against PLO targets in Lebanon. Begin addressed the cabinet:

> *Gentlemen, after this assassination attempt, as far as our operation*
> *is concerned we must be prepared for the maximum. It is unfeasible*

*that these scoundrels can bring such harm to our ambassadors. I must
add that we have to be prepared for further developments . . . there is
firm reason to believe that our villages in the north will come under fire
. . . and if this happens, we will not be able to sit on our hands . . .
We must take into account that next week it may be incumbent upon
us to launch the larger mission and try to demolish them . . . We may
sum up by saying that the proposals were unanimously agreed upon.
Chief of Staff—good luck. If it can be carried out today—carry it
out . . . try to do it as soon as possible. Tomorrow evening there will
be a cabinet meeting. Thank you, the meeting is closed.*

The air force attacked, but war was not yet inevitable. The
PLO's response would dictate that. If they responded by shelling
the villages and towns of the Galilee, there would be war. If they
refrained, it would be avoided. No one was surprised when, at five
thirty in the afternoon, PLO rocket fire and artillery shells began
to fall on the towns of the north. The next day, June 5, we boarded
an air force Boeing 707 in Bucharest. The pilots let me sit in the
cockpit. I remember that after takeoff an enormous cloud loomed
in front of us, and the pilot explained to me that it was better to fly
around it than through it. Upon landing, my father drove straight
to army headquarters and from there to the cabinet meeting in
Jerusalem.

At that June 5 meeting, Begin asked for cabinet approval.

The resolution that was passed, with fourteen ministers in favor, two abstaining, and none opposed, declared:

a. To charge the Israel Defense Forces with the mission of
removing all of the towns and settled areas of the Galilee from
the range of the terrorists, who are deployed, in their bases
and their headquarters, in Lebanon.

b. The name of the operation: Peace for Galilee

c. The Syrian army is not to be attacked while carrying out
this mission, unless it attacks our forces.

The IDF advanced north on three main routes—the Lebanese coastal route, the central sector, and the Bekaa Valley in Lebanon in the east. Progress along the coastal route was swift, unlike that in the Mount Lebanon and Bekaa areas, where the PLO terrorists were gathered within the Syrian army's lines. The Syrians increased their presence in the area, advancing to the south and west and shelling IDF units even though they had not been attacked. Israel did not want a full-scale war with Syria by any means, and it did not want another front opened on the Golan Heights. In the meantime the terrorists fired freely from within the Syrian-occupied areas, striking villages in the Galilee. On June 6, the first day of the war, my father depicted the current situation at that evening's cabinet meeting: "I'd like to report to you the summary of the meeting I held with the prime minister, the chief of staff, and the head of military intelligence regarding our options in dealing with the problem of the Syrians. There are two options: one, to forgo the notion that they are not involved in the war and to attack them directly. And two, to proceed farther north and try to pressure them, approaching them from the rear, with the understanding that the threat of encirclement would induce them to retreat. Our recommendation was to try and advance farther north and attempt to threaten them with encirclement, not to attack them directly. Perhaps, in the face of the danger of being surrounded, they will move, and that is what the prime minister asked me to present to the cabinet." One of the ministers asked whether this advance would take us north of the thirty-mile mark. "Yes, we will cross the thirty-mile mark."

The cabinet authorized the flanking maneuver, to be carried out by the force on the coastal route, which would reach the Beirut-Damascus highway and advance east along the road toward the rear of the Syrian positions. Earlier that day, speaking before the Knesset Committee for Foreign and Security Affairs, my father raised the possibility that the IDF would reach the Beirut-Damascus highway.

On the morning of June 7, the UN Security Council passed a

resolution calling for Israel to cease all offensive action in Lebanon and to withdraw to the international border. President Reagan sent his envoy, Philip Habib, to the area. The cabinet decided to inform the Syrians, via Philip Habib, that they should push the terrorists in their midst farther than thirty miles from our border." All that was asked of the Syrians was that they remain in the positions they had held since 1976 and that the terrorists be kept out of range of Israel. Had they accepted these terms, there would have been no need for a confrontation or a flanking maneuver, nor would these have taken place. That day the Syrians increased their troop levels in the Bekaa and in the central region, and the terrorists continued to shell the towns of the north from within Syrian lines.

During the June 8 cabinet meeting, Begin said,

After the cabinet meeting I will invite the heads of the Labor Party in for a briefing. This morning I let both of them know that something grave had happened. After I spoke with them, everything that was said was published in Haaretz *and the* Jerusalem Post—*and that is a serious matter. This is translated into Arabic . . . it is heard in Damascus and in Beirut and in all Arab countries; does the enemy need to know that there is a debate among us about whether or not to fight the Syrians? That was simply information given to the enemy, both of them apologized profusely. Rabin said: I was really shocked. Peres also said that he really can't understand how this could have happened. How did it happen—I don't know. All I know is that I told no one, so one of them must have relayed the information. It was just the three of us in the room, and they promised to keep the matter confidential. They apologized, and we made our peace. I will invite them in after the cabinet meeting. Arik has to go back up north because the campaign is at its height, but I will ask Yehoshua [Saguy], the head of military intelligence, to stay for twenty more minutes in order to brief the heads of Labor, it's important that we receive from them what's known as good will.*

At the June 9 cabinet meeting Begin reported back about his meeting with Philip Habib. "I have an urgent message for you," Begin told him.

> *Please pass it on to President Assad from me that:*
>
> a. *We do not want war with your army.*
>
> b. *Instruct your army not to fire on our soldiers. If our soldiers are not hit, they will not attack your army.*
>
> c. *Have your army withdraw from west to east and from south to north to the starting point where it was positioned before the campaign began.*
>
> d. *Instruct the terrorists to retreat 16 miles to the north. If this is done, then the military stage of the operation will have been concluded and the diplomatic stage can commence.*

The Syrians had brought in six additional SA-6 missile batteries into the Bekaa. This was a striking violation of the status quo between Israel and Syria. The Syrian missiles put Israeli jets in peril when they took to the air against PLO targets, and they endangered Israel's reconnaissance flights, needed for aerial photos in Lebanon and the deep view into Syria, which was an important tool in assessing whether Syria was gearing up for war. The air force was confident in its ability to destroy the missile batteries. In fact, the cabinet had already authorized a strike against those missile sites, more than a year earlier, back on April 30, 1981, but bad weather had delayed the strike, and later the American envoy Philip Habib had come to the region to try to resolve the issue diplomatically, and the operation had been delayed yet again. My father was happy for the delay, because the timing conflicted with something more important: the strike against the Iraqi nuclear reactor, the destruction of which was an existential necessity. An attack on the surface-to-air missile batteries would have alerted the Arab air forces in the region and put the more pressing operation in jeopardy. This is why the Syrian surface-to-air missile batteries remained in Lebanon until Operation Peace for Galilee. During the fighting, the Syrians brought in additional missile batteries.

At the June 9 cabinet meeting, Prime Minister Begin informed those present, "I've called Lewis [U.S. ambassador to Israel Samuel Lewis] that he inform Assad that he must immediately withdraw the new missile systems. He must withdraw them by 5:00 a.m. The meaning of this is that if he does not respond, we will act."

The vote was unanimous: the cabinet authorized the defense minister's suggestion to attack and destroy nineteen Syrian surface-to-air missile batteries.

By noon on June 9 there was still no response from the Syrians. At 2 p.m. an attack was launched, and the missile batteries were destroyed. The thirty Syrian MiGs that rose to engage the Israeli jets were all downed, and the planes remained untouched. Only after that blow had been inflicted did President Assad agree to meet with the American envoy, Habib, who had been left waiting in Damascus. After the strike, the Soviet Union began to press the United States for a cease-fire. The Soviets were rattled, too, for this aerial attack had proved quite conclusively that their air defenses were exceedingly vulnerable to attack; they had been destroyed in a short time. The Soviet pressure on the United States caused the Americans to exert pressure on Israel. On June 10, at two in the morning, Begin received a personal message from President Reagan, demanding a cease-fire effective at 6 a.m.

The international pressure on Israel intensified. It took one more day for the cease-fire to go into effect. The IDF stopped thirty miles from the border, in the Bekaa; it could have continued to advance, but since the Syrians had respected the cease-fire in that region, so too had the IDF. This would probably have been the end of the war had the terrorists in the Beirut region respected the cease-fire. Instead, they chose to violate it time and again.

The fact that the IDF reached the Beirut-Damascus highway was a result of the flanking maneuver and the terrorists' ongoing violations of the cease-fire. As my father pointed out a few years later, "Our policy was to respond to the attacks, not at the time or convenience of the terrorists, but in accordance with our goals." That said, reaching the highway and linking up with the Christian forces had tremendous advantages. There was a reason the government had set that destination as a goal and opposition leaders such as Yitzhak Rabin had pushed to see it achieved. The key to Lebanon was control of that road. It cut Beirut off from Syria, and it gave Israel a military and diplomatic advantage.

On June 24 the cabinet ordered the IDF to expand its hold on the highway, because the corridor that cut off Beirut and was linked to the Christians was too narrow. In addition, the topography was inferior and under terrorist attacks, and the IDF corridor would be cut. Begin concluded the meeting, "Based on the suggestions heard, there was not so much as a single opinion opposed to the suggestion by the defense minister and the chief of staff. A very serious and very important discussion took place. I believe I can sum up the matter of our occupation of the Beirut-Damascus highway by saying that the operation has been authorized, and may we have good luck."

The following day the mission was completed. The key point for control of Lebanon was in the IDF's hands. Capturing that point is what led, in the end, to the deportation of the terrorists from Lebanon and the destruction of the PLO's terror state, which had threatened Israel. The advantages of linking up with the Christian forces was not the goal of the operation, but it was a welcome by-product. This stirred hopes of establishing a relationship with Lebanon, a country with which Israel had no territorial dispute. Unfortunately, this never happened.

Back in 1982, with my father serving as minister of defense, it took the IDF a mere seven days to reach the Beirut-Damascus road, removing the threat of rocket fire from northern Israel. This advance was accomplished while combating the Syrian army that provided the terrorists with cover.

OPERATION PEACE FOR
GALILEE, 1982

*Mediterranean
Sea*

LEBANON

Beirut

BEIRUT-DAMASCUS HIGHWAY

Demour

SHOUF
MOUNTAINS

AMPHIBIOUS
UNIT

Sidon

Damascus

Tyre

Beaufort Castle

Kiryat Shmona

UPPER
GALILEE

NO MAN'S LAND

SYRIA

Nahariya

GOLAN
HEIGHTS

Haifa

*Sea of
Galilee*

ISRAEL

N

Jordan River

JORDAN

0   20 km

SIXTEEN

With Rabin before flying to the outskirts of Beirut
during the 1982 Lebanese war. (Uri Dan)

I srael's ties to the Lebanese Christians began during Rabin's
first term in office, circa 1976. These Christians were by far
the most sympathetic element in Lebanon on all matters pertaining to Israel. They were at war with our enemies, so just as we
assisted the Kurds and the Yemenis, so too did it seem natural to
come to the aid of the Christians in Lebanon. This was in Israel's
national interest, and it was clearly part of the Israeli consensus.

Rabin, in his memoirs, explained why Israel assisted the Christians in Lebanon during his first term in office: "Thousands of Palestinian terrorists are fleeing to South Lebanon and taking it over, all along Israel's northern border. The terrorists are entrenching themselves. Amassing arms. Setting up a logistical infrastructure. Even were this not the case, Israel could not allow itself to sit by idly, swathed in its own neutrality, in light of what is going on in Lebanon. A religious minority [the Christians] fighting for their lives against the forces of fanatical Muslim nationalism deserves to be helped by Israel, and Israel has no right to turn them away empty-handed . . . I was convinced that it was our duty to aid the Christians."

Chaim Bar-Lev, the former chief of staff and a member of the Labor Party, further explained in April 1981: "We should help the northern Christians, not only because we are neighbors and cannot remain impartial to such annihilation by the Syrians . . . What happens there has direct relevance for us not only from the humanitarian aspect . . . If the Christians in the north collapse, there will also be direct implications for the situation in the south."

Begin's government did not alter this policy. In January 1982, five months before the outbreak of the war, my father flew to the Christian enclave in Lebanon to assess the situation for himself.

In darkness, the helicopter lifted off from an air base in Tel Aviv and headed north along the coast, turning out to sea to skirt the PLO-ridden southern coast and Beirut, and then back to the port city of Junia. My father looked around and was surprised to see fully lit ships bobbing in the sea. "Those are ships, cargo ships," said Bachir Gemayel, the leader of the Christian Phalangists, who greeted him upon arrival. "But there's a war on. How can you have ships out there like that?" my father asked. "Yes," Gemayel said, "there is a war; but war is one thing, business is something else entirely."

After dinner they set out for Beirut. Buildings were pockmarked by bullet holes, the city was divided everywhere by roadblocks, and yet the cafés were filled, bars and restaurants churned with excitement beneath neon lights, and fashionable women streaked past in some of Europe's finest automobiles. This in a country where, over six years of civil war, 100,000 lives had been lost and a quarter of a million people had been wounded, among a population of only 3 million. But this is Lebanon, especially Christian Lebanon, where, beyond the usual duality throughout the Arab world, war and peace are accompanied by French chic.

In Lebanon with Bachir Gemayel (in sunglasses) before the war.
(Uri Dan)

The following day my father visited the headquarters of the Lebanese Christian forces in the Beirut port, along the seam line with West Beirut, which was in Syrian and PLO hands. From there they drove to the mountains, surveying the capital city from above, and from there to the villages and the outskirts of the Christian enclave in the mountains, and to Bachir's village of Bikafiya.

They visited the cemetery, where Bachir's daughter was buried. She'd been killed by an explosive charge attached to the family car. Suleiman Faranjia's family, a rival Christian clan that had ties to the Syrians, were the perpetrators of that act, revenge for the explosive charge that the Gemayels had planted, killing Faranjia's son and daughter-in-law. This was the reality in Lebanon, where everyone killed everyone else and then went out afterward, until the next murder. Nine months later, Bachir was buried in that same cemetery.

Back at Bachir's house the heads of the Christian forces were interested in hearing about the likelihood of an Israeli invasion in Lebanon. If the Syrians increased their involvement, they said, they would not be able to face them. My father said that Israel was trying to avoid war, but if the terror continued, we would have to take action. "If we do come in, we would come in in order to defend our northern borders. But as a result of that you might have a chance to restore a normal life to Lebanon."

When my father arrived home late that night, my mother was waiting for him. "What was your impression of the Lebanese?" she asked.

"They are people who kiss ladies' hands—and murder."

———

There was a wide national consensus when the war began. In the Knesset vote on June 8, 1982, ninety-four MKs supported the government; only three voted against it.

While reporting to the Knesset Committee for Foreign and Security Affairs about the war, my father received full support from the heads of the Labor Party, which was in the opposition. During the committee's first meeting during the war, Rabin, Peres, Bar-Lev, and others all supported the goals of the operation. On June 9, a mere three days after the beginning of the war, Rabin voiced his strong support in favor of occupying the Beirut-Damascus highway so that Israel could begin negotiations from a position of power.

Motta Gur, a former chief of staff serving as a Labor MK, suggested the option of invading Beirut to strike at the terrorists' infrastructure and roots. The Beirut-Damascus highway is north of the thirty-mile belt that for some reason became a mantra to those from the opposition that later attacked the government over the war, even though the leaders of the Labor Party had pushed, during the early days of the war, to reach the Beirut-Damascus road.

On June 10, Motta Gur, in responding to my father's briefing before the Knesset Committee for Foreign and Security Affairs, said that to his mind it was a grave situation if the PLO leadership, remaining in Beirut, could restore its stature immediately as long as Israel did not enter Beirut. "This will not solve the political, strategic problem," said Gur, who felt that the IDF had to enter Beirut and was more aggressive than the government. My father responded that an invasion of Beirut was not in accordance with cabinet decisions at that time. Yitzhak Rabin voiced unequivocal support for the cabinet decisions during the same meeting.

Several days later, Rabin and my father visited Lebanon together. They stood on a lookout point above Beirut, and Rabin suggested tightening the siege around the city.

Principally, then, there were no differences of opinion between the Likud and Labor, nor should there have been. Israel had gone significantly farther for less, in the 1956 Sinai War and during the Six-Day War, too.

To the Brother Hadj Ismail [commander of PLO forces, Region of Southern Lebanon]: Greetings and Blessings of the Revolution Be upon You.

The resolution of the General Military Committee is to concentrate on destroying Kiryat Shemona, Metulla, Dan, Sh'ar Yeshuv, Nahariya, and its suburbs.

Kiryat Shemona: All elements of the Revolution will take part in hitting this objective. Kiryat Shemona is to be bombarded by "improved Grad" [missiles].

Metulla will be shelled by the Palestine Liberation Front and by Saika, using 160mm mortars.

Nahariya and its suburbs will be shelled by the First Battalion, using 130mm artillery.

Dan and Sh'ar Yeshuv will be dealt with by the Eastern Region.
I yet again bring to your attention the matter of the ground patrols.

Yasser Arafat
July 18, 1981

This is one of hundreds of documents seized during the war, and it, like the rest of them, reveals that the objective was the killing of Jewish civilians. The PLO operatives wanted to shed civilian blood.

International pressure on Israel against the war was enormous. After the successful strike against the Syrian forces and the siege on the terrorists and Syrian soldiers in West Beirut, the pressure only intensified.

The proposed American cease-fire included the point that Israel would withdraw from the Beirut-Damascus highway and, in essence, allow Arafat and the thousands of terrorists under his command to perpetuate the terror state they had erected in Lebanon. The government could not in good conscience agree to those terms. The siege around West Beirut was tightened, and the air force increased the intensity of its strikes. The terrorists were on the verge of collapse. Peres, with his insatiable ambition, his drive to be prime minister, was not willing to see the government's achievements as a national success. This is Shimon Peres at a meeting of Labor Party leaders during the siege on West Beirut: "Contrary to our previous fears, the war is a big success. It is about to reach its most important objectives. In a few days—and one cannot escape the facts—an Israeli-Lebanese peace treaty will be signed. This will be their [the Likud's] second peace treaty with an Arab country. They will also succeed in sending Arafat and his terrorists to hell as well as in breaking the PLO."

"They," he said, as though the objectives of the war were a matter of party politics, something that primarily stood between him and the prime minister's office.

Rabin and Dayan fully understood Peres's agenda and, as the former wrote in his book: "It was completely clear to me that Shimon Peres believed that he and the prime ministership were utterly worthy of one another and that they should not be separated . . . the passion was boundless. An urge that respected no borders. All was permissible, leaks and a constant digging under the foundations of the prime minister's position, a weakening of the government's status—all according to the well-known Bolshevik tenet: the worse it gets, the better it will be for Peres's people and their leader."

In addition, Dayan wrote this note to my father on February 14, 1979, while they were dining along with Peres and others at the American ambassador's residence in Israel: "This lust for office is deranging them [the Labor Party]. Shimon [Peres] is simply dangerous in his assertions. Imagine what he says to the Americans when others are not present."

Labor Party members and the press began to attack the government, focusing on my father and Begin. Antiwar protests started to surge around the country. It was, to a certain degree, a mimicry of what had happened in America during the Vietnam War. Like so many other imported imitations, this one came ten years late and in no way resembled the original. The United States had not been attacked by Vietnam, and the Vietcong forces had not set the goal of destroying the United States and founding a Vietnamese state on its ashes. But these details did not matter. Antiwar protests, guitars, and long hair were always an easy sell. Nor were there any Katyusha rockets falling on Tel Aviv, as they were in the north.

Arafat and his men were encouraged by the domestic and international pressure being brought to bear on Israel. Documents seized in Beirut after the deportation of the terrorists told the story. In a speech my father gave in the Knesset on September 22, 1982, he quoted from documents seized during the Lebanon War: "The most important thing is to increase the demonstra-

{ 274 }   SHARON

tions all over Israel. To do that, we have to mobilize all possible means. If the demonstration in Tel Aviv is not repeated, all its effect will be lost." The material taken in Beirut also provided reports of Abba Eban's and Peres's meetings with Egyptian diplomats, in which the two Israelis ruthlessly attacked Begin and his government.

The antiwar sentiments and actions encouraged the terrorists and delayed their eventual deportation. They were not worthy of pity. Those who had shown no pity on the children of Avivim, Misgav Am, and Maalot were not worthy of Israeli sympathy, especially when all that was being asked of them was to board a ship and leave.

The American position delayed the deportation and generally made it more difficult. On June 15 my father presented Israeli demands to Philip Habib that all PLO terrorists be expelled and all foreign forces evacuated from Lebanon.

"The withdrawal of external forces cannot be symmetrical," Habib said.

"What do you mean it can't be symmetrical?"

"Well, the Syrians have security interests in Lebanon," the American diplomat said.

"What security interests do they have in Lebanon? Did Lebanon ever attack Syria? Did they ever threaten Syria? Has Syria suffered from any terror activities coming from Lebanon?" my father asked.

The answer to all of those questions was a resounding no.

Habib tried a different tack, asking if "perhaps instead of expulsion the PLO could be made into a 'political' body."

"They have to leave," my father said.

The State Department officials were under the impression that Lebanon could serve as a lever to solve other problems in the Middle East. They believed that the PLO's dire position in Beirut could help solve the problems of the Palestinians in the West Bank, and they thought they could curry favor with the stricken Syrian regime by advocating for their demands vis-à-vis Israel.

The government of Israel did not accept this: Lebanon had been so war-torn, its internal problems so entrenched, that it would be a great accomplishment to solve its problems. Using it as a lever to solve other problems was an unattainable aspiration.

The military pressure worked; the air force bombs and the threat that Israel might invade their enclaves forced, on August 21, eleven weeks after the beginning of Operation Peace for Galilee, some 9,000 terrorists and 6,000 Syrians from Beirut. They were deported to eight different countries: Syria, South Yemen, Tunisia, Yemen, Algeria, Sudan, Jordan, and Iraq.

The deportation took ten days, marking the end of the PLO state in Lebanon, and the communities of the north knew years of quiet.

Terrorists being deported from Lebanon, August 1982.
(© Micky Sepharti, courtesy of the IDF
and Defense Establishment Archive)

Two days after the beginning of the deportations, on August 23, 1982, Bachir Gemayel was elected president of Lebanon. This gave hope that Lebanon's civil war might end and that the country's relationship to Israel might change. On September 12 my father met with Bachir in his village, Bikfaya. The two discussed the need to ensure that no terrorists remained in West Beirut, a mission best handled by Lebanese forces. An estimated 2,000–2,500 terrorists remained in West Beirut, and there was concern that they might reestablish their PLO terror state.

With Uri Dan and Bachir Gemayel in Lebanon, September 1982.
(Uri Dan)

They spoke of the difficulties awaiting Bachir, who, as president, would be asked to rule over areas like Tyre and Sidon, places that he could not so much as visit

The matter of diplomatic relations with Israel arose as well, and the two men agreed that a peace treaty could be signed. They determined that in three days' time there'd be another meeting, attended by Foreign Minister Yitzhak Shamir. That meeting was never held. Two days later, Bachir Gemayel was killed by an explosive charge planted by a Christian who belonged to a pro-Syrian party. The Syrians did not want him sworn in to office. That would weaken their position in Lebanon.

This gave rise to a concern that in the wake of the murder and the political uncertainty in Lebanon, the terrorists remaining in West Beirut would reassemble and attempt to reestablish their terror state. Prime Minister Begin, the chief of staff, and my father met to discuss the developments. Some eight hours after the murder of Gemayel, after midnight on September 15, it was decided that the IDF would move into West Beirut, taking key positions and junctions, denying the terrorists the opportunity to assemble their forces, but would not enter the neighborhoods. The Christian forces would be in charge of that. They were better at discerning between terrorist and civilian; they spoke Arabic and could tell in an instant whether someone was local or foreign; and at the end of the day this was their capital city. They wanted it united and under their control, which meant that they had to shoulder some of the burden, too.

In Israel, there had been considerable complaining about the passive role of the Christian forces from the early stages of the war. Israel's action, all taken in the country's own self-interest, served the Christians well. Victor Shem-Tov, an MK and secretary general of the opposition left-wing Mapam Party, said about three weeks after the war broke out, "In the name of God, what are our Phalangist Christian allies tasked with in this matter? We were told that at some point the Christians would go into Beirut. In the end, it is their capital, and they too must do something in order to liberate their homeland. What is their role in this war? Is the full extent of their role to shower our soldiers with rice and to hand out flowers while we shed blood? After all, we paved the path for the Christians when we were in office."

As far back as June 15 the cabinet decided that the Christian forces had to take a more active role in the fighting in Beirut.

My father arrived in Beirut on the morning of September 15. At division commander Amos Yaron's headquarters, in a tall building southwest of Shatila, Chief of Staff Rafael Eitan informed my father that he had coordinated with the Christian commanders their entrance into the neighborhoods of Sabra and Shatila. The details of their mission were to be coordinated with the commander of the northern front and Yaron. My father authorized the decision and spoke with Begin, reporting back to him and discussing the political situation in Lebanon before going to pay condolences at the Gemayel family home.

From Yaron's headquarters my father went to the Phalangists' headquarters near the port, where he had a brief discussion about the new political situation. There too it was settled that the IDF would occupy certain central positions in West Beirut, but the Phalangist forces would be the ones to operate. From there my father continued on to Bikafiya to pay a condolence call. He met Pierre Gemayel, Bachir's father, and other family members. My father expressed his sorrow in the name of the prime minister and the state of Israel. Pierre Gemayel emotionally thanked Israel for all it had done for Lebanon's Christian community, which had been forsaken by the rest of the world. The two also spoke of Lebanon's political future. My father left for Israel.

The following day, September 16, the chief of staff briefed my father on the IDF moves in West Beirut. At around the same time, the Phalangist forces, who were to be going into Sabra and Shatila, were at northern front commander Amir Drori's headquarters, finalizing their coordination and completing their preparations. They met with Amos Yaron and were instructed that the operation was against PLO terrorists alone. Civilians were not to be harmed. In the evening the Phalangists entered Sabra and Shatila, at the same time the cabinet in Israel convened to discuss the new political situation in Lebanon.

Senior military and intelligence officials attended the meeting, along with the country's attorney general. My father described the IDF operation to take key points in West Beirut, and as he spoke, he received a note informing him that the Christian forces had entered the neighborhoods. He let those present know of the developments. No one in the room voiced any reservations.

The next day was the eve of Rosh Hashanah. As always, we went to the memorial service for Gur. From there my father continued on to Jerusalem for a meeting at the foreign ministry. At nine in the evening, back at the farm, he got a call from the chief of staff. Eitan, back from Beirut, said that there had been some problems with the Phalangists' actions, that they had harmed civilians. "Gone too far," was how the chief of staff put it. In light of this, the commander of the northern front had stopped the operation. The chief of staff highlighted three points: the Phalangists' operation had been stopped; no further forces were entering; and they were organizing their withdrawal. They will complete the withdrawal by 5 a.m.

An hour or so later, the situation officer at the defense ministry was on the line. He reported that Israeli paratroopers had clashed with soldiers from Sa'ad Haddad's southern Christian forces. (Major Haddad's forces had worked in concert with the Israelis for years in South Lebanon against the terrorists there.)

According to the reports, the paratroopers had opened fire on Haddad's men, who were shooting civilians. One of Haddad's men was killed, and two were taken prisoner. The information confirmed what my father had heard from the chief of staff. If IDF forces were shooting at, and killing, Haddad's men, who had been our allies for years, then in fact drastic measures had been taken to stop the violence against civilians.

An hour and a half later, at approximately 11:30 p.m., my father received a call from an Israeli journalist named Ron Ben-Yishai, who said he had heard from Israeli officers that the Phalangists were killing civilians. This was a development my father was aware of at this point, and it was also the reason that their operation had been halted, that they had been instructed to withdraw, and that Haddad's men had been shot. Awful crimes were committed by the Phalangists, and the IDF had taken swift and concrete action to stop this.

Those are the facts. Sad and shocking, but such is Lebanon. This was not the first, nor would it be the last, massacre in Lebanon. The twentieth century was riddled with ethnic massacres in Lebanon, especially during the civil war, which started in 1975. In December of that year, PLO terrorists murdered dozens of Christians in Beirut. On January 18, 1976, Christian forces massacred about one thousand Palestinians in Karantina, Beirut. Two days later Palestinian forces gunned down hundreds of Christians in Damur. In July of that year, PLO terrorists killed 120 Christians in the north Lebanon city of Shaka. In August, the Christian forces murdered some 3,000 Palestinians in Tel a-Zaatar, on the outskirts of Beirut. In October 1976, Palestinian terrorists and local Muslims murdered dozens of Christians in the town of Aiyshiye in South Lebanon. In July 1980, the Phalangists struck against a rival Christian force. In 1983 Syrian forces murdered many Palestinians in the camps near Tripoli, and in 1985 Shiite forces massacred Palestinians in Beirut. This is a brief and incomplete history of atrocities in Lebanon. It does not even touch on the assassinations—for example, the February 2005 assassination of Rafic al-Hariri, the former prime minister of Lebanon. Hezbollah and Syria have staunchly denied any involvement in this murder, which claimed the lives of twenty other people, despite the findings of the Special Tribunal for Lebanon.

The IDF entrance into Lebanon on June 6, 1982, brought the Lebanese forces into line, and for most of the war they acted, under IDF guidance, as a proper military force, following reasonable military decorum.

Lebanese forces operated around Reichan, at the science college, and took Jamhur Junction, east of Beirut. They were sent in to clean the route from Aley to Suk al-Arab on the way to Dar-mur. They operated in Beit Adin in the Druze region, in the Shuf Mountains, where Christians had been murdered by Druze. And in all of these places they had acted properly. Atrocities such as the massacre in Sabra and Shatila had not happened under the guidance of the IDF.

With Begin in the Knesset.
(© David Rubinger, Yedioth Ahronoth)

SEVENTEEN

Over the years some have tried to portray Prime Minister Begin as someone who was divorced from reality during the war, as though he was incapacitated and unaware of what was happening. He has been depicted as someone who was pulled along blindly, with no control over the government he headed. To portray Begin in that light is to do him a great injustice. I remember Begin pounding the table during an interview with Yaakov Achimeir and Elimelech Ram. This was on June 15, 1982, during the war, while speaking on the Israel Broadcasting Authority's *Moked* program. At the time, the media had been raising the issue of whether or not the prime minister was being led astray. Begin decided to go on air to refute those claims.

BEGIN: *One of the strangest cases of misunderstanding that I've ever encountered. On account of the range of the terrorists' cannons and Katyushas we set a range of forty, actually it should be between forty-three and forty-eight kilometers, because the terrorists have in their possession a Soviet cannon, albeit outdated, from World War II, 180mm. The important thing that must first be attained, is making it to that line and that it be a line of peace and not a line of war, so first we had to push them north, and here, there was a situation in the central and eastern sectors, we faced them to the south, they were positioned eighteen kilometers from Kiryat Shmonah. Had we remained in place, we would have solved nothing.*

However, this is a frontal assault, and in frontal assaults the attacker takes many losses, and for us each loss is a world onto itself.

You must know that we lost 170 lives. For us this is an open wound for all the days of our lives. It is an awful sacrifice of blood for the nation of Israel, but we had no choice. Because they did not allow us to live, to breathe, did not allow our children in the Galilee to go to school, and they sat day and night in the shelters. Even there it was hard to breathe. We had no choice, and we have found out that for the nation of Israel to live in the land of Israel there must, at times, be a willingness to give one's life, otherwise it will not be able to exist, for it is surrounded by enemies.

However, we must be sparing with our blood, and so, had we launched a frontal assault, we would have suffered more severe losses. What did we do, we decided to flank the enemy . . . and therefore we took another route, the king's highway, and we moved north, in order to take the enemy from two sides, from the north, too. We reached a point, say, sixty kilometers north of our border, and from that point we moved south in order to take the enemy from both sides.

ACHIMEIR: *You explained, sir, about the cabinet meetings, and you said that it has not faced fait accompli situations. You are perhaps aware that in certain parts of the public, of which I cannot measure the scope, there is the suspicion or the fear that the defense minister, Ariel Sharon, is the one who dragged the cabinet into moves that strayed beyond the original forecast for the mission.*

BEGIN: *Every matter was determined by decision. No one pulled the cabinet, no one would pull the cabinet, and why would the defense minister, a man learned in battle, a true patriot, devoted heart and soul to the nation, pull the cabinet along behind its back and so forth? It simply never happened in any shape, way, or form, and I really would like to take this opportunity to call on these journalists, perhaps you should start checking your facts, perhaps put your fingers aside. You have sources, why suck it all from your fingers? On the contrary, be in touch with us, we will give you information. Of course of the type that can be given while the enemy listens. There was no trickery and no tugging along; the things were done by government decision.*

At the cabinet meeting on Saturday night, June 5, 1982, one day before the war began, Begin said, "I would like to say that the government will keep its finger on the pulse throughout the duration of the mission. If the need to take Beirut arises, the government will decide on it. Nothing will simply roll along and take shape as has happened in previous Israeli governments."

In the June 27 cabinet meeting, while the siege on West Beirut was ongoing, the matter of pressuring the terrorists was raised. Begin said, "Time is not on our side, internationally speaking as well . . . what do our members, who reject this mission, suggest? What they suggest in essence is a war of attrition. If we do not act, they will act . . . This nation will not be able to stand it. Our time is limited, we do not have unlimited time, we do not have weeks, days—yes . . . We are wasting the fruits of our victory. We require such a plan, because that is the way to push the scoundrels off balance and to force them to flee Beirut. If we take the airport and advance an additional 1,500 meters, I think we will pressure them hard."

In the cabinet meeting of August 1, Begin said, "If there is no other choice, then we will enter Beirut one of these days, I shall recommend it. We mustn't say, we will not enter there. That would do very grave damage. There is a limit, time is passing. We will plan and we will decide. He who says today that we mustn't enter West Beirut is helping the PLO remain there."

Is this the sound of a prime minister who is divorced from reality, unknowingly led into conflict? Not at all. Begin led and presided over his government at the time. Nor did he have any reason to be surprised by any of the developments. In September 1979, by order of Defense Minister Ezer Weizman, the IDF drafted a plan that stated, under the line "Intent," "The IDF will advance through south Lebanon to the Junieh-Zahkleh line, destroying Syrian and Lebanese forces as necessary to complete the mission."

That is farther than the line that the IDF actually reached in the field in 1982.

After the terror attack in Kibbutz Misgav-Am, Defense Minister Weizman said, on April 8, 1980, "In a war against the terrorists, serious operations will have to be carried out . . . In the end we will have to reach a situation in which we've taken control of a large part of Lebanon . . . from the Beaufort to the Zaharani and Beirut."

In 1980 my father, as minister of agriculture in Begin's government, heard a representative of the IDF general staff say in a meeting with other ministers, "Assuming that there could be a situation that requires us to take all of South Lebanon up to the Beirut line in order to link up with the northern Christian enclave . . . there is a plan based on standing army with reserves reinforcements."

Yechiel Kadishai, perhaps the man closest to Begin, who stood by his side for dozens of years, his aide for half a century, his chief of staff, the man who cared for him all the way till his dying day, wrote on June 12, 1996, "I've been asked, did Ariel Sharon lie to Menachem Begin, may his memory be blessed, and to that I have answered, then as now: 'Ariel Sharon did not lie to him.'"

On February 10, 1997, Kadishai wrote, "I never once heard a statement of any kind from Menachem Begin, may his memory be blessed, never saw a sign or indication of any kind that he felt or believed that Defense Minister Ariel Sharon lied to him; based on my deep, fundamental, and years-long acquaintance with Menachem Begin, with his opinions, positions, and responses, and taking into account the close and tight relationship between myself and Menachem Begin, I have determined that had Menachem Begin felt that Defense Minister Ariel Sharon had lied to him, I would have known of it."

In a Channel 2 interview on November 7, 1997, Kadishai said, "The fact of the matter is that insofar as the Lebanon War is concerned, Begin never said that Ariel Sharon lied to him, and in my opinion, he did not lie to him."

Begin's military liaison officer, a man who never had my father's best interests in mind, gave an interview thirty years after
those events and suddenly "remembered" that my father would
call Begin at night, and it would be draining to the prime minister. But even he, when asked directly in the same interview if
"Begin would ever, even privately, speak out against Sharon?"
answered, "Never."

In 1982 Begin's government demonstrated its responsibility
toward the citizens of Israel, setting clear goals for the war
and attaining them. As Begin said, "There is no country that
runs its wars like this, with the entire cabinet making decisions.
Throughout the Falkland Islands campaign, Mrs. Thatcher
did not convene the cabinet once." And in fact throughout the
course of Operation Peace for Galilee the cabinet convened every day, at times more than once a day, collectively deciding the
next steps. This was in sharp contrast to previous wars, where,
for instance, the Golan Heights were taken and the Suez Canal
reached with nothing more than a directive from the minister
of defense.

In praise of Begin's government, in which my father served
as minister of defense, one can say that it was unwilling to accept a situation whereby the communities of the north would
live under fire, held hostage by the terror organizations, for if
Israel responded to the terror attacks they launched, the terrorists would retaliate against the residents of the north with rockets
and artillery.

From today's perspective, after the IDF spent a month in Lebanon in the summer of 2006 during the Second Lebanon
War—in which I served as a reserves officer in the region—and
managed to advance essentially nowhere, and after the cessation
of rocket fire, one of the main goals of the war, was not achieved

in the field of battle but rather through a cease-fire agreement with Hezbollah, one can appreciate the achievements of 1982.

The nearly monthlong "Operation Cast Lead" that began in December 2008 and was intended to put an end to the rocket fire from Gaza into Israel further underscores the impotence of the Israeli leadership. In fact, the goals of the operation were stated in such a way as to remain amorphous so that the government would not be caught in abject failure, repeating the debacle of the Second Lebanon War, where the goals went unrealized. The result of this was that after nearly a month of combat, the IDF had seized a few neighborhoods on the outskirts of Gaza City. There was no clear and decisive policy, and as a result there were no clear orders and the rocket fire continued throughout the operation and beyond. Defense Minister Ehud Barak boasted at the end of the operation, "The IDF has fundamentally changed the unbearable reality for [the communities] surrounding Gaza."

The terrorists in Gaza apparently see things differently.

In 1982 the government enjoyed scant international leeway. From the onset of the campaign, phone calls from President Reagan made clear that the timetable for military operations was limited. In 2006, and then again, if to a lesser extent, in 2009, the government was given ample latitude, but Ehud Olmert's government failed to grasp that the license given by the international community would not extend forever, nor would the patience of the citizenry, who were under fire for the duration of the Second Lebanon War in 2006 and "Operation Cast Lead" in 2008–2009. If pressed into war, Israel must prevail swiftly and decisively. Large swaths of our very small country cannot be left under fire for long durations of time.

On July 17, 2006, Prime Minister Olmert spoke before the Knesset:

We can all see how the majority of the international community supports our battle against the terror organizations and our efforts to remove this threat from the Middle East.

We intend to do this. We will continue to operate in full force until we achieve this. On the Palestinian front, we will conduct a relentless battle until terror ceases, Gilad Shalit is returned home safely, and the shooting of Qassam rockets stops.

And in Lebanon, we will insist on compliance with the terms stipulated long ago by the international community, as unequivocally expressed only yesterday in the resolution by the eight leading countries of the world. We will not stop our actions until this is achieved.

The return of the hostages, Ehud Goldwasser and Eldad Regev;

A complete cease-fire;

Deployment of the Lebanese army in all of southern Lebanon;

Expulsion of Hezbollah from the area, and fulfillment of United Nations Resolution 1559.

Inspirational words—the trouble is that if you have no intention of adhering to them, they're best left unspoken, so that your warnings are not yet again defanged, dismissed as not credible.

The bodies of the soldiers, whose abduction triggered the war in 2006, were returned two years later, as part of a prisoner exchange; the fire did not stop throughout the duration of the war, with rockets striking Israel up until the initiation of the cease-fire; Hezbollah remains in the area and has bolstered its firepower much more than before the war, and we have ceased our activities.

In the summer of 2008, Olmert, this time with Ehud Barak as his defense minister, reached a semiofficial cease-fire understanding with Hamas—without the return of Gilad Shalit, who was kidnapped from Israeli territory, and in the face of ongoing rocket fire toward the western Negev. The result: Hamas lessened the frequency of the attacks, but never fully ceased. After six months of this lopsided cease-fire arrangement, during which Hamas continued to smuggle arms into Gaza through the tunnels connecting the Strip to Egypt, the terror organization decided to increase the intensity of its rocket offensive to force Israel to open its checkpoints into Gaza. This led to an Israeli operation. On December 28, 2008, a day after the Cast Lead operation began, the government secretary quoted Prime Minister Olmert: "The state of Israel launched a military operation yesterday in the Gaza Strip region, in order to restore peace and quiet to the residents of the south, who for many years have suffered the continuous fire of rockets, mortars and terror activity aimed at disrupting normal, tranquil, quiet life, as any region in any state deserves and is entitled to." At the beginning of the government meeting on February 1, 2009, some ten days after the cessation of "Operation Cast Lead" and after that morning's rocket fire on Israel, Olmert said, "The response to the [rocket] fire will be disproportionate."

Beyond the feebleness of the threat lies a more fundamental lack of understanding of terrorism and the fight against it—what does a "disproportionate response" mean? What would be deemed proportionate: a rocket-for-rocket exchange? The proportionate response to the violation of Israel's sovereignty and the attempts to kill its citizenry should be sharp and severe, and should not be dictated by the enemy. An appropriate response should simply halt the terror activity, and not, in the name of some skewed proportionality, strive to match the terror attack in severity.

EIGHTEEN

My father in New York after his libel trial
against Time *magazine, January 1985.*
(© David Rubinger, Yedioth Ahronoth)

After the events of Sabra and Shatila, the world press pounced on Israel. The Israeli media followed suit, attacking relentlessly, particularly my father and Begin. There was scrawled graffiti everywhere proclaiming "Begin is a murderer" and "Sharon is a murderer." The roads all the way from Tel Aviv to our farm were filled with that filth. Some of those signs stayed in place for years.

I am sure that many of the Israeli protesters really did grieve for the victims of those murders, but it was Arab Christians who murdered Arab Muslims. The Jews self-flagellated themselves for this, and the international community, which has shown little interest in slaughtered Jews, Armenians, and Africans, and in atrocities in other places around the world, now reveled in the spilt blood, heaping their contempt on Israel. Peres and his people in the Labor Party did not intend to let this opportunity slip away, attacking with a potent blend of self-righteousness and venom. Protests were held at the gates to our farm. Placards with the words "Sharon Is a Murderer" were held aloft. Members of the nearby kibbutzim, including classmates of mine, were in attendance. I also remember there were girls in my class who thought it was not right to protest outside my house.

My school, Shaar HaNegev (Gate of the Negev), was for years a small and homogeneous rural school. It served the children of the nearby kibbutzim. The children of the villages went to a different school, and the children from the town of Sderot were rarely accepted. Our farm is situated in the district of the Shaar HaNegev regional council, though, and since 1978, when we moved, we attended the local schools.

My brother was in ninth grade and I was in seventh when we enrolled in the school. We stood out starkly against the local landscape. The school was fully political, with the teachers and students all unabashed supporters of the Labor Party. Students were sent, on regional council buses, to attend Labor Party political rallies in Tel Aviv, which were extremely popular with the kids because the trip to the big city included Cokes and fries and occasionally a hamburger, too.

We were not from the party, did not support the party, and did not hide that fact or bend our opinions to fit in with the rest. We argued a lot. We may have been a bit combative about it. During elections and after the Lebanon War, tempers flared, but in general the atmosphere was pleasant, the lawns large and green, and the studies not very intense. Excelling in school was not considered particularly important. "You city slicker son of a bitch" was a very popular curse in those days. One of my friends finished eleventh grade with ten failing grades on his report card, but soon afterward he was given the honor of operating the giant cotton picker. A short while later he dropped out of school altogether and began working the picker for the kibbutz. During the final matriculation exams in sociology and economics, I was the only person in the room. The rest of the class had decided to skip the tests.

In November 1982, my brother was drafted. During paratroopers basic training in Sanur he bore the brunt of the hostility toward our father. When the rest of his platoon finally went to sleep, he remained outside and endured many extra sessions of running. He took it like a man, running and not complaining.

To the army.
From left: My mother, Omri, my dad, and Vera, at Omri's
swearing-in ceremony at the Western Wall, 1982.

After the murders in Sabra and Shatila, our friend Uri Dan was the first to see the rising wave of resentment. He begged my father to appoint a military commission of inquiry. There were, on occasions, prophetic sides to Uri. He knew they were out for

my father's head, and that the media storm and the public pressure would eventually hit him.

My father was concerned that such a commission might harm
the army or make it seem as though he were hiding behind the army's skirts, and he was not willing to do that. This was a mistake.

Begin, who had shown real leadership during the war and had
stood firm during the siege and the lead-up to the PLO deportations, began showing signs of weakness. I recall my father saying
how one time, during the siege, Begin had gotten a call from President Reagan, who was furious about the aerial strikes in Beirut.
My father watched as Begin rose from his chair and stood by the
side of his desk. "Yes, Mr. President," he said repeatedly, like a soldier accepting orders. It left him sad, seeing a proud Jewish prime
minister diminish himself before the American president. But at
the July 18 cabinet meetings Begin was bolder. "Do we really want
to harm the civilian population? But to say that if we harm several
civilians then we should cease all activity—where would that kind
of policy lead us? If we do not enter Beirut, the PLO will be victorious. President Reagan promised President Assad that he would
keep in mind the suffering of the Palestinian people, not the suffering of the Jewish people at the hands of terrorists." Speaking
with Reagan, two weeks later, he was less of a hero. Perhaps the
"Begin Is a Murderer" signs being waved outside his office and
the especially grave placards stating the number of Israeli dead
since the start of the war took their toll. On September 28, 1982,
Begin bowed to the political, public, and media pressure, passing
through the cabinet the decision to establish the Kahan Commission of Inquiry into the Events at the Refugee Camps in Beirut.

My father, despite his vast experience, was naïve. "I have nothing to hide," he said. He did not grasp that it didn't matter that
he had nothing to hide. The antiwar protests, the attacks from the
left and the media, they all demanded a victim. Someone would
have to pay with his head. It didn't matter that the cabinet had
made all of the decisions during the war. It didn't matter that
the opposition had supported the stated goals of the war, and it

didn't matter that we did not murder and we did not know that the Christians were going to do so.

Not long afterward, in November 1982, the prime minister's wife, Aliza Begin, died. The funeral was held on a cold wintry day on Jerusalem's Mount of Olives. During the procession, my father turned around and saw two figures behind him—they wore black hats, black ties, and black coats and looked at him darkly. After the funeral my father went to the Knesset to deliver a speech. Standing at the podium he saw, in the guest section, the same two figures, both staring at him hard, like a pair of black ravens. The two men were justices Yitzhak Kahan and Aharon Barak, both members of the three-person commission of inquiry. Their stares left no room for doubt.

My father denied his lawyers the right to question the chief of staff, so that it would not seem as though he was trying to blame or hide behind the army. His lawyer, Dubi Weissglass, tried to explain to him the difference between the military echelon, whose role, he said, "was to weigh the facts, take all matters under consideration and reach an operational decision, in light of all the different considerations, and between the authorizational echelon, which is not supposed to go through the same process but to simply assess its reasonability."

"Mr. Defense Minister," Dubi said, "you keep on saying I instructed them, but that is not the way it was. You authorized the chief of staff's instructions."

"I come from a different culture," my father responded. "In the culture in which I was raised, someone under my command presents a plan to me and I authorize it and from that moment on, it is my plan and my order. And I will not hide behind these distinctions."

The Kahan Commission finished its report on February 7, 1983. The following morning, my father found Dubi Weissglass and Oded Shamir in his office, smiling as they read through the report. The committee found that the Phalangists bore direct responsibility for their murderous acts, and "in having the Phalangists enter the camps, no intention existed on the part of anyone who acted on the part of Israel to harm the noncombatant population."

"You better skip through to the conclusions," my father said. They did, and they stopped smiling.

The head of the Mossad, the commission ruled, did not foresee or warn against the dangers of a massacre. The commander of military intelligence did not sound the warning bell, either. None of them did. And yet my father was found to bear an "indirect responsibility" for what had happened. He should have foreseen the potential for atrocities, they ruled. Their recommendation was that he step down from his position as minister of defense.

This is the one and only instance in which an Israeli official has been found to bear indirect responsibility for someone else's heinous acts. In fact, such indirect responsibility is unprecedented in Israel and across the world. Nor has it happened since then.

Prime Minister Begin and Foreign Minister Shamir were also found to bear a degree of responsibility, but no personal recommendations were filed against them. The commission found fault with Chief of Staff Eitan's behavior, but since he was close to the end of his term, they refrained from recommending his termination.

Yehoshua Saguy, the commander of military intelligence, was ousted from his position, while the head of the Mossad, whose behavior was also found to be lacking, was not.

Two days after the February 8, 1983, release of the report, the cabinet met to consider its next steps. My father couldn't make it to Jerusalem on time: there was a demonstration on the road outside our farm, and the protesters were unrelenting. Chants of "Sharon is a murderer" followed him as he left the farm.

An hour later, he reached the prime minister's office. There was a mass demonstration outside—this time, of his supporters. Although he had as many supporters as detractors, the detractors were far more vocal, and they were buoyed by the press, which presented, one can say, a unified front against him. There were no anemic opinions about my father. Either you admired and loved him fiercely, or you hated him. No one was indifferent.

My father stopped for a moment alongside his supporters. Countless hands reached out to shake his, and the crowd shouted calls of encouragement and support. Those calls were mixed with the shouts of an approaching demonstration. "Sharon is a murderer, Sharon is a murderer." The supporters called back, "A-rik, A-rik."

The calls followed him into the cabinet meeting. The government secretary, I think, was the one who shut the windows. The crowd circled the building, and the "A-rik, A-rik" calls came in from the other side of the building. The secretary shut those windows, too, but the chants were still heard inside the chamber.

His supporters had no way of knowing that the louder and more passionate their chants grew, the quicker his already small chances dwindled. The chants of support had more of an effect than those calling for his head. His colleagues hated the disproportionate support he was given from within the Likud camp.

Toward the end of the cabinet meeting, the government secretary received a note reminding him to take into account the votes of two absent ministers, Simcha Ehrlich, who was against my father, and Professor Yuval Neeman, who opposed the commission's recommendations. "Better sixteen to one than seventeen to two," was the response. And so it was: the cabinet decided to accept the commission's recommendations by a sixteen-to-one vote.

On February 14, 1983, the Ministry of Defense held a farewell ceremony for my father. As he walked through the parade grounds and inspected the honor guard outside the ministry, he thought of his father, Samuil, who had told him thirty-seven years earlier, in the orchard in Kfar Malal, "Never turn Jews over. Never do that." And now, those who had been the victims, the underground fighters turned in by their Jewish brothers, were the ones who handed him over to the mob.

A few months later, sitting in Begin's office, my father told him what he had been feeling on that day. "Menachem," he said, "it was you who handed me over to them. You are the one who did it."

All things considered, I believe Begin was an honest man. Perhaps his conscience was troubled by the fact that he had chosen to hide behind my father, to sacrifice him to keep his own position. Perhaps this is also one of the reasons that, a short time later, he stepped down and cloistered himself in his own home until the day he died.

My father's career seemed over and done. This is why Uri Dan's pronouncement, on the same day my father left the Ministry of Defense—"Those who don't want Sharon as defense minister, will get him as prime minister"—was openly ridiculed. But as I said, Urileh, as I sometimes called him, was a bit of a prophet.

———————

The local and international press had a field day at my father's expense. *Time* magazine took things the furthest. In February 1983 the magazine published an article asserting that the Kahan Commission's report, in the classified section of the ruling, Appendix B, found that my father, when in Bikfaya on his condolence call, had urged the Phalangists to take revenge on the Palestinians for the murder of Bachir Gemayel.

This was an outright lie. It's possible that the magazine thought they were dealing with a beaten man, about whom you could write whatever you wanted. It's also possible that the publication's institutional arrogance—according to Dubi Wiessglass, they prided themselves on the fact that during their sixty years of publication they had never apologized for a news item; never published a correction; and never paid any compensation. *Time* published a lie, and they most certainly underestimated the person they lied about. One can't say my father was in high spirits at the time, but he was far from beaten. Angry might be a more apt description of his mood. He had left office and he felt he was being pelted with stones as he left, but this one particular stone was harder than the rest, because it charged him with incitement to murder, and that was a charge he would not accept. As soon as it went to print, he decided to turn and fight.

I was with my parents in New York on the day the trial began and through later hearings, until I returned to Israel for induction into the army. I remember how awful the first day was. Representing the magazine was a battery of lawyers, led by a red-faced, self-satisfied attorney. The *Time* magazine attorneys screened all nature of trash and slurs about my father from publications around the world in order to prove that their article had not unduly defamed him. They contended that my father was libel-proof—that his reputation was so severely defamed that he could no longer file for libel.

At the end of the first day, sitting with our lawyers, I was encouraged—"They've put all of their trash on display already, and now we can get on with it," I said—but the looks I got indicated that not everyone agreed with my not-so-learned opinion.

Among the jurors was one large-framed former military policeman. "I would talk about how the military police directed the forces, and each time I'd talk about a junction, it seemed like he was about to leap out of his seat and start directing the traffic," my father would recall with a smile.

One of the women on the jury had a boyfriend in the army reserves, and my father went out of his way to praise the contribution of our reserve troops during the war. Even in court he kept his sense of humor.

When my father missed some of the words from the bench or didn't understand Justice Abraham Sofaer's ruling, he would look toward the rows where the Israeli journalists sat. "If they looked crestfallen, I knew he had ruled in my favor; if they looked joyous, I knew he had ruled against me," my father would say.

On January 25, 1985, the jury in New York ruled that *Time* had published an article that was both false and defamatory. They could not say whether the magazine had published the article with actual malicious intent, but they did rule that "certain *Time* employees had acted negligently and carelessly."

In Israel, where a plaintiff docs not need to prove malicious intent along with lying and defamation to win a libel suit, Time Inc. settled the case and paid compensation for the lie it published. The suit sent tremors through the magazine, forcing a change in the upper management of the magazine and the ousting of the reporter who filed the story in the first place.

After being forced from the Ministry of Defense, my father's natural inclination was to go back home to the farm and tell everyone to get lost. Uri Dan urged him to stay in the game. I think it was our friend Reuven Adler who coined the saying "So long as you're on the wheel, you're still spinning." In other words, so long as you remain in politics, you remain relevant and you may rise again. My father kept the position of minister without portfolio, serving in an office situated in a desolate and depressing building in East Jerusalem.

In 1984, an election year, my father decided to run against Yitzhak Shamir, who had replaced Begin as prime minister, for the position of Likud leader and prime ministerial candidate. He did not think he had much of a chance of beating Shamir, who was backed by the ministers and the party establishment, but the minister without portfolio appointment, along with the unceremonious manner in which his colleagues had disposed of him, was enough of an inducement. He decided to fight.

He, too, was surprised when the final tally revealed that he had received over 42 percent of the vote in a contest against a sitting prime minister. The excommunication to the far reaches of government had clearly come to an end. As usual, support for my father came from the grass roots, the people, and bypassed the ministers and the party establishment.

The national elections ended in a type of stalemate. Labor had received more votes but was unable to assemble a coalition government. My father met with Shimon Peres secretly.

They agreed on a national unity government with a rotation

at the top between Peres and Shamir. This was the first but not the last time that my father advocated in favor of a national unity government. Having seen the polarization of Israeli society during the Lebanon War and the way in which security matters were twisted into political affairs, he felt a wide consensus at the top was the best way to govern the country.

He was appointed minister of industry and trade and made a member of the security cabinet.

At the close of the 1988 elections, Yitzhak Shamir formed a national unity government, this time without a Labor-Likud rotation at the top. Once again, my father was appointed minister of industry and trade. Fourteen months later he resigned, unsatisfied with the government's response to the outbreak of terror in Judea, Samaria, and Gaza one year earlier (the First Intifada). He felt that Rabin, the defense minister, did not wage the struggle against terror properly, and the prime minister, Shamir, did not do much of anything at all, although he was fastidious about his afternoon naps.

In March 1990 Shamir's government fell to a no-confidence vote. Several days later, the Likud Central Committee convened. Shamir had been weakened, and those closest to him were planning a coup. The plan apparently was to replace Shamir with his defense minister, Moshe Arens. Help came from an unexpected source, my father. He spoke before the central committee and called on everyone to support Shamir against the threat of Peres, who was seeking the prime ministership. As problematic as Shamir was, my father felt the alternative was worse, and so, in order to foil Peres's plan, my father used all of his connections with the ultra-Orthodox leaders, including the Lubavitcher Rebbe.

My father's efforts took him to Shaar HaGay for a nighttime rendezvous with Agudat Israel power-broker Avraham Shapira. The two drove to nearby woods, where they could speak confidentially. "He looked like a scared Jewish boy," my father said;

"the sounds of the night, the darkness, he wasn't used to any of those things, he was afraid of forest animals." Shapira's driver produced a sheet of apricot fruit leather and a few boxes of sugar-coated almonds, Shapira's standard snacks in case he got hungry during the forty-minute drive from Jerusalem to Tel Aviv. My father was downright gaunt in comparison to the Agudat Israel man. "It's a little bit scary here at night," Shapira said. My father responded, "It would be a lot scarier if you were on your way through the sands of Rishon LeZion with Yitzhak Shamir." This was in reference to the night when Shamir, the commander of the underground group Lehi at the time, took a trip with his deputy to the dunes and returned alone.

On a fine April day, Peres and his handpicked cabinet appointees arrived in the Knesset in pressed shirts, prepared to display their new government and have it authorized by the Knesset. Soon enough, it became clear that Eliezer Mizrahi and Avraham Verdiger, both members of the ultra-Orthodox party, had disappeared from the hall and could not be found. There was no government formed on that day. This was perhaps one of the worst blows that Peres suffered in his long and bruising political career. "I hid Mizrahi in the orchards between Yavne and Ashdod, and I put Verdiger in the caves of Qumran," my father used to say, laughing.

In June 1990 Shamir formed a new government, this time without the Labor Party. My father was, predictably, not rewarded. The fact that he had saved Shamir from an in-house coup and foiled Peres's plot got him nothing in return, nor had he expected anything. This was Shamir. But that, too, was for the best. My father was appointed minister of housing and construction and chairman of the Ministerial Committee for Immigration and Absorption. The Soviet Union was collapsing at the time, and a blessed torrent of immigrants started to move to Israel. During the 1990s, one million immigrants left the crumbling empire and moved to Israel, adding about 20 percent to the population. This was as if the United States suddenly had 60 million new immigrants. A Herculean effort was needed to build hundreds of thousands of apartments for the new immigrants on short notice. My father threw himself into the task. He would visit the different construction sites, urging the contractors to get the job done. Incentives were given for apartment buildings that were completed quickly. Caravan sites were assembled. During ordinary times, trailer parks would be far from optimal housing, but what mattered at the time was that each immigrant that arrived was given a roof over his head. After they'd been settled, the permanent housing could be completed. This temporary solution also served to keep the rent and the prices of the apartments from skyrocketing.

My father would receive constant updates, in real time, about the number of immigrants that had arrived, the number of sites under construction, and the number of apartments ready for habitation. He spent days and nights on the road, visiting construction sites all across the country, meeting mayors, contractors, Israel Lands Administration workers, treasury employees, bankers, and army officers, whose bases were even used to house some of the immigrants. He met with all relevant parties, anyone who he thought could help. He pushed them all, shouted, encouraged, and patted whoever deserved it on the back.

All told, after two years as minister of housing and construction, my father had presided over the commencement of construction of more than 170,000 housing units and the planning of hundreds of thousands more apartments. The wave of immigration afforded Israel a once-in-a-generation chance, if not more, to inject vitality into and make growth possible in the towns and cities of the periphery, places that the state had long neglected.

Rabin, Peres, and my father on a tour of the Golan Heights.
(© Yaacov Saar, Israeli Government Press Office)

In June 1992 the Labor Party, headed by Rabin, defeated the Likud, led by Shamir, in a landslide. Slightly over a year after the elections, the Oslo Accords were signed. My father vehemently opposed the agreements. They were not exactly what Rabin wanted, either: he opposed giving Jericho over to Arafat but was outmaneuvered by Peres, who had orchestrated the agreements.

"What was Peres up to last night?" was Rabin's staff's standard question in the morning during his second term as prime minister. Peres tried to pull similar stunts with my father, but got nowhere.

Rabin and my father spoke before Rabin's trip to Washington, D.C.: "He told me that handing over Jericho was unnecessary. The fact that Israel could no longer act in certain territories worried him as well. He would go for a Palestinian state in Gaza, and he would happily, as he said to me, 'let things drag on another twenty years in Judea and Samaria.'"

On September 13, 1993, Israel and the PLO, with great fanfare, signed the Oslo Accords on the White House lawn. The main idea, aside from mutual recognition, was a commitment to resolving the conflict peacefully, with a total cessation of terror. That was the heart of the agreement. In 1994 Arafat arrived in the territories along with thousands of his armed men, members of his organization, all charged with keeping the peace. In addition, he smuggled four men into Gaza in the trunk of his Mercedes. All had been denied entry. One was linked to the murder of the Israelis at the Olympic Games in Munich, another to the slaughter of the children at the school in Maalot. The other two were of a similar ilk. My father was resolutely opposed to allowing Arafat into the area. He viewed him as a murderer and his men as a band of terrorists.

In an article published in *Yediot Ahronoth* on September 4, 1994, my father wrote,

> *Israel has serious security problems. Some of them have already been dangerously heightened by the Oslo Accords. The handling of Palestinian terror has been given over by the government to a certain Palestinian terror organization, the Fatah, under the assumption that it will take action against the other terror organizations, which continue and will continue to operate. In the meantime, the operational arm of Fatah, the Fatah Hawks, has resumed its operations against us. This assumption, which lies at the heart of the Oslo Accords, has already proven to be fallacious . . .*
>
> *Once the government has taken the fighting of terror out of the IDF's hands in the autonomous areas, we may foresee a scenario whereby Palestinian terror increases, triggering sharp Israeli responses, and the Arab states, unable to stand by idly, take a series of steps that are not war but that force Israel into tough dilemmas.*
>
> *Terror and national security should not be divided. They have always been linked. That is why the Oslo Accords, which have taken the fighting of terror out of Israeli hands, contain the seed of the next war.*

And in terms of the peripheral states Iran and Iraq, we're talking
about a true threat within several years, one that has no connection
or affiliation with the peace accords with our neighboring states. The
matter can only be taken care of by an international coalition under
the leadership of the United States.

But his admonishments did not help. I believe that had Yitzhak
Rabin known how many Israelis would pay with their lives on
account of the Oslo Accords and how many Palestinians who
were not involved in any hostile actions would suffer as a result of
the war on terror, he would perhaps have had greater doubt about
the merit of those agreements. But that we will never know.

One day before the signing of the accords, on September 12,
1993, three IDF soldiers were shot and killed by Palestinian gunmen in Gaza. In Hebron, a terrorist wounded a bus driver, and
on the Ashdod-Ashkelon road a terrorist managed to murder a
bus driver.

Ten days after the festive signing in Washington, Yigal Vaknin
was murdered by terrorists in the village of Batzra, not far from
Kfar Malal. He was stabbed to death. On October 4 a suicidal
terrorist blew himself up alongside a bus near Beit El, wounding
some thirty people. Five days later, two Israelis, Aran Bachar and
Dror Forer, both in their twenties, were murdered while hiking in
the beautiful Wadi Kelt. The two were shot and stabbed to death
on October 9, 1993. The Palestinians certainly showed in which
direction they were heading.

Kochava Harush from Sderot worked at our house on the farm
for decades. Her son Kobi was my father's driver for many years.
We considered them to be part of the family, and we knew her
family that lived nearby, in the village of Yoshiveha. On the night
of February 9, 1994, Ilan Sudri, Kochava's twenty-three-year-old
nephew, stopped his taxi on the way from Ashdod to Ashkelon for
what appeared to be a few ultra-Orthodox men. The Palestinian
terrorists he allowed into his taxi killed him. His body was found

two days later. And now, over fifteen years later, I see his mother, Simi, at family events, and her face still looks stricken with grief.

On January 22, 1995, on Sunday morning, two terrorists blew themselves up at the Beit Lid junction. Both my brother and I used to come through that junction on Sunday morning on our way back to basic training. It was crowded with soldiers returning to the nearby base after a weekend at home. All told, twenty-two people were killed in the two blasts—many were wounded.

The dozens of murdered and the hundreds of maimed in the terror attacks that took place after the signing of the Oslo Accords ate away at public support for Yitzhak Rabin and his government, driving their popularity rating into the ground. The contrast between what had been promised by the Nobel laureates—Rabin and Peres together with their dubious partner Yasser Arafat—and the terror and death spreading through the streets enraged the public. They felt cheated. After all, the cornerstone of the agreement had been the resolution of the conflict by peaceful means and the cessation of Palestinian terror in exchange for Israeli recognition of the PLO and the transfer of control of part of the territories to the Palestinians.

On November 4, 1995, Yitzhak Rabin was assassinated by Yigal Amir, a fellow Jew. Rabin's murder sent shock waves through Israeli society and altered the public discourse. Although the failure of the agreements was underscored by each new and ghastly terror attack, the trauma of the assassination caused even legitimate opposition to the accords, to be viewed at times as incitement. There was much understandable attention devoted to the matter of incitement, but far too little coolheaded analysis of what the accords had wrought.

In 1994 Rabin signed a peace treaty with King Hussein of Jordan. My father abstained when the peace treaty was voted on in Knesset. He of course supported the peace, but opposed the article in the agreement that granted the Jordanians special status vis-à-vis Jerusalem. Why in the world should the Jordanians have special privileges regarding Jerusalem? Because they captured the eastern part of the city in 1948, destroyed the Jewish Quarter and all of its synagogues, held the territory for nineteen years, and during all that time, never allowed Jews to enter the Old City?

———

The suicide bombings of 1994–95 created a dreadful feeling. The public had been fed a false vision of what Peres loved to call the New Middle East, but what ensued was the same old Middle East we knew, only worse.

On a personal level, my father and Rabin were on good terms and had been for years. As my father told Uri Dan,

> Our meetings touched on many military matters and he would listen to what I had to say. He also used to speak with me and consult, as though the two of us were not on the opposite sides of the political spectrum. [We would discuss] matters of state, matters pertaining to Iran and Iraq, the Middle East and others. It took place on a different plane, not political. There was talk about maps, security zones, it certainly had an impact. There was of course another matter that we discussed more than once. I tried and I explained to him and he understood that a peace process is built on compromise. It is very difficult to attain peace without widespread political support. That's why I always thought that a great effort needs to be made to attain national unity, a widespread national consensus.
>
> I was very very pained by Rabin's murder. I appreciated him and I was friendly with him—or maybe I should say that we had mutual respect for one another, which endured through the years.

After the assassination there were those who tried to turn Rabin into a left-wing personality more willing to make concessions than he really was, and to use his name to further their own overly concessionary agendas. In 2000, during Ehud Barak's term as prime minister, my father, the head of the opposition at the time, called Rabin's dying wife, Leah. At the time, Barak seemed ready to offer Arafat overwhelming concessions. Leah Rabin said to my father "Yitzhak would never have given up Jerusalem? Right?" She was stating a fact more than she was asking a question.

After the murder, Peres took over as prime minister. Shortly thereafter I drove my father to a late-night secret meeting with Peres at his Jerusalem apartment. He had not yet moved into the prime minister's official residence. The telephone wasn't working, and Peres fumbled with the receiver and the cord. "Why doesn't the minister of communications [Shulamit Aloni] come straight down here with a pair of pliers and take care of this?" my father asked with a smile. Sonia, Peres's wife, was in the kitchen, and some drinks were poured. On a personal level my father got along very well with both of them. I heard my father tell Peres that it was critical that they reach some type of national unity or consensus in order to lower the tension level. "True, no one would have gotten exactly what they wanted, but it would have led to some sort of agreement."

That was his message—to soothe and unify the people so that we could better face the demands of enemies and friends.

"May I speak in front of Gilad?" Peres asked.

"Speak freely," my father said.

They spoke, and it was clear that Peres was not interested in a unity government. He wanted to beat Likud head Netanyahu in face-to-face elections and rise to the post of prime minister, just as Rabin had done to Shamir in the '92 elections. Peres moved up the elections to May 1996. He was sure of victory, as were the members of his party and the vast majority of the media.

The disastrous terror attacks of February and March 1996, including the two bombings of the Number 18 bus in Jerusalem and the attacks in the Dizengoff Center in Tel Aviv and in Ashkelon, resulting in the deaths of dozens of people, shook a large swath of the voting population free of the trauma of the Rabin assassination. Many began to discern between their contempt for the murder of Rabin and their opposition to the Oslo Accords and the terror that those agreements had brought into Israel.

My father worked hard for Netanyahu's campaign. He pulled every string he could in the ultra-Orthodox world to ensure their support. In addition, since he knew that it would be a close contest and that each vote would count, he managed to induce Rafael Eitan and his party, Tzomet, and David Levy, who had left the Likud and formed the Gesher party, to run on the Likud ticket. Having them join the Likud gave the impression of a wide and united front.

"There is a danger that our candidate is going to be elected," he would say ironically when he came home in the small hours of the night from yet another function or meeting on behalf of the Netanyahu campaign.

Netanyahu beat Peres by less than 1 percent of the vote. Peres had gone to sleep thinking he had won, as the polls had indicated, and had woken up as head of the opposition, tacking yet another loss to his overall record.

Netanyahu was elected prime minister, and soon enough the key traits of his character were put on full display. In those days the aspect that defined him above all others was the man's loyalty, or lack thereof. Netanyahu, who had promised my father one of the three main ministerial posts before the elections, decided not to offer my father any position at all and to exclude him from the government entirely. One Friday night, not long after it became clear that Netanyahu had reneged on his promise, I escorted my father to the Israel Broadcasting Authority studios in Jerusalem for an interview. I went into the studio, sat off to the side, and watched him. He told the interviewer, "I hope he keeps his promises to the voters more than the ones he made to me." Truth be told, the betrayal was not a surprise. One night during the campaign, Netanyahu had come into a room where our friend Reuven Adler, the campaign's creative director, had been sitting and said, "What a great defense minister Arik will be, eh?" Adler had called my father and told him the news. "He won't abide by a single thing he said," my father responded. Neither my father nor my mother believed Netanyahu. They felt he was traitorous and a liar, and in that they were not mistaken.

Soon enough it became clear to Netanyahu that his decision not to give my father any role in his government was not worth the trouble it caused. Netanyahu faced pressure across the board, and his swearing-in ceremony in the Knesset had the atmosphere of a Turkish bazaar. Three weeks after Netanyahu's inauguration, my father was appointed minister of national infrastructure. The post was created for him and included authority over Israel's energy sources, electricity, water, gas, and others.

In the summer of 1997, Netanyahu pushed finance minister Dan Meridor out of the government and then turned to my father to see if he would be willing to fill the post. He agreed. Time passed, and nothing happened. This surprised no one. Yet another Netanyahu promise that he had no intention of fulfilling. Netanyahu summoned my father to a meeting in his office. Standing at the entrance to the room and putting an end to the shortest meeting in the history of the prime minister's office, my father said to Netanyahu, "A liar you were and a liar you have remained."

*My father with King Hussein. Prime Minister
Benjamin Netanyahu is in the background, August 1997.
(© Amos Ben Gershom, Israeli Government Press Office)*

NINETEEN

On September 7, 1997, Inbal gave birth to our first child, a boy, whom we called Rotem. He is my parents' first grandchild, and the two of them were ecstatic.

Shortly thereafter, on September 25, 1997, the Mossad tried to assassinate a Hamas man by the name of Khaled Mashal. Mashal was exposed to a deadly poison and his condition quickly became critical. The operation went awry, and the two agents carrying out the mission were caught and placed in a Jordanian jail. Israel was forced to provide the antidote to save the dying terrorist. King Hussein was livid and refused to meet with Netanyahu. My father was the only member of government with whom he would talk. My father's sensitivity to Israelis held in Arab countries was not something new, and he began taking swift action to free the men.

My father had first forged ties with the Jordanians in 1996, when a Jordanian delegation arrived at the Ministry of Infrastructure to discuss water rights and joint projects. The Jordanian delegation was shocked to find that one of my father's close advisers was Lieutenant Colonel Majalli Wahabi, a Druze Arab. This contradicted the notion of my father in Jordan as "an enemy of all Arabs." They were also pleasantly surprised by my father's attitude toward the dispute over water. Although the Israel-Jordan peace treaty is not completely clear on the matter of the division of and payment for the transfer of water to the Jordanians, my father announced that "agreements must be honored," and the agreed-upon amount of water would be given to them. "I will not accept a reality where my flock has water to drink and yours does not," he told the Jordanians. That was the backdrop to the warm relations between King Hussein and General Sharon, as he was known in Jordan. At the time, seventeen projects had been launched, and most of them had been completed, among them a free trade zone. In addition, to streamline the process, an informal principals' committee was founded. The king and his secretary, General Ali Shukri, and my father and Majalli Wahabi sat on that committee and together managed to skirt much of the red tape. They oversaw the completion of projects and strengthened the strategic pact between Israel and Jordan, which rests on the stability of the Hashemite Kingdom as well as on security cooperation.

On the day of the assassination attempt on Mashal, Majalli was in Jordan on a mission for my father. He was scheduled to meet with King Hussein. Neither he nor my father had any knowledge of the Mossad operation.

Majalli felt something was not quite right when the major from the king's palace who ordinarily picked him up did not show. He waited for hours and was then not taken to the king's usual palace, but to a secret one he had never before seen. The place was not visible from the outside, and from the inside one couldn't see out.

Majalli waited by the pool and sipped the coffee that he had been served. The king's convoy arrived. The king embraced him, as usual, but the customary smile was gone. "Sit down, my son," the king said. "How are you? How is General Sharon?" Majalli discussed with the king the joint projects and relayed the messages my father had asked him to pass on to the king. All told, they met for half an hour, less than usual. Hussein apparently came away with the impression that Majalli did not know anything about the Mossad operation. He ended the meeting. On his way home, Majalli decided to stop in at the Israeli Embassy and visit the ambassador. When he arrived, he saw that the building was encircled by Jordanian troops in armored personnel carriers. A Jordanian officer, who had not realized that Majalli was an Israeli, stood in his way when he tried to enter the building, and said, "Are you crazy?" Majalli got to the Allenby Bridge, one of the crossings between Israel and Jordan, called my father, and told him "something strange" was afoot.

On the evening of Sunday, September 28, three days after the assassination attempt, a helicopter took off from Jerusalem to Jordan. On board were Prime Minister Netanyahu, Defense Minister Yitzhak Mordechai, my father, and others. Since the king refused to meet with Netanyahu, the delegation was met by the crown prince, Hassan, and the head of Jordan's General Intelligence Department, General Samih Battikhi. The likelihood of freeing the Mossad men from Jordanian prison did not seem to be anywhere on the horizon.

Several days later Majalli called Ali Shukri, the king's secretary, who told him that His Majesty did not believe Netanyahu and did not want to hear from him. General Sharon should take this assignment on himself. Majalli and my father set out for the prime minister's residence.

"Let me take care of this," my father told Netanyahu.

"How can I let you do that?" Netanyahu asked.

"Majalli is the only one who can talk to them."

"Get me the king," Netanyahu told Majalli.

When Majalli tried to reach Hussein, he was told, "The king is asleep. He has had two rough days. He does not wish to speak with the prime minister."

Netanyahu agreed to let my father take care of it.

Majalli and Shukri set up the visit. The Israeli secret service spoke with their Jordanian counterparts, but too many people were in the loop. The Jordanians got cold feet and canceled the visit.

The two went back to the phones and set up a clandestine meeting. The prime minister, the head of the Shabak, two secret service agents, and my father's driver were perhaps the only others who knew about the meeting.

The Jordanians did not want the helicopter taking off from Israeli soil. They wanted the Israelis to cross the Allenby Bridge and to take off from there, near the bridge, facing Jericho. My father changed into a suit that was brought from the farm while Majalli went across to Jordanian soil to get the king's vehicle, driven by a Circassian officer, an aide to Ali Shukri. Once they'd joined up with the waiting Jordanian convoy, they drove a short distance to a heliport among banana groves and boarded the waiting helicopter.

They landed at the palace in Amman and were met by the crown prince, General Battikhi, and Ali Shukri. The king's secretary approached my father and asked him how he wanted to conduct the meeting—four eyes, six eyes, eight eyes? In essence he was inquiring whether my father wanted Majalli present or not. "With Majalli," was the response my father gave. The two Israelis and the three Jordanians went into the room, while the Israeli secret service agents stayed with their Jordanian peers. Battikhi looked sour. My father didn't like the look on his face and told Majalli, "He's going to get it from me soon enough."

"You know why I'm here," my father began.

Coffee was served.

"How could you do such a thing to us?" Battikhi said, still sour and unpleasant.

My father was prepared to answer, but at that moment Prince Hassan announced the king's arrival.

King Hussein entered, and after the customary greetings my father said, "I will speak Hebrew, and Majalli will translate; you speak Arabic, and he will translate back to me, so that all is clear and every word is understood.

"Our intention was not to harm you or the sovereignty of Jordan. I am not shirking responsibility. I am part of the government. What happened, happened. I would like to tell you, Your Majesty, that you should have taken greater care that a murderer like him would not sprout in your yard. You should have taken care of him. I am not only in favor of eliminating these types of terrorists, but I am angry with our men for not doing a better job."

Ali Shukri signaled to Majalli to step aside. Standing by the door, he told him that in effect his boss was criticizing the king. "If he wants to speak critically of the king or the kingdom, he should not do so in this forum, but in a face-to-face private meeting."

Majalli returned to my father's side and asked him in a whisper to soften his tone. My father carried on, "If we could have finished the job properly, we would have helped you and ourselves. We do not make light of Jordanian sovereignty, but Mashal's actions warranted this type of operation. I also would like to tell you, Your Majesty, that as someone who knows the customary ways, we are willing to pay a ransom for this mistake. If you give me back those two men, let them get on board the helicopter with me, I'm going to tie them up like lambs, you know what I mean, Your Majesty, I am going to tie them up and I am going to sit on them, and as you see, I've got some weight on me. Not because they're not good, but because they failed in completing their mission."

"Softer, softer," Majalli pleaded in Hebrew.

Battikhi went on the offensive. "This is not how peace is made, this is not proper behavior." The king cut him off. "I would like to respond to General Sharon. I protest the fact that this was perpetrated on my soil. I see that as an insult. It is good that we are speaking; we must find a solution, and it has a price. We passed on the names of the people we want released."

The king had a point. My father, who had not known of the mission but still represented the state of Israel, said, "I have a request: give [the agents] over to me and I will recommend that the people you've listed will be released."

"We've given you a list of Jordanians and Palestinians that we want released," the king said.

"As far as the Jordanians are concerned, I am in favor; as far as the Palestinians are concerned, I need to check."

"You did something wrong on our soil, and now you're arguing about the ransom?" Battikhi said.

"The Jordanians that we have, we will let go; in terms of the Palestinians, some of whom have considerable blood on their hands and were already tried for their actions, I cannot commit.

Please do not ask that which I cannot commit to. The number of Jordanians suffices, especially for that terrorist—remember who this Mashal is."

"What will Hamas get?" Shukri asked twice. His question revealed the Jordanian fear of in any way changing the invisible balance in the region, a balance that was key to the survival of the kingdom and the king, who lived his entire life on a tightrope, attuned to the slightest tremors, fearing that one of the forces in the region could at any point cut him from his throne.

Israel had agreed to release the head of Hamas, Ahmed Yassin.

"They got Yassin. That's more than they ever dreamed of. The only place they thought they might see him was in a coffin," my father said.

My father's tone and body language set the Jordanians on edge. Battikhi continued to push, and the conversation began to worsen. The king looked displeased. Shukri called Majalli over to the door again and said, "Let's not go into the list. I'm sure we'll work something out. Change the direction of the conversation— don't you see that this Battikhi is trying to blow the whole thing up? Don't give him that pleasure."

"We keep seeing trays of coffee and juice," my father said to Majalli; "when are we going to see some food?"

Majalli told the king about our farm. He told how whenever a calf is born, my father is told about the event. He really would have notes slipped into meetings informing him of the gender of the calf, or the amount of rain that had fallen that day. That kind of information energized him, gave him strength to carry on. My father always hoped for female calves so that the herd would grow. I had a bit of a preference for bull calves, so that we would have something to sell. On our endless drives around the farm, with me behind the wheel, I'd say to him, "Dad, if it was up to you we'd open an old-age home for cattle and sheep that neither give birth nor milk." He had a hard time parting with the animals, even when they were well beyond their prime. "Give her another chance," he'd say.

When the king heard from Majalli that my father would stop meetings in order to be updated about births on the farm, he got excited. "Really?" he asked.

My father confirmed it and told him about the herd.

"I like horses and camels," the king said. "I have a farm, and it's my daughter, Hayyeh, who takes care of them. I passed that love on down to her."

My father told the king about the Arab shepherds and cow-hands who used to work on the farm and how they would call him to help with the births because his palm was not too large. He told the king about how lovely he always thought it was, the way the Arab workers would sit down to lunch together. Each would take out what he had—pita bread, onion, hummus—and place it in the middle of the circle, and they would all eat together. I, too, have heard him praise this custom many times. "Not everyone takes their own food and goes off to a corner and eats it," he'd always say in appreciation.

"I've come to the conclusion that the flock does far better if they're given cold water," my father told the king, who was shocked and moved by the fact that he was speaking to a real man of the earth, a raiser of cattle and sheep.

"I like my sheep over hot coals," Majalli said. Prince Hassan got the hint and hit a bell. It was after midnight, but in the nearby dining room one could hear the plates flying. "Let's settle this and go eat," Hassan said. "If we go eat, then what General Sharon and I decide is what will be. We are in agreement regarding the Jordanians, and insofar as the Palestinians are concerned, we shall negotiate," the king said.

"Had you let us know in advance, we would have prepared properly," Prince Hassan said when they sat down to eat. His statement, of course, was ridiculous. They literally had a king's banquet, with lamb and spiced rice and all the rest.

The king ate quickly and sparingly and told Majalli, "Tell General Sharon to help himself. I eat like a Bedouin. Fast and in small portions. Always alert. That's why I'm so small."

"Tell him not to worry," my father responded. "I am hungry and I am planning to eat."

The meal lessened the tension. At two in the morning, the king got up and bid farewell. "Majalli and Ali Shukri will be in touch and will nail down the details. Do not worry, we will reach agreements."

My father said, "I will go back and meet with the prime minister. You will have authorization on the lists by tomorrow." The meeting had lasted for close to four hours.

On the way out of the meeting, my father, who always took care of his security guards, asked them, "Did you get something to eat?" They had been well taken care of. They got back on the helicopter and flew to the banana grove heliport near the Allenby Bridge. My father thanked the Israeli commander of the bridge for keeping his entry a secret.

Back on the Israeli side of the bridge, he called Netanyahu, who asked to see them at once. They arrived at three thirty in the morning. Netanyahu offered them coffee, which he prepared himself.

By morning, my father, Netanyahu, and Majalli had been joined by the prime minister's military attaché and the Mossad representative. They reviewed the list of prisoners, and Majalli worked on the phone with Shukri. The Jordanians refused to work through any other channels.

At six in the morning, twelve days after the two Mossad agents had been imprisoned, an air force helicopter waited for Majalli at the Sde Dov airfield. The previous night the head of the Mossad had called and told him, "I know you're off tomorrow. I don't believe it's going to work, but I was told to offer assistance. Can I be of any assistance?"

"You could get me a shirt and tie. I've been wearing the same outfit for days, and I am going to meet with some senior people there."

"Where am I going to get that now?" the head of the Mossad said.

In the morning there was a Mossad man waiting by the helicopter with a shirt and tie. Majalli, ever elegant, did not approve of his new clothing, but he had no other option.

The pilots seemed surprised. "Who are you? What's going on?" they asked.

"What does it matter? I'm Majalli."

At seven thirty in the morning he landed in Amman. The Circassian major drove him to Shukri's office in the palace. "Let's have breakfast," Shukri said. He looked at Majalli's shirt and tie and immediately sent someone to bring back more suitable clothing. Then they drove together to the offices of the General Intelligence Department, where Battikhi waited for them.

"You want to see the two guys?" Shukri asked before the meeting began. "What use is it seeing them?" Majalli answered. "I want to free them."

Battikhi, his assistant, Shukri, and two other generals all greeted Majalli at Battikhi's headquarters. "Just me against all of you?" Majalli asked. "Why so many?"

"Each person has his own role," he was told.

The meeting began at nine thirty in the morning. Majalli had come with the agreed-upon list, but Battikhi started down the same old road, demanding that more names be added to the list, more Hamas terrorists, that a terrorist by the name of Mousa Abu Marzuk be allowed to fly to Gaza with Ahmed Yassin, and making other assorted demands. All the things he hadn't dared to ask for in front of my father, he now tried to squeeze out of Majalli, who felt his self-confidence begin to weaken. He walked into another room and put a call through to my father. It was clear that the phones were being monitored. My father, though, could hear in Majalli's voice that he was being pushed against the wall.

"You could have heard Arik yelling through that phone even if you weren't anywhere near the telephone," he recalled. " 'Don't forget who you represent. Tell them that is what was agreed upon, and that's it. If they aren't willing, leave!'" Majalli hung up the phone with his confidence restored. He walked back into the room and started to collect his papers.

"What are you doing?" the officers asked.

"I'm leaving. I spoke with Ariel Sharon. He told me that this is what was agreed upon, and that if you aren't willing, then I should leave." The Jordanians were stunned.

"Hold on," Shukri said, "let's have some lunch first."

"What kind of lunch? It's only eleven a.m."

"So what?"

Majalli stayed put at the headquarters, and Shukri hurried over to the palace, a forty-minute drive. Majalli got a call from the palace: the king wants to make sure that General Sharon will be able to deliver all that he has promised, Shukri told him.

"You have the word of Arik Sharon," Majalli told him. The king agreed.

Shukri returned, and they sat back down to finalize the details. "Yassin will land here, and then we will release the two prisoners," Battikhi said. "Absolutely not," Majalli responded. "As soon as Yassin boards a Jordanian helicopter in Israel, our guys board an Israeli helicopter in Jordan."

Later in the day, Majalli was taken to the waiting helicopter. He met the two Mossad operatives there, both of whom had been described by the media as using the cover of Canadians. "One was large and the other was a small Yemenite," said Majalli, who remembers thinking, What kind of Canadians are these? On board the helicopter, they had tears in their eyes. "We screwed up. We never thought we'd get out this fast. Thank you," they said. This was after twelve days in a Jordanian prison. "You should thank Ariel Sharon," he responded.

Back at the Dov airfield, Majalli called my father and handed the phone over to the two Mossad men so they could talk to him. Then my father told Majalli, "Have the pilots take you to Jerusalem, to the cabinet meeting."

On this leg of the trip the pilots did not ask him anything.

King Hussein was impressed with several things. My father would not commit to what he was not certain he could deliver; he bargained hard, but kept his promises; and he was especially won over by my father's love of the land, of farming and animals.

Several years later, Battikhi was ousted from office on corruption charges and imprisoned.

Khaled Mashal, the Hamas man, was deported from the Kingdom of Jordan several years later, but in February 1999, during King Hussein's funeral, my father and the rest of the Israelis present saw him in attendance. As far as I know, my father was the only world leader to go and visit the dying king in the Mayo Clinic. He went along with my mother.

Minister Sharon giving Governor George W. Bush
a tour of the country, late 1998. (Uri Dan)

TWENTY

In late 1998, the popular young governor of Texas, George W. Bush, came to Israel for a visit. My father took him on an aerial tour of the country in a helicopter. Flying over Israel with my father is an unforgettable experience, one I have had more than once. His knowledge of the lay of the land is absolute. He knows the name of every village and town, is familiar with the curve of every ridge and canyon. He's walked the entirety of the country. My father showed the governor the strategic challenges inherent in Israel's position, flying him over the country's narrow waist, a less-than-ten-mile-wide stretch of land between the Green Line of 1967 and the sea in Netanya. "The average farm in Texas has a longer road than that," Bush wrote in his memoir *Decision Points*. "This was my first trip to the Holy Land. My most vivid memory from that trip was the helicopter tour that Sharon led for us. We flew over Israel and he showed me where he had fought, where he had established settlements. On that ride I became convinced that it was our responsibility to preserve our relationship with Israel and to strengthen it."

In October 1998, my father took over as minister of foreign affairs, in addition to his role as minister of national infrastructure. Once again Washington, D.C., was open to him, after the years of the first President Bush's administration, during which he was somewhat of a persona non grata.

Upon taking on the role of foreign minister, my father learned that Netanyahu was negotiating with the Syrians, via the prime

minister's personal friend Ron Lauder. Netanyahu, a polling ad-
dict who pores over the results and treats them with oracular
reverence, must have seen somewhere, in one of the many polls,
that the public is not that concerned with the Golan Heights. The
reason for this result was likely that the public did not feel that
the Heights were in jeopardy with a Likud government in power
and therefore was less interested in it, but no matter: as far as Netanyahu was concerned, he was willing to give the land to Assad
senior.

Assad demanded to see a map. Netanyahu, in typical smart-
aleck fashion, suggested drawing the new borderline in a thick
pencil, so that it would remain open to interpretation. My father
was not willing to turn over the Heights, and the fat-pencil tricks
were not to his liking. He managed to sever those negotiations.

Netanyahu's coalition was shaky. On December 21, 1998, a
majority of eighty-one Knesset members voted to move elections
up to May 1999.

In the internal Likud elections my father finished in the eighth
spot. It was a jolting blow. I recall sitting in the living room of
our house on the farm one Shabbat and saying to my father, "We
aren't doing this professionally. If we're in this all the way, then we
have to be professional."

Intraparty politics never really interested my father. He found
them distasteful. After the slide in the Likud primaries, my brother
Omri decided to take matters into his own hands. As my father's
political strategist, he took over a spot that had previously been
unoccupied. My father always had his staff focused on the matters
of his ministry. My brother dove deep into the political pool and
stayed put for close to seven years. Omri is a warm and friendly
guy, and the political activists received him with open arms. He
knew each and every one of them, their families, and what they
felt most strongly about. He attended every bar mitzah and wedding. There wasn't a wedding hall in the country he hadn't been
to. He wasn't playacting; he really is interested in people, and
many of them are still in touch with him today, years after he

has left the political arena. He took on, with stunning success, an entire side of our father's political career that had been neglected and without which it is impossible to lead a large political party.

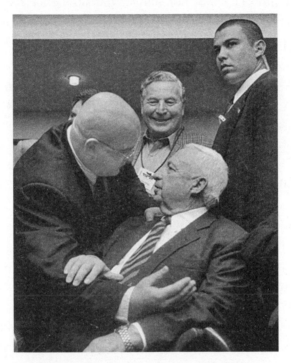

With my brother Omri and Cyril Kern,
after winning the election for prime minister, February
2001. (© David Rubinger, Yedioth Ahronoth)

I realized one day how little I understood of the way things work in politics, when I was with Omri in his pickup truck as he spoke with one of the political activists on the speakerphone. "What kind of tone of voice is that?" I asked him when the conversation was over. "Why do you even talk to that guy?" He had sounded aggressive and hostile.

"What are you talking about?" my brother said. "That guy's one of our supporters."

In May 1999, Israel went to elections. It seemed clear that Barak was going to beat Netanyahu by a significant margin, but my father worked with energy and enthusiasm as though victory were near at hand. This was not par for the course for other senior Likud Party members. On May 17, Netanyahu delivered his concession speech, thanking my father for all of his hard work. He recognized the fact that even though he hadn't wanted to appoint him to a cabinet-level position, and despite their personal differences, my father had remained loyal and fought for a doomed campaign to the last.

Aside from Netanyahu's loss to Barak in head-to-head elections, the Likud also managed a meager nineteen seats in the knesset, its lowest tally to that point.

That night, Netanyahu announced his resignation as chairman of the Likud. Leadership of the Likud seemed an attainable goal for my father. He was appointed temporary chairman of the party—and a temporary position is a good thing to have in this country if you're looking for something permanent. Elections for the position of chairman of the Likud were set for September 2, 1999. The major figures in the party seemed to want my father in the top spot, knowing that he would be able to rehabilitate the party and ready it for the next cycle of general elections; they did not want him heading the Likud in national elections. Some had prime ministerial aspirations of their own, and some feared he would never be able to win a general election. My father offered them a solution: elections now for chairman, and then again before the country voted for prime minister. This solution dovetailed well with what I heard my father say so many times before: "If you have two problems, one right in front of you and one far away, deal with the close one first." The faraway elections really were a problem, but first he had to get himself elected as chairman of the Likud for the first time.

On September 2 my father won handily and rose to the position of chairman of the Likud and head of the opposition. Consistent in his ways, he advocated for a national unity government and met with Prime Minister–elect Barak even before my father had been voted in as chairman. Barak, though, preferred the far-left Meretz and the Sephardic ultra-Orthodox Shas party as coalition members.

My father began working to rehabilitate the Likud, even though he was going through a difficult time. Several months earlier my mother had been diagnosed with cancer. She endured the treatments heroically, more troubled by her concerns for my father than by her own suffering.

My father and mother on the farm, 1977.
(The Times, NI Syndication)

TWENTY-ONE

My mother was born in Romania in 1937, a fact you were unlikely to hear from her lips. The city in Transylvania where she was born, Braşov, was Hungarian until the end of World War I, and that is how she always viewed it. She could speak Romanian, but, proud of her Hungarian heritage, she never did. Instead she spoke Hungarian with her sisters and parents. Her father died a few months before I was born, and her mother passed away when I was three. My only memory of her is from a visit to Tel Hashomer Hospital. What I recall is standing outside a long pavilion with its rounded roof, and not my grandmother Ida, known as Anioka ("Mama" in Hungarian), who was dying on the other side of the window.

She was her parents' beloved youngest daughter. They spent part of the war years in a small village outside Braşov called Zizin. Jews were not forced to wear the yellow Star of David, but they were prohibited from going to the market until ten in the morning, by which time there was nothing left to buy. The Gentile father of my mother's friend Milika would shop for the family; my mother managed to live through the war without experiencing serious trauma. In 1950 she moved to Israel along with her parents. Her three sisters, Olga, Margalit, and Yaffa, and two brothers, Eliezer and Yitzhak, had already come to the country.

Her father, Leopold, or Rabbi Yekutiel Yehuda, as we refer to him in memorials, was a rabbi. His father had instructed his sons, "You can be whatever you wish, but first you must be ordained as a rabbi." My grandfather was an erudite scholar of the Bible, a book he knew by heart. But he reviled the rabbis and the religious establishment, saying, "The bigger the beard, the bigger the lie."

In Israel, they lived sparsely, in a drab apartment in Kiryat Motzkin near Haifa. My mother was sent to a state boarding school in Petach Tikva. When she came home from school, she would help her mother mop other people's apartments.

After the Haifa rabbinate learned of my grandfather's extensive Bible knowledge, a rabbi approached him: "Grow a beard and come work for the municipality," he said. To which my grandfather responded, "I will not grow a beard, and I will not work for the municipality." Instead he earned his wages by cleaning a supermarket and working as a night watchman.

In the 1920s in Europe he was well-to-do, the owner of a small hotel in Braşov. Over time he lost his assets, but never his sense of humor. "You're going to be late for synagogue!" my pious grandmother would wail on Friday evenings. "Don't worry," he'd respond, "I promise I'll leave along with everyone else." This prompted tears. In general, she cried a lot: happiness, sadness, it didn't matter. The important thing was to cry.

My mother's parents were very religious, and they did not even imagine that their youngest daughter would enlist in the army. "I have to go to the recruitment base to settle the matter of my exemption," she told her innocent parents on the day that she signed up. She did not for a moment consider bypassing military service like some of the other religious girls, who were granted an automatic exemption if they so desired. She saw military service as the essence of being an Israeli, the epitome of belonging, and being Israeli was important to her, as evidenced by the differing accounts of her age upon immigration to Israel. Five was the number I once heard, but she was thirteen when she reached these shores.

She was tall and beautiful, a black braid dangling down her back. She excelled in basic training and was chosen to demonstrate to the prime minister of Burma at the time, U Nu, how Israeli female soldiers throw a grenade. Upon completion of her training she was posted to the military intelligence department of the Paratroopers battalion, which my father, her brother-in-law at the time, commanded. "I couldn't figure out what she liked about him," my mother once told me, referring to her older sister's decision to marry the tough regiment commander.

My mother in the Paratroopers
Intelligence Unit, at Tel Nof, 1956.

My mother was given the task of draftswoman, drawing the battle diagrams. She was a skinny girl, her uniform hanging off her frame, and the men were understandably terrified of approaching her, since their commander was her brother-in-law. So much so that Kokla, who spent time with my mother's roommate in the girls' quarters, jumped out the window as she approached. Years later we still laughed with Kokla, a brave warrior, about his atypically fearful retreat.

After her army service my mother got a job drawing police sketches. She had studied art at the Avni Institute of Art and was living in Tel Aviv. When my brother Gur was born, my mother lived not far from my father and Margalit's house in the Tzahala neighborhood of Tel Aviv. She would come over often and babysit for him. The two of them were very close. In May 1962, Gur's mother, Margalit, was killed in a car accident. Gur, five at the time, became my mother's life project. My father was a thirty-four-year-old widower, a field officer who had taken a grave hit on the home front. The family nearly crumbled. Then my mother stepped in, for Gur. "In the beginning I did not love him," my mother said of my father. It took time. At first they were just partners in a joint project, then they became friends, then love grew.

In 1963 they got married. In 1964 my brother Omri was born, two years after that I was born, and eleven months after that, Gur was killed.

———————————

She devoted herself to the family with everything she had. I remember one time early in the 1970s, back when my father was the southern front commander and we lived in Beersheba, the principal of the elementary school grabbed Omri by force. I don't think such things were particularly unusual then. My mother filed a complaint with the police. The principal begged for my furious mother's forgiveness. In all matters pertaining to us she had the instincts of a lioness. I was young and not yet a good swimmer when one day I spent my time jumping into the shallow end of the pool. She stood outside, fully clothed, talking with a friend. Once, running back toward the pool, I found the shallow end occupied, so I sprinted over to the deep end, jumped in, and promptly began to sink. She came charging in, fully clothed. I knew you'd come, I mumbled.

My parents, circa early 1970s.

The food she prepared was fantastic and plentiful, the finest of Hungarian cuisine. She had a magical palate. All it took was one taste in a restaurant or in someone else's home, and she knew how to reproduce a dish, and how to make it even better.

I imagine that at some point we'll publish a cookbook of hers, which incidentally is no small undertaking since she never measured quantities. She was a natural; she cooked by feel. Luckily my wife, Inbal, spent twelve years with my mother, some of the time in the kitchen. The tastes have stayed exactly the same. "A good cook is judged by the meat," my mother would say with great seriousness—and the slow-cooked lamb, the ghoulash with the *csipetke* noodles, Vera's roast, the stuffed fish (put in a little more sugar, my father would urge my mother when he tasted it, and he continued to urge the same from my wife), the calf's-foot jelly, the special Passover noodles, all those Inbal learned to make from my mother, along with the Rosh Hashanah duck and the Hungarian specialties: *szilvásgombóc,* a type of dumpling filled with pear, coated with bread crumbs, and drowned in a sea of melted butter; *turosgombóc*, cheese dumplings; noodles with sugar, melted butter, and either crushed pecans or freshly ground poppy seed; *kaposztás taszta*, square noodles with cabbage that's boiled or fried or God knows what till it comes out so delicious. Those dishes and many others have stayed with us exactly as they once were. The Hungarian kitchen has four pillars, I would say to my mother, laughing with her a little and at her a little—paprika, poppy seed, melted butter, and goose fat.

A passionate people, the Hungarians. You could hear it in the way they talked, everything dramatic in the extreme. Two Hungarians talk to one another and it sounds as though one is eager to slit the other's throat, when in fact they are merely greeting each other or discussing the weather. If you were not born to it, there is no chance of acquiring the language. Hungarian and Finnish are part of the same linguistic family. Apparently they were ancient tribes with some connection. Spies in Hungary must be locals. There is no teaching someone that tongue. For many long years I heard my mother talk to her sisters in Hungarian, and aside from an *igen* here and an *igen* there, I understood nothing. Worse, I couldn't even tell where one word ended and the next one began. All I knew was that sometimes they paused in order to breathe.

My brother and I did not miss a lot of opportunities to laugh at the Hungarian penchant for passion, the way everything was emotional and irrational with them. Relations with other human beings were especially intense. "A rottweiler lets go of the kid eventually," my brother used to say, "a Jewish mother never does."

She was smart, with a fine sense of humor, a sharp and quick tongue. She never left the last word unsaid. "Why must you always have the last word?" a teacher asked her when she was a little girl, she told me. "Please," my mother said, "go ahead." She was given what's known as a solid European education and had a terrific grasp of languages. Aside from Hebrew and Hungarian she also spoke perfect German, a language she could read, too, and French, Italian, Yiddish, and Russian—I have no idea how, perhaps from Vera—as well as English and Romanian, which she did not admit to knowing.

She had great taste. The house was always spotless. The gardens bloomed; she knew each flower personally. "A good gardener knows when to uproot," she would say, "no great shakes knowing only to plant." Some of the roses on the farm are the ones she planted nearly forty years ago, and they still bloom. Wherever she sat or spoke on the phone you could find small drawings she had left on notes and pieces of paper—little birds, butterflies, flowers, and the faces of little girls.

She was always well dressed. I never saw her lounging around the house in sweatpants. Avraham Yoffe, my father's close friend, who would come to the house early in the morning before they left for an excursion, would say, "Even at six in the morning Lili looks fresh, she never looks like she just rolled out of bed." And it's true; you would not find her with rollers in her hair and a face puffy with sleep. He also said, "Arik is the only man I know who still wants his wife."

My parents, on a visit to Egypt, early 1980s.

My mother loved music, and for decades my parents would go to the philharmonic in Tel Aviv. My father used to say that in the beginning they sat in the back of the Mann Auditorium, in the far corner behind a pillar. During the course of the years they moved more toward the front and center till they reached row 5, seats 24 and 25, where they stayed. That changed when my father was elected prime minister, at which point he was relegated, out of security concerns, to the rear. "I'm back behind the pillar," he would say.

She used her own innate sensibilities to design our house, then friends' houses, and later, once we'd grown up, she turned her hobby into a profession, designing homes, apartments, offices, and occasionally sets for television. She did not concern herself with measurements—someone else did that job—she simply designed the way she cooked, with a lot of feeling and good taste. Often she would walk into a space, get everyone up on their feet—the couch needs to be there, the bookshelf there—and within minutes the place would look better, to the delight of the owner.

Her many talents did not prevent her from devoting the majority of her time to my father. No one coerced her into it, nor did my father expect it of her. He respected her and relied on her. She loved him, but she did not by any means hold back her feelings. In fact, she always told him exactly what she thought. She did not throw the full force of her being behind him out of weakness or a lack of alternatives. She was strong enough to decide that that was what she wanted. My father, the family, the house—those were the major endeavors of her life. She never forgave those who spoke badly of my father, even when he had long ago let bygones be bygones. I am not sure it was advantageous politically or with the journalists, but that's just the way she was. It was not a show. She was convinced with all her heart that her husband was a great man, better than the others and worthy of serving as the head of state. He was appointed prime minister on her birthday, a year after her death.

My mother accepted my father as he was, except for, perhaps, his diet. She always wanted him to go on a diet, although at the same time the house was always full of baking, broiling, and simmering dishes based on the Hungarian pillars of cooking. This didn't really help his diet.

She knew him well, down to the very depth of his soul—"He can't stay alone," she told me before she died. "He likes it when someone from the family is around; it gives him strength." "His eyes get greener when he's aggravated," she told me several years earlier. She had been with him for long enough, she knew what he had been through, she had seen him off on missions, waited for him during wars, witnessed the triumphs and the disasters, knew what he loved and where his weak spots lay, knew what kind of lion she had roaming around the house and did not try to tame him.

My mother accompanied my father to each and every television appearance, fixing his collar, straightening his tie. She sat in the front row when he spoke, and he would always look for her loving and supportive face in a crowd that was not always nearly as supportive.

My parents at the Knesset, January 1997.
(© David Rubinger, Yedioth Ahronoth)

She was both impulsive and intuitive. In our house there would often be six pots going on the stovetop and a cake in the oven while my mother found the time to assemble stunning flower decorations. She'd juggle those things like a whirlwind, the cake coming out of the oven just as the guests arrived, poppy seed or date or one of the many other delicacies she made.

My mother had sharp intuition. What do you think, doctor? Professor Goldman would ask her when one of the family members was ill. He knew she always hit the nail on the head.

She also knew exactly how she would die. Years before she passed away, she used to say that cancer would kill her; "You'll be sorry you didn't behave better when I'm gone." "Enough with the threats," I'd reply, or joke, "Oh, yeah, well we're going to bury you in this or that cemetery." And she'd say, "What, are you crazy? That'd kill me." Then she'd tell me the punch line of the joke about the dying woman who, in her last wish, asked that her husband walk arm in arm with her sisters during the funeral procession. "That's one funeral he's not going to enjoy," she'd say.

In 1999 she was diagnosed with lung cancer. She was not surprised; she'd been expecting it. And she was heroic, maintaining her dignity and fighting nobly to the end.

On December 19, 1999, our house on the farm went up in flames. A spark from the chimney lit the dovecotes between the ceiling and the roof, and from there the fire spread fast. My parents were not home. I took the picture of Gur out of the black dresser and stashed it somewhere safe. Then I tried to fight the fire until the firemen showed up. The top floor was completely ruined, and the bottom was flooded with water. My parents came home, and my mother told a cluster of reporters outside the house, "Now everyone knows we have a warm home," my brother said. "The bad news is that the house burned down, the good news is that Mom is going to have to buy a whole new wardrobe."

The sight of the house in the morning was awful. Lingering smoke and smoldering books, the charred hull of a home . . . somehow Bibles had emerged unscathed from the fire even though the books nearby had gone up in flames.

Our house on fire, December 1999.
(© Danny Salomon, Yedioth Ahronoth)

My parents were calm and collected. They were raised up to the second floor in a citrus fruit picker placed in a tractor, since the steps were closed off. My mother looked around their room, but there was not much to see. Some of the workers on the farm had tears in their eyes, and I was in an efficient, emotionless state— the house had burned down; we had to clear it out, take care of everything, build it back up and that was that.

She spent her final eighteen days in the hospital, entering on March 7, her sixty-third birthday. My father was devastated; it seemed he was angry with her for leaving him.

I'd wake up every morning at five o'clock bathed in sweat, call the hospital, and ask how she was doing. Afterward I would come up and sit with her. "You have to stop with this," she said. She was worried about me.

I was told that a lung transplant was out of the question because the cancer had spread and she would never be able to endure it, but I was unwilling to give up on my mother. I made sure she received omega-3 fatty acids and medicines. My parents were apparently worried about me and sent Professor Goldman—our friend Bolek, at that time the director of the Tel Hashomer Hospital, where she was being treated—to talk to me. "Let's go get some coffee," Bolek said. We went, and I listened to him delicately explain the situation to me. "I understand the situation," I said, "but still I want to do what I can." I was driven mad by the notion, which still torments me, that a remote control vehicle can be made to land on Mars, collect soil samples, and send pictures back to earth, while the problem of cells multiplying uncontrollably, eventually killing the body, cannot be solved. The dying, day after day.

For my part I tried to improve my mother's mood. Hoping to lift her depression, I brought her sister's son, Chanoch, for a visit. Dr. Chanoch Miodownik, a psychiatrist, was my mother's first nephew, and they had a special bond.

"I want you to make her happy," I told him. He went into the room, but emerged shortly after. Even if happiness was within reach, he wasn't the right person for the job. He looked up to her, which surely didn't help, and anyway she was right, and he knew it: she was young, she was going to die, and there was nothing anyone could do. There was nothing to be happy about.

The pious die on Shabbat, she had told me years before. "Both of my parents died on Shabbat," she added.

On Friday, as evening approached, she asked, "What time is it?" I knew why she was asking. I didn't want her to die, but I also didn't want her to be anxious, I wanted her to be calm. "Shabbat has already come," I answered.

"It's deterministic," she said without enthusiasm, perhaps the day before, when they discussed increasing the dosage of her morphine. They raised it. Night fell, she lay in bed, my wife, Inbal, to her left and Omri's girlfriend, Efrat, to her right. We walked around outside the room, peeking in every once in a while.

At dawn my father went into the room. The three of them were asleep—my mother's left hand in Inbal's hand and Efrat's head on my mother's right arm. The picture is engraved in his mind. I have heard him talk about it. The intervals between breaths increased. She died at seven in the morning, on Shabbat, March 25, 2000. The soothing sound of the next breath never came. The girls woke up and cried. My father had just left a few minutes before to wash his face. I went to go get him. "Dad, come," I said. His bodyguard shook his hand and offered his condolences. We went to her room; my father entered, touched her face gently with his fingers. "She's still warm," he said softly. Saturday morning, I drove him back to the farm.

My father threw himself into his work. On weekends he'd finish his tours of the farm on the hillside where she is buried, sitting on the wooden bench and looking out at the stunning view, which changes every day and with each passing season.

At my mother's funeral.
From the left: Maydanchik (with the hat and beard), Efrat
Yoffe, me, my father, and my brother Omri.
(© Danny Salomon, Yedioth Ahronoth)

My father on the Temple Mount.
(AP Photo / Eyal Warshavsky)

TWENTY-TWO

On September 28, 2000, my father went up to the Temple Mount. I accompanied him. We drove from the farm to the secret service offices in Jerusalem and exchanged cars there. Our adopted brother, Roni Schayak, from Jerusalem, joined us.

I met Schayak in reserves. He was under my command, and ever since our first meeting, in an eleven-man field tent on the far reaches of the Tze'elim base in the desert, he has been family. I brought him home to the farm for a visit, and the rest of the family adopted him, too. It has been over twenty years. Roni still comes to us for all of the holidays. We are all in close constant contact.

We entered the Temple Mount through the Mughrabi Gate, accompanied by several other Likud MKs and a guide from the Israel Antiquities Authority. Hundreds of policemen stood guard. From my perspective their presence was more of a show of weakness than of strength. What type of sovereignty do we have over the place if in order to visit the courtyard, without even stepping foot in the mosques, we need this type of security force? This is Jerusalem, our home, not some distant colony.

After leaving the Temple Mount I shared these feelings with my father, and he agreed. It seems that Dayan's position on the Temple Mount, expressed after the Six-Day War—"What do we need this whole Vatican for?"—was not attuned to the feelings of the Jews. He was knowledgeable, apparently, about the historical side of archaeology, but less so about its implications for the present and the future.

The Temple Mount, the site of the two destroyed temples, is not some insignificant religious headache. It is of tremendous significance to Jews. After 1967 Jews were allowed, for the first time in hundreds of years, to tread on that hallowed ground. Barak's pledge to turn the Temple Mount over to the Palestinians during the second Camp David and Taba talks infuriated my father and many other Israelis. Neither the public nor the Knesset agreed with Barak's concessions, as evidenced by the huge majority my father received in the 2001 elections, resulting in Barak's defeat.

The visit to the Temple Mount was an expression of my father's opposition to Barak's readiness to give the site up. It was not his first time up on the Temple Mount; he and many other Jews had visited it on other occasions since 1967. The Mitchell Report of 2001 came to the conclusion that my father's visit did not cause the Intifada. But it was used by the Palestinians as an excuse to launch a wave of terror attacks that had been prepared in advance, orchestrated and encouraged by Yasser Arafat.

In July 2000, Prime Minister Ehud Barak met with Arafat and President Bill Clinton at Camp David. Despite the far-reaching concessions Barak offered, Arafat refused. President Clinton blamed Arafat for the failure. Shortly thereafter an orchestrated terror campaign was launched, worse than anything Israel had known.

———

On September 30, 2000, an Israeli border policeman was shot by a Palestinian policeman while on a joint patrol. These joint patrols were an unnecessary relic from the good old Oslo days. On several occasions Palestinian policemen had drawn their rifles and threatened their Israeli peers, events that passed without a fitting Israeli response. Chief Inspector Yosef Tabja paid with his life.

The Jerusalem neighborhood of Gilo, near the Palestinian neighborhood of Beit Jallah, became a firing zone.

On October 1, Madhat Yusuf was killed at Joseph's Tomb in Nablus. Yusuf was part of a small force guarding Joseph's Tomb.

A Palestinian crowd stormed the tomb, which was, according to the agreements, an Israeli "island" in the midst of the Palestinian-controlled city of Nablus. Palestinian gunfire wounded Yusuf with a shot in the neck. IDF officers contacted Jibril Rajoub, the head of the Palestinian Preventative Security Force in the West Bank (a rather long name for something that provided such scant security), in hopes that his men would save the bleeding soldier. For close to five hours Yusuf slowly bled to death in the compound surrounding the tomb. Throughout, IDF troops were stationed several hundred yards away. Among them were some very senior officers, and yet, rather than storming the tomb and saving the soldier, they waited for Rajoub to rouse himself from his beauty sleep or whatever it was that he was doing and come to the soldier's aid, which of course he did not do.

One year later, on October 4, 2001, with my father already prime minister, two Israelis entered the Palestinian village of Jaljilia in Samaria and were attacked by an armed mob, which besieged the house they sought safety in and threatened to lynch them. This time, though, things ended differently. Within a short period of time an IDF force entered the village and extracted the two Israelis.

On October 7, 2000, four and a half months after the IDF's withdrawal from Lebanon, Hezbollah abducted three Israeli soldiers from Israeli territory. The three, Benny Avraham, Adi Avitan, and Umar Suweid, were attacked on Dov Mountain, on the Israeli side of the internationally recognized border. The prime minister and defense minister at the time, Ehud Barak, did not respond to this attack. This decision contradicted his statement given not long before, as Israel withdrew from Lebanon: "The prime minister and defense minister has made clear that from this morning on, responsibility for all that happens in south Lebanon rests with the Lebanese government and with Syria, and he warns that Israel will respond decisively and forcefully to all attacks on its sovereignty, citizenry and soldiers, as is required of it for self-defense." Truly a worthy statement, but unfortunately when it came time to follow through on it, Barak had second thoughts. It's possible that had he taken decisive action at the time, all of the other subsequent conflagrations would have been avoided. For all intents and purposes, Barak led Hezbollah leaders to believe that Israel's retaliatory capabilities were feeble, and there would be no real price to pay for attacks on her sovereignty and citizenry. Later, once Hezbollah had placed thousands of rockets along the border with Israel, the meaning of setting the "norm" was to go to war.

Several days before the abduction of the three soldiers, an Israeli citizen, Elhanan Tannenbaum, was kidnapped abroad and taken to Lebanon. Hezbollah also continued to sporadically fire at Israel.

The territories, meanwhile, were on fire with mass unrest and a wave of terror attacks. Israeli Arabs, citizens of Israel, joined in the demonstrations. Nahal Iron, which is a major artery to the north of Israel, was blocked entirely for some time, and rock throwing and other violent forms of protest also occurred in parts of Jaffa, just south of Tel Aviv. One of the worst incidents was the Ramallah lynching. On October 12, 2000, two reservists, Yossi Avrahami and Vadim Nurzhitz, mistakenly drove their car into Ramallah. A Palestinian mob blocked their way and began pelting them with stones. Palestinian police officers led them to the police station, where the officers and the feverish mob beat them, stabbed them, and then threw them from the second-floor window into the hands of the mob below. Their bodies were mutilated and dragged through the city. A television crew from RAI, the national Italian state-owned channel, filmed the events and broadcast them all around the world. One and all could see the delight of the murderers as they displayed their bloodied hands. The TV crew was subsequently threatened with their lives by the Palestinian authorities, and the representative of the Italian TV station here in Israel, Ricardo Christiano, apologized in writing to the Palestinian Authority for filming the lynching. It was published in the daily *Al-Hayat al-Jadida*.

Barak's response was to warn the Palestinians, and then, once they'd left the building, to bomb the empty police station, settling the score with real estate.

After my father took office, he ordered the security services to catch the perpetrators of the lynching. It took seven years, until the last one was apprehended.

The Palestinian lynchers showing their bloodied hands, October 2000. (BBC News World Edition, http://news.bbc.co.uk)

None of these events deterred Ehud Barak from arriving in Sharm el-Sheikh on October 16, 2000, for a meeting with Arafat, Clinton, and Mubarak. Barak, his staff, and the Israeli journalists were treated in a particularly degrading manner by the Egyptians, and of course nothing came of the meeting.

(Their humiliating treatment of Barak was so extraordinary that during Barak's trip to Sharm el-Sheikh, his chief secretary,

Marit Danon, was locked in a hotel room. The Egyptian security officer refused to let her out. This was a traumatic event for her, as she herself later related to me. As a further insult, Israeli flags were not raised at the summit, and Barak's team was searched, with all cell phones confiscated, including Barak's own.)

Despite the raging terror—perpetrated in part by members of the Palestinian Authority—Arafat and other PA leaders continued to travel freely, and Barak steadfastly tried to coax Arafat into "calming the area."

Shimon Peres's late-night meeting with Yasser Arafat on November 1 was part of that hopeless initiative. Barak, in a conscious act of humiliation, had named Peres minister of regional cooperation, a superfluous position created specifically for Peres. He was sent to the meeting so that "senior sources in Jerusalem" could say, at the close of the meeting, "We're optimistic that in the future Arafat can be negotiated with to calm the area." It semed like Palestinians ridiculed the way the Israelis courted them.

On November 18 a Fatah man, a Palestinian police officer, snuck into Kfar Darom and killed two soldiers. The terrorist, who was killed, was posthumously promoted in rank by the Fatah. (Commemorating terrorists is common in the PA. There are 100 cases in which places or events in the PA are named after forty-six different terrorists. Among them are schools, kindergartens, summer camps, soccer tournaments, community centers, a sports team, a square, a street, a military unit, and a TV series. Among those commemorated terrorists one can find suicide bombers, a terrorist connected to the murders at the Munich Olympics, and others of the same kind, according to Palestinian Media Watch.

On November 20 terrorists lay near Kissufim Junction and watched as an explosive charge ripped through a bus carrying schoolchildren, which they had specifically targeted. Two of the passengers were killed, and nine others were injured.

Two days later, terrorists blew up a car alongside the Number 7 bus in Hadera, killing two people and wounding sixty more.

The next day, Ehud Barak went to visit the injured at Hillel Yaffe Hospital in Hadera. "These barbaric attacks will not deter us. We are on the way to making peace and we will make peace," he said from the hospital. A giant wave of terror had begun to descend on Israel, and Barak, completely divorced from the reality on the ground, simply failed to grasp what was happening. He thought he was on the verge of making peace.

—————

Despite the conciliatory stance of the Barak government, Oman, Tunisia, Morocco, and Qatar all severed their ties with Israel. The designated Jordanian ambassador never arrived, and Egypt recalled Ambassador Muhammad Basyouni to Cairo. Basyouni had served as ambassador in Israel for many years, during which he had been the object of fawning, and sickening, attention from the upper classes. His schedule was so full of dinner and cocktail engagements that his wife used to complain that at times they were invited to places with only a few months' advance notice, and their schedules were simply too full to allow them to attend. Yet while Basyouni was wined and dined in Israel, the role of ambassador to Egypt in Cairo has long been considered a punishment to whoever holds the title. It is a lonely diplomatic post, subject to total segregation and isolation.

On the personal level, Basyouni's departure was not a great blow. During his years of service in Israel he was not known for his heroic efforts to strengthen the ties between the two countries, but rather for the sex crime of which he was accused. Barring the crude intervention of the ministry of foreign affairs in the legal proceedings, he would perhaps have been convicted and placed in prison with other sexual offenders. I can only imagine what would have happened had our ambassador to Egypt been accused of rape.

In Egypt, members of Parliament called for a boycott against Israel. There were also calls to boycott the detergent Ariel because it had the same name as Sharon.

In Jordan, the deputy counsel was shot and wounded by terrorists. That was on November 19. Two weeks later, an Israeli embassy worker was shot in Amman.

On December 10, 2000, Ehud Barak stepped down as prime minister after a mere one and a half years in office. Elections were called for sixty days later. Despite the fact that he held a minority in Knesset and elections were a few weeks away, Barak embarked on talks with the Palestinians on January 21, 2001, in Taba. The degrading treatment he had received at the hands of the Egyptians in Sharm el-Sheikh did not deter him from returning to Egypt for another round of negotiations. He did not understand the behavioral code of our region, where the weak are held in contempt, an area where if you lack honor, you lack everything. National honor is an integral element of national security here, but he failed to grasp this. Since he had resigned his post and set the course for elections, he had neither a mandate from the people nor the legitimacy to conduct negotiations with the Palestinians. Terror continued to burn through the country—shootings, explosive devices, bombers—and Barak's government continued to meet with the Palestinians. Throughout December and January he sought to reach an agreement, straining against the time constraints of Bill Clinton's term in office, which was to end on January 20.

The Taba talks began sixteen days before the elections, driven by an obsession to seal a deal before the nation went to the polls. Israeli television audiences were treated to footage of a senior Israeli officer, dressed in a tank top, leaning over a map in a hotel room in Taba. It was embarrassing. My father watched this with disgust. In the meantime, the streets were on fire and Israelis were being murdered.

Ehud Barak is known as a man who can disassemble watches and break them down into their smallest components. Perhaps he is even capable of putting some of the parts back. But he can't tell the time, is completely incapable of looking beyond the details and seeing the larger picture. By conducting talks under the fire of the murderous terror campaign, he in essence allowed the Palestinians to use terror as a tool during negotiations. Using terror for leverage, they pushed Israel into more and more concessions. Peace talks and terror do not go hand in hand. All this was compounded, of course, by the fact that he had already resigned.

Barak was willing to concede more than Israel could afford, and more than the public and the Knesset were willing to sanction. Perhaps he felt that an agreement, if reached, could alter the results of the upcoming elections. Otherwise, why didn't he take the obvious course of action, and put the negotiations in the hands of the person soon to be elected by the public? Clearly negotiations on such a tight deadline—of which the Palestinians were aware—made it harder to hold out for what is right, and contributed to the weakening of Israel's national interest.

It also revealed an underlying misconception: that the conflict was rooted in some type of misunderstanding or territorial dispute. I recall this arrogant statement, attributed to Barak: Give me fifteen minutes in a closed room with Yasser Arafat, and I'll seal a deal with him. As though the whole problem were the result of a misunderstanding and now the great brainiac would show up and explain it all away.

In the end, as usual, we could simply have counted on the Arabs to save the day. They refused what had been placed on the table— everything. Much more than what Israel's security needs permitted.

In January 2001, while I was serving as a captain in the reserves, our battalion was called up for active duty in the Ramallah region. During pre-deployment training I was rather surprised to see that the main briefing was not given by someone from the Intelligence Corps or the Department of Operations but a liaison officer with the Joint Israeli-Palestinian Liaison Committee. Rather than receiving information about the enemy and the nature of our missions, this officer explained how, during Palestinian riots, he calls one of his friends, a Palestinian merchant and storekeeper in the center of town, who then relays to the officer what is happening. Our orders regarding the use of live fire seemed utterly ridiculous: an armed Palestinian in Area A whose weapon is not aimed directly at you should be treated like this; an armed Palestinian in Area B, moving in the direction of Area A, should be treated as such.

In order to follow those orders, each soldier would need a personal lawyer and cartographer. These instructions were not military orders. They were unclear and illogical. You cannot send a soldier to active duty and call the clashes he is sent to deal with "low-intensity confrontations," yet another one of the stupid and unnecessary terms coined at the time. For Madhat Yusuf and the other downed soldiers, there was nothing low-intensity about it. It was live fire, and they were thoroughly dead.

That was the shape the army was in back then. It was clear that the military and the legislative branch that controlled it had not internalized the fact that we were at war—a harsh war against a particularly nefarious strain of terror.

From July 2000 on, Barak's governing coalition had consisted of 32 MKs out of 120. Once the Gesher party defected, too, his coalition was reduced even further, and yet he sought, with such a slim minority of the people behind him, to decide on matters of far-reaching national importance. The date for Likud primaries was set for December 2000. My father had committed to that over a year earlier.

At around the same time, Netanyahu decided to return to politics. His resignation after the last elections had somehow managed to purge him of his sins in the eyes of the Likud faithful. The manner in which Barak's coalition crumbled, the way the people closest to Barak distanced themselves from him like the plague, the far-reaching concessions he made to the Palestinians that were greeted with more fire and terror—all these mitigated Netanyahu's faults in the eyes of Likud members, and he was now considered the likely winner of the primaries.

My father had no intention of conceding, but the numbers showed a grim picture. According to the figures we had, Netanyahu was slated to take 80 percent of the votes in the December 2000 primaries. But Netanyahu had a condition: he demanded the elections be for both Knesset and prime minister, not only for the prime minister. He felt that the nineteen seats that he left the Likud would make it impossible to govern effectively. Barak liked his chances against my father, and the Knesset itself was none too eager to go back to elections, so it was decided that the elections would be for prime minister alone.

Netanyahu announced that if there was not a general election for Knesset, he would not run. When it became clear that that was not going to happen, he decided, after some deliberation, to keep his word. Perhaps he was afraid to renege on his very first pledge since returning to politics; perhaps he really did feel it would be impossible to govern at the head of such a small party. Either way, on the morning of the primaries, Netanyahu rescinded his candidacy, and my father became the Likud candidate for prime minister.

(In November 2002, during the Likud's next primaries, after two years in office, things were already very different, and my father beat Netanyahu by a wide margin, after governing with a nineteen-mandate party, forming a national unity government, handling state affairs with great authority, and striking a debilitating blow to terror.)

In 2001, though, my father faced Barak in a head-to-head election for prime minister. Barak's weakness and the swirling terror brought Moshe Dayan and later Uri Dan's more explicit prophecy to life: "When there's a catastrophe, they'll turn to you." The public saw my father as the national authority on security, and they wanted him in power.

Nonetheless, there were many hurdles to clear. In one focus group, a woman from Kiryat Motzkin said she had never seen my father so much as kiss his kids, and that he seemed "inhuman" to her. The public had to be shown his less well known characteristics. In terms of his security bona fides, no legwork was needed.

PR people are not magicians. They cannot turn someone into something he is not. The public is perceptive, and it can sense a forgery. But the best in the business can certainly emphasize a candidate's strong suits and shade his weaker ones. Our friend Reuven Adler is an exceptionally bright person. He ran the campaign. After the victory, he was asked what was it like working with Ariel Sharon. "I felt like a doctor operating on a family member," he said. Adler wanted to soften his tough image. He drafted the slogan "Only Sharon Will Bring Peace."

He has the security side locked up, anyway, he said, so he thought it would be right to run with a message of hope. Everyone wants peace. My father drove the point home by speaking of the "painful concessions" he was willing to make for the sake of peace. The commercials he ran showed him at home on the farm with the grandchildren, planting a tree, walking through the fields, and on the couch in the guestroom. He was far more relaxed than when he was asked to speak directly into the camera. His tone was naturally warm and welcoming.

Although the campaign made sure the camera showed the less-known warm side of my father, he had also truly softened some. "It took me seventy years to learn that," he would say to us sometimes when we were impatient about something. His years of experience and the weight of responsibility had made him more amenable. My father was in no way less energetic or determined, but he started to listen with more patience, and at times he conquered the urge to "not suffer fools" even if they deserved it.

On February 6, 2001, he was elected prime minister, beating Barak almost 2 to 1. He arrived to the post of prime minister mature and well prepared, a responsible adult. Love and admiration for him crossed over traditional party lines.

And yet Marit Danon, the chief secretary in the prime minister's office, said to Barak after he had lost to my father in the elections, "With that man, I will not work." She, like many others, had a very low opinion of my father. Barak got angry and slammed his hand down on the table. "That man is not who you think he is. He is a sensitive man, well read, a lover of music; you'll see that he's not who you think," he said. "You must stay on." She thought to herself, Okay, I'll give it a chance.

It didn't take long for her to change her mind about the new prime minister. Several months into his first term, a woman who had served tea to the visitors walked out of the prime minister's room. My father called Marit over and said quietly, "Follow her, see what's going on, she has sad eyes." Marit followed and spoke with the woman. It turned out that she was going through some very difficult times at home. "Amid all the commotion, he noticed people, and that was true throughout his time in office," Marit told me.

Three months into his first term, Marit's conscience got to her, and she went into his office to come clean. "Mr. Prime Minister, I have a confession. I have to beg your pardon. I was part of the flock, and all my life I believed some bad things about you, and I have to ask for your forgiveness." He was surprised and moved.

From then on, whenever a negative column would appear in the paper or something would upset him at work, he'd say to her, "Now I don't want you going back to those old opinions you had in the beginning."

"You know that's irreversible," she would say.

Five years later, when he could no longer serve as prime minister, Marit, who had worked with five prime ministers, no longer wanted to keep her position. She didn't want to work with anyone else.

Marit speaks eight languages, perhaps more. She got used to my father teasing her about her Italian studies, and their discussions about the differences between a trough and a water basin and the nature of awns and other parts of the barley. He would write her,

Marit,
    Please check if it's raining [on the farm]. I have to cheer up.
                                     Thanks, Arik

Her response is written on the same piece of paper:

It has been dripping all day long, no serious rain.
    And please cheer up!!!

June 28, 2005

Prime Minister,
    Gilad reports: 203 cows are pregnant, 21 are not.
    Is this grounds for significant improvement of mood?

                                                          Marit

During the days leading up to the elections, I was in the reserves. I came home just in time for the birth of our twins, Yoav and Uri, on January 30, 2001. (Eight and a half years later, our daughter Talya was born. The scope of this book does not include that period, but I have no intention of leaving her out.) Election Day was eight days after the birth of Yoav and Uri. The mohel would arrive at the farm in the morning. Before the circumcision, Inbal and I drove down to Sderot to cast our votes for my father. On the way back to the farm, we found that the gate was swarming with journalists and cameramen. We decided to avoid them and to cut back to the house through the fields, but it was winter and muddy, and the car got stuck. There was a morning sun after the rain, all around us it was green and quiet, and at home we had twins, one mohel, and my father, who had a wildly hectic day ahead of him.

We made it back home, the bris ceremony was performed, and my father headed out into the chaos of Election Day.

———

"Did you believe this would ever happen?" he asked me after the elections. There was something in his voice that indicated that he hadn't. I was on the stairs, facing him, and he was looking up at me from the ground floor of our farmhouse.

"I never had the slightest doubt," I told him.

PART IV

2001–2006

*In our living room at the farm during the
campaign for reelection, January 2003.
(Courtesy of Ziv Koren)*

TWENTY-THREE

My father's attitude toward individual European states was always shaded by their treatment of Jews. He attached great significance to this matter, examining Europe's treatment of Jews both over the long course of history and especially during World War II, when a third of the Jewish people was obliterated without too many outward signs of sorrow on the continent, even among those who played no active role in the obliteration.

A pragmatic man, one who constantly had Israel's best interest at heart, as a minister in government and especially as prime minister, he was quite capable of discerning between past, present, and future. He believed in keeping a fastidious historical record but was also committed to combating the still-simmering European anti-Semitism of today and maintaining Israel's foreign relations and international standing.

European hostility toward Israel, he always believed, was not just linked to the anti-Semitism of old, which has been recycled as anti-Israel and anti-Zionist opinions, but was also a means of clearing the consciences of many nations. If the Jews, the victims, had turned out to be so dreadful, these cruel and crushing occupiers of the poor Palestinians, then perhaps the sins that these nations had visited on the Jews of Europe were not so bad after all.

My father always believed that the one-sided anti-Israel sentiment heard all across Europe is rooted in ancient guilt and bolstered by economic concerns; that is, the European sense of justice had been further skewed by a taste for oil and a desire for a maximal market for their goods.

Another reason for the anti-Israel policies in Europe is a recent development: millions of Muslim immigrants have quietly, resolutely made their way to the continent and become an influential factor. The modern wave of immigration is threatening—without a single battle—to achieve what the Moors attempted via Spain in the eighth century and what the Ottomans longed for from the fifteenth century to the seventeenth century, until, in 1683, they were finally stopped at the gates of Vienna.

Today the lifestyle and character of many European nations is under threat, and while in the past the threat came from afar, today it comes from within. Indeed, the terror acts that took place in London and Madrid, and even those of 9/11, are connected to the Islamic communities in Europe.

Muslim immigrants arrive in Europe from their native lands with a fully cultivated hatred of Jews and Israel, and elements of that population have committed many anti-Semitic crimes and perpetrated most of the attacks on Jews solely on account of their being Jews.

This minority's loud and sometimes violent pressure has its effects on the governments of Europe, too. It is entirely possible that these governments have tried to buy domestic silence by adopting a pro-Arab line, opposing and condemning Israel, and even turning a blind eye toward armed Arab terrorists within their borders. These practices were given public backing when Francesco Cossiga, the president of Italy from 1985 to 1992, went on record with the *Corriere della Sera* in describing the leeway Italy granted to Palestinian terror organizations during the 1970s and '80s. His remarks were underscored by Bassam Abu-Sharif, a leader in George Habash's terror organization, the Popular Front for the Liberation of Palestine, who, speaking to the same paper, asserted that in fact his organization and other terror organizations were given free rein when operating in Italy.

Muslim immigration is not, above all, the result of a shared allegiance to the ideals of humanism and liberalism that Europe sees as the bedrock of its existence; it is born of economic disparity and the gaping difference in quality of life between the immigrants' home countries and Europe.

There is a degree of historic justice in that countries that for hundreds of years used their colonies in Asia and Africa to bolster themselves, their economies, and their power are now also receiving, after the flow of pillaged goods, their people. It will be interesting to see how they fare during this challenge.

A s a government minister, before serving as head of state, my father was free to take foreign relations less formally. Thus in 1988, my father and my mother and our good friend Professor Boleslav Goldman were in Poland on an official visit. A native of that country, "Bolek," as his friends call him, was a child during the Holocaust. Poles murdered his father, grandfather, and uncles while they were scavenging for food. The rest of the family hid in a hole in the ground deep in the forest. Bolek, who had been fortunate enough to obtain forged identity papers, and his grandmother survived. This was his first trip back to the country he thought he had left forever. For him it was an emotional visit, a full circle—you tried to kill me, and yet I've returned, a professor, the director of a hospital, and I have come on official business, along with my friend, the minister of industry and trade, General Sharon.

My parents were also moved, particularly when visiting the Warsaw Ghetto. My father didn't miss a single opportunity to remind his hosts that Polish soil had just recently become home to the world's largest Jewish graveyard. The Jewish pride that had always been an important part of his personality was strengthened in Poland, and the man the Poles saw before them was certainly a stark departure from their notion of a Jew.

Upon arrival at their hotel in Warsaw, Bolek went straight to his room, but after a few minutes he heard a knock at the door. "Arik would like to see you right away," an Israeli security guard told him. Bolek came immediately.

"Come stand over here," my father said, placing him so that he faced the ceiling. "Now please translate this into Polish," he said.

"The president of Poland, a visionary leader, is in fact so visionary that he has placed a camera in my room. Luckily for him, Lily has come along with me on this trip, making the viewing all the more interesting." Stunned, Bolek translated verbatim.

Similarly, on a trip to Nicolae Ceauşescu's Romania, my father spoke into the microphones and praised his hosts but added, "I'm out of soap and would be happy if more could be brought"—a request that was filled within minutes.

On a trip to Germany in January 1999, when he served as foreign minister, a German helicopter pilot asked if he had any special requests during their flight from Frankfurt to Bonn. "I'd like it if you could fly at the level of the cows' horns," my father said. The Germans, not particularly known for their sense of humor, fulfilled his request to the letter. It was a great experience for him. In general, when on official business he would often ask to see as many cows and other animals and as few people as possible.

In April 1999, during my father's term as foreign minister, he and my mother paid an official visit to Pope John Paul II. When my father invited the pope to come see the Holy Land, the pope, revealing his scriptural erudition, responded by saying that the Holy Land is for everyone, but the Promised Land is for the Jews alone. My father told the pope, as I heard him say countless times before, "When you hold the Bible in your hands in Israel you don't need a guidebook of any kind. The names are preserved as they once were, all is written down. Jerusalem is Jerusalem, Hebron is Hebron, Beth-El is Beth-El, Mount of Grizim is Mount of Grizim, Mount Carmel is Mount Carmel . . . ," and the pope himself continued, "Mount Tabor is Mount Tabor, Mount Gilboa is Mount Gilboa, the Jordan River is the Jordan River."

With my mother and Pope John Paul II, April 1999.
(© Fotografia Felici)

A s prime minister, my father found himself in a complex situation. On the one hand, he was elected to office in an era when an onslaught of Palestinian terror struck relentlessly at Israeli civilians, a particularly intensive and cruel assault. It was larger in scale than any of the familiar offensives over the more than one hundred years of Palestinian terrorism against Jews in the land of Israel. On the other hand, there was the diplomatic front, particularly vis-à-vis Europe, where democracy and liberalism are the cornerstones of society and a citizen's rights to peace and serenity are deeply entrenched. The countries of that continent, for the most part, held one-sidedly pro-Palestinian views regarding Israel's war against Palestinian terror, ignoring the right of the Israelis to the same peace and security. The Europeans refused to see what was plain to the eye: that the Palestinian Authority, and particularly its chairman, Yasser Arafat, were deeply involved in perpetuating terrorism, and that Arafat habitually spoke out of both sides of his mouth. Not only did the PA refrain from combating terror, but the majority of terror attacks were conducted by organizations under Arafat's control and funded by his administration.

Thus, while waging war against terrorism, my father also took pains to expose Arafat and the PA's true face to European and American leaders. On Israel's public relations front, he had to contend with the joined ranks of bias, hypocrisy, and double standard.

This Sisyphean battle for public opinion turned, slowly, into a successful campaign, thanks to my father's steadfast efforts and to the ongoing barbarism of the Palestinian terror attacks and their refusal, despite all political or economic opportunities, to rise above their desire to kill Jews. Later in his career, after initiating and executing the withdrawal from Gaza, despite internal contention in Israel, my father proved to the world that Israel's intentions were peaceful and that he was willing to go to great lengths to achieve peace, even evacuating Jewish settlements in the Gaza Strip. But in no way was he willing to compromise the safety and security of the state of Israel and its citizens: "As someone who participated in all of Israel's wars, someone who has seen the horrors of war, was severely wounded twice and has witnessed his friends fall in battle, I had to make life-and-death decisions regarding the fate of others and my fate—and so I understand the importance of peace and strive towards peace. But this peace must be the kind of peace that gives Israel security."

Conversations with world leaders were an important tactic on the diplomatic front. My father had many of these phone conversations at home, and I heard some of them; others he told me about, and many of them I learned about from the verbatim copies of conversations that were in his papers. (Although many of these conversations were held in English, the documents were written in Hebrew and later translated into English for this book.)

Usually the leaders of countries regularly discuss trade agreements, aviation agreements, and perhaps the nature of global warming. Israeli leaders, when speaking with their foreign counterparts, must touch on these matters but are also obliged to discuss terrorism, the dangers facing Israel, the endless negotiations with the Palestinians, anti-Semitism, and the constant struggle to prevent the UN and other international organizations from passing resolutions hostile to Israel. There are other nations that indeed do commit atrocities: there are massacres in Africa, abolition of human rights in the Arab dictatorships for many years and in some of the countries in Asia. But these do not draw anywhere near the ceaseless scrutiny of Israel and its conflict with its neighbors.

A February 2001 poll in Britain, organized and financed by the Jewish community there, revealed that the majority of the British public harbored negative feelings toward Israel. A strong majority of the representative sample of 1,000 people perceived the state as aggressive, unduly forceful, militant, and fanatical. Over 40 percent felt that the term *terrorist* was an apt description of Israel; less than 40 percent disagreed. In all likelihood, the situation was no different in other European countries.

The Palestinians were not seen in a better light, but Israel, a Western-style democracy facing a terror onslaught of exploding buses, bombs, shooting sprees, stabbings, and more, was abysmally misunderstood in terms of its right to defend its own citizens.

This was before 9/11, before the bus and train bombs in London in 2005, and before the foiled plan to blow up several airplanes en route to the United States in 2006, which was fortunately prevented.

After the bus and train attacks, a wave of violence against Muslims swept through Britain. More than a hundred attacks occurred, and many Muslims feared for their personal safety. In Israel, no such sprees of violence against Muslims have occurred even after years of terrorism, at times perpetrated with the aid and assistance of Israeli Arabs. Although there have been cases of Jewish terror against Arabs, they happen rarely, are roundly condemned by the government and the public, and do not in any way reflect the prevailing mood in Israel.

This was the background of the meeting between my father and a delegation of high-ranking European Union officials that took place on March 13, 2001, less than a week after he had assumed office. The delegation arrived in Jerusalem straight from discussions with Arafat in his Ramallah headquarters. There they assessed, among other things, the financial situation in the Palestinian Authority in advance of the upcoming European foreign ministers meeting, a forum that would decide on the extent of European aid to the PA. My father gave the delegation, which included the Swedish foreign minister, the United Kingdom's EU commissioner for external affairs, and the Spanish EU special envoy to the Middle East, his view of the current situation. "The Palestinians," he told them, "persist with their terror, violence, and incitement.

"Terror," he explained, "is not a tactical weapon. It is a strategic one. It is used for political purposes . . . The role of the European Union . . . should be to pressure Arafat to cease and desist from terror-related activity . . . We demand the following," he said. "For him to (a) call for an end to terror, (b) stop incitement, (c) begin counterterror actions, and (d) coordinate security [with Israel]."

This was part of a *tour d'horizon* he conducted, which included Iran's arming of Hezbollah, the Iranian development of weapons of mass destruction, the need to keep Saddam Hussein under close watch, and other areas of deep Israeli concern. The Clinton proposal at Camp David, which Ehud Barak had agreed to, he said, was "null and void." Beyond that, he made it clear that Israel would simply not negotiate under fire, that he would not allow terrorism to be used for political leverage, and that Israel would never capitulate to it.

The EU delegates may or may not have been impressed by my father's description of Iran's arming of Hezbollah or about the dangers of that country's WMD programs. But they were adamant about the need for Israel to change its policies. They insisted that Israel reopen its borders to Palestinians, which had been closed back in February under Barak to prevent the infiltration of terrorists. They demanded the release of tax moneys to the Palestinian Authority, which Israel had earlier begun withholding in response to the continuing Palestinian violence. And they objected strongly to what they called "extrajudicial killings"—that is, to Israel taking preventive action against individuals who were about to launch terror attacks, killing those who were about to commit terror. Their position was the moral equivalent of terror.

O n April 16, 2001, after a rocket from Gaza—the first—was fired into the Israeli city of Sderot, an order was given to take military action that included a limited advance of Israeli troops toward the site of the rocket launch in the northeast corner of the Gaza Strip, in the town of Beit Hanoun and in another area close to Kibbutz Nahal Oz. This type of action would seem logical, if not totally inadequate, for any other country with citizens under foreign fire. But for Israel at that time, about a month after my father took office, the situation was fraught with controversy, and the operation grabbed headlines all over the world, reaping harsh condemnations.

My father had to put pressure on the IDF, which had grown stale with inactivity over the years, relying heavily on the Palestinian Authority forces to prevent terror, an action the PA obviously would not do. He also had to launch an international diplomatic campaign, explaining the legitimacy of our actions. Later, as the terrorism against Israel escalated, the IDF responses increased and the public relations campaign was able to reveal the true face of the PA and its chairman, Arafat. By then the IDF's increasingly extensive actions were seen as more understandable in the court of world opinion. That, however, is not to say that Israel's preventative actions were supported but merely that an understanding of Israel's need to protect itself arose—a change that took more than a year to take shape, during which many Israeli lives were lost to terrorism.

But in April 2001, a few short weeks after my father formed his government, this IDF action in Gaza elicited a massive media response, triggering a political memo from his diplomatic adviser, Danny Ayalon. Dated April 19, 2001, the memo details the diplomatic onslaught from other nations undertaken in light of the army's advance, which took several hours and covered less than two miles.

The memo further adds that Paul Simons of the American embassy in Israel, along with David Ivri, Israel ambassador to the United States, the political advisers of Chancellor Gerhard Schröder of Germany and President Jacques Chirac of France, the British and Swedish ambassadors, and EU envoy Miguel Moratinos, had all been notified about the rocket fire on Sderot and Israel's military response.

That evening Paul Simons called Ayalon, with whom he had spoken a few hours prior. He was reportedly shocked by Israel's response to the rocket attack, of all things, and requested further clarifications. Finally he announced that the State Department was set to issue a harsh statement. Secretary of State Colin Powell's condemnation was indeed published.

This incident brought home the realization that it was imperative to establish an open line between the prime minister's office and the offices of world leaders. In fact, on Ayalon's April 19 memo, my father scribbled a note saying that "Secretary of State Powell should have been directly informed." He recognized the importance of having an open channel of communication in order to pass messages in real time and avoid misunderstandings. In addition, he would send heads of the intelligence community abroad to brief world leaders, arming them with irrefutable intelligence information when possible.

On April 22, 2001, my father met Foreign Minister Louis Michel of Belgium. At the time, Belgium was set to replace France as president of the EU, and Michel asserted that it was hard for him to defend Israel's action in the Belgian parliament, and that, to put it mildly, our policies were not always understood in Europe.

My father asserted: "Our policy is—to make things easier for the civilian population and to combat terror."

The Belgian foreign minister wanted answers and explanations.

"Under normal circumstances you would not ask that question. It is hard for me to understand why I need to explain that, as a Jew, I have the right to protect myself," was my father's answer.

As the Palestinian violence intensified, Israel's responses did as well. Alarmed by the deteriorating situation, former American senator George Mitchell, who had negotiated the Northern Ireland peace agreement a number of years earlier, was sent to the area by President Bill Clinton to head a committee tasked to examine the situation and make recommendations. The Mitchell Report was completed on April 30 and formally issued on May 20. Among the report's stipulations was the following:

> *The PA should make clear through concrete action to Palestinians and Israelis alike that terrorism is reprehensible and unacceptable, and that the PA will make a 100 percent effort to prevent terrorist operations and to punish perpetrators. This effort should include immediate steps to apprehend and incarcerate terrorists operating within the PA's jurisdiction.*

Israel accepted the report's recommendations, as did Arafat, and on May 22 my father announced that Israel was declaring a unilateral cease-fire.

Ten days later, on June 1, a Palestinian suicide bomber walked into a crowd of young Israelis waiting to get into the Dolphinarium, a Tel Aviv nightclub, and blew himself up. Twenty-one teenagers were killed, and a further 120 were wounded. After visiting with the wounded at Ichilov Hospital, my father made a statement to the media. "Restraint is also a type of strength," he said. It was not an easy thing to say, especially not for him, after seeing the teenage boys and girls in their torment in the hospital, and especially not for him, a man conditioned to defend them. Moreover, public sentiment, both in the streets and politically, was very much in favor of a strong military response. I felt similarly. Truthfully, I think the Palestinians and the international community expected it. For another leader, restraint may have come more naturally, but the public would have had a harder time accepting it. For my father it meant conquering his natural inclinations, and everyone knew it. His good judgment served Israel's national interest well. Later, when he began to fight Palestinian terror in earnest, everyone knew, in Israel and across the world, that he had stretched Israeli restraint to the maximum.

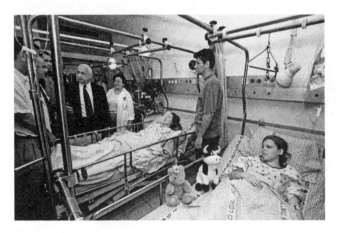

Visiting victims of the Dolphinarium bombing, June 3, 2001. (© Avi Ohayon, Israeli Government Press Office)

Joschka Fischer, Germany's foreign minister and vice chancellor, was on an official visit to Israel around the time of the Dolphinarium bombing. "Since we've declared a unilateral cease-fire," my father announced at a joint news conference held on June 3, two days after the bombing, "there were over 140 terror attacks that caused many casualties of Israeli citizens. When will Arafat quit his whirlwind tours around the world and declare an end to violence? Israel is committed to peace, security and regional stability—but our utmost obligation is the safety of our citizens . . . we cannot restrain ourselves forever—there will be no negotiations under fire."

During this press conference, my father called on the Europeans to assist the Palestinians with tangible projects and not cash. That way there was a chance that aid would reach those in need and not just line the pockets of senior Palestinian officials, primarily Chairman Arafat.

With Germany's Foreign Minister Joschka Fischer, December 2004. (© Amos Ben Gershom, Israeli Government Press Office)

Furthermore, in light of President Bashar al-Assad of Syria's coming official visit to Germany and his country's possible nomination for a spot on the UN Security Council, my father asserted that Hezbollah, with Syrian and Iranian backing, was turning South Lebanon into a terrorism stronghold. Later, Syria was indeed given a seat at the Security Council, serving from 2002 to 2003. This underscored the truly absurd nature of many UN decisions, as Syria is host to many known terror organizations, including Hezbollah, Hamas, Palestinian Islamic Jihad, the Popular Front for the Liberation of Palestine, the Democratic Front for the Liberation of Palestine, and others. Perhaps, in light of the daily massacre of Syrian civilians that began in the spring of 2011, it is clear to everyone, even the UN, that Syria's membership in the security council was indeed absurd.

In the wake of the Dolphinarium murders, President Bush sent CIA director George Tenet to Israel to propel the Mitchell initiatives forward and implement procedures that would calm the situation and lead back to political negotiations. Tenet's emphasis, as was Mitchell's, was on the cessation of violence and the initiation of measures to counter terror.

One concrete development from the Tenet plan was the reinstitution of security meetings between the Israeli and Palestinian security services, held jointly with CIA representatives. The intent was to start sharing intelligence and building cooperation to apprehend terrorists before incidents actually happened. Israel would turn over information on terrorist planning and operations to Arafat, and Arafat would make the arrests. The latter never happened.

Israel's diplomatic ties to the United States of America were and will continue to be at the heart of its foreign policy. The United States is its most significant and important ally. The shared values of democracy and freedom, the former campaign against Soviet expansionism, and the current battle against Islamic extremism have forged close ties and a firm alliance, a sense of intertwined fate. The 9/11 attacks tightened the ties and the feelings of allegiance. That was how my father saw it. His excellent rapport with President Bush, the mutual sense of trust between the two leaders, further reinforced the bond between the two nations. In the beginning of my father's term as prime minister, it was Arie Genger, a friend who lived in the U.S., whom he engaged as his personal envoy to the White House.

On July 30, 2008, President Bush spoke to me in the Oval Office about my father:

"Well, first of all, I met your father in a very unusual way. He was a minister and I was a governor during the Netanyahu government in 1998. And I was getting briefed about—I was there with three other governors, so we were getting briefed about Israeli politics and Israeli policy. And one of the ministers said, 'Would you like to take a helicopter ride of the West Bank?' I said, 'I'd love to.' That minister was your father. And so in my—had to have been my second—or first full day in Israel, I was given a helicopter ride over the West Bank by Ariel Sharon . . . He knows everything. This is where—and so it started off with, 'I was a young commander here, and there, in this village here.' And so that's my first introduction. And then of course we had a lot of meetings—at Crawford and here in the White House. And then over time, we grew to trust each other as people who, when they gave their word, kept their word. And to me a man's word is very important, and your father always kept his word. If he said he was going to do something, he did, and if he said he couldn't do it, he explained why."

"How much, if at all—the interaction, the personal relations—how much does it affect relations between countries' policies, if at all?" I asked the president.

"I think it affects them, because I think that if you earn somebody's trust it makes it easier to listen and it makes it easier to exchange ideas. If it's not trustworthy, if people feel like you don't care what they think or if you think that you don't share the same goal, then it's really hard to have a constructive dialogue. And so your dad and I worked hard to develop a mutual trust. Secondly, what really helped define the relationship, of course, was the evidence of extremism on our own soil. The attacks of September 11 were very defining, and it put most Americans squarely on the side of other countries that were fighting extremists, like Israel. In other words, it became very evident they [the Americans] were part of the same front against people who murder innocent people to achieve their political objectives. And I think your father appreciated the fact that I saw that so clearly and stated it so absolutely, that the United States would fight extremists and we would help our friends in fighting extremists and that we would promote democracy as the alternative to the vision that these people had. And I think a lot of people in Israel at this time were wondering whether the United States fully understands their security risks. And I think my statements made it clear I did understand them, and I think your dad appreciated that. So it was—we not only had a good personal relationship, but we shared the same kind of philosophy of how to conduct yourself in a very dangerous world.

". . . I admired him. He's a tough old tank driver. That's how I looked at him . . . I felt his decision on Gaza was a very bold and strong decision, and I admired him a lot for that."

"How did—standing in front of the Old City, Jerusalem, how did it affect you? They say that site affects especially people who believe in God."

". . . We got there and it was late, so we didn't see anything. We drove in from the airport. And I had this beautiful room, and I opened up the curtains and the sun had just come up, and the Old City was gold; it was golden. It was an amazing experience. I said [to Laura] you're not going to believe this. She had to put her contacts on first. And then I fired up the coffee. And so we drank our coffee watching the city literally turn gold. And it was an amazing experience. I've never forgotten it . . . it was awe-inspiring. I'm not sure if I would view it as a religious experience, necessarily. I think the whole getting in touch with much of the sites of the Bible and the history of the Bible is important for a person's religious experience. I've spent time in the Sea of Galilee and I've been to Bethlehem. The great thing about Israel and the Middle East is so much of what Westerners—at least Americans—study in their religious life, no matter what your religion, that it becomes evident and real there . . . I went to the Wailing Wall and left a prayer . . .

it was a fantastic experience and it affected me. I remember your dad flying down the line there [the Green Line] and saying, '67, look how close it is between this hill and this strategic site—city."

"I wanted to ask you, how can we be sure that the world, including the United States, will not stand on the side while the life of Jews are being threatened, especially by Iran, like it was in World War II?"

"That is to have leaders who understand the threats, leaders willing to have strong relations and listen to our friends and allies. You don't have to worry about the next six months," the president said, laughing, referring to his last months in office.

". . . Look, I think any American president is going to take the Iranian threat seriously. The question is, how does he deal with it?"

Before parting, the president said to me, "I had great respect for your dad, and I found him to be a courageous man."

O n May 23, 2001, President Bush called my father.
"Hello, Ariel, I hope all is well with you."

"Thank you, George, I'd like to thank you for your clear call for an immediate and unconditional cessation of terror and violence and your rejection of the linkage between terror and settlements. Israel has initiated a unilateral cease-fire. It is most important now to pressure Arafat to hold his fire too. In the meanwhile there is no change in Arafat's actions and only half an hour ago two Israeli students were injured in [the city of] Ariel from Palestinian fire, one of them critically. We will be happy to work with your representative Bill Burns, we will make an effort to move forward and we will use restraint in the coming days in order to give Arafat a chance to examine his actions. Up until now he has not done a thing. We will wait a few days, during which we will not initiate, in order not to give him an excuse not to declare a cease-fire."

"Thank you for operating in this manner," Bush said. "The entire world appreciates your leadership. We will continue to demand of Arafat that he lay down his arms and we understand that there will not be any confidence-building measures without quiet and a cessation of violence. Thank you for your cooperation with Burns and for the steps you have taken. The world sees that as an important step and I called simply to thank you for your leadership."

My father said, "There is a danger in the border with Lebanon. Israel is making every effort to keep the border quiet and to insure stability. Hizballah is planning a terror attack with a green light from Syria. We have already warned of the repercussions and serious consequences of such an act. It is important to maintain the pressure on Bashar Assad and the leaders of Lebanon. We already discussed this with your people the day before yesterday. They are planning something to commemorate the first anniversary of our departure from Lebanon. We want to prevent escalation there because it will cause all of us to confront a serious problem."

President Bush answered, "I appreciate that. I am aware of what is going on in the North—the cooperation between us is close. We've demanded of the Syrians that they not allow and encourage violence against Israel and will continue with those messages. Thanks again for your leadership."

In June 2001, some three months after my father formed his government, the BBC aired a documentary on *Panorama*, its investigative current affairs program. The central theme of this program was to prove that my father was responsible for the murder of Palestinian Muslim Arabs at the hands of Christian Arabs in the Sabra and Shatila neighborhoods of Beirut nineteen years prior.

This false claim stands in stark opposition to the findings of both the Kahan Commission and the findings of courts in the United States and Israel.

Reactions came swiftly on the heels of the screening, ushering in a wave of claims regarding supposed war crimes from all corners of Europe, including Spain, Denmark, and Norway, and even one from the American side of the Atlantic.

But Belgium trumped them all—a nation notorious for the genocide it perpetrated in the Congo, a nation that, as I reminded my father in a memo I wrote at his request on February 2, 2003, provided two volunteer divisions to the ranks of the SS, the Flemish Division 27 (the Langemarck) and Division 28 (Wallonien), along with the Flemish Black Brigade, which predated the other two divisions and was made up of Belgian volunteers. And we haven't forgotten the Hitler Youth. The very same Belgians, riding on the gusts of anti-Israel sentiment blowing through Europe's street corners and governments, buoyed by the attempts to demonize my father in mass media, assumed the role of global arbiters of justice and bearers of the flag of morality—a somewhat surprising stance from a nation known primarily for its fine chocolate and its questionable treatment of children.

At no point did they consider trying Arafat, whose hands were covered with the blood of the innocent, or any other Arab leader, many of whom had killed vast numbers of their own countrymen and neighbors.

Before long, though, the situation began to spin out of control for the Belgians. For some reason they were offended by the insinuation that their actions spoke of anti-Semitism. On February 25, 2003, a Belgian foreign ministry official spelled this out in a meeting with an Israeli diplomat in Brussels. The Belgians were also very concerned about the claims Iraqis were filing against American officials, including George H. W. Bush, Dick Cheney, Norman Schwarzkopf, Colin Powell, and, within three months' time, General Tommy Franks and Colonel Bryan McCoy.

On February 18, 2003, in response to a question, Secretary of State Powell said, "It's a serious problem. The Belgian legislature continues to pass laws and modify them over time, which permits these kinds of suits, and it's the same kind of law that affected Prime Minister Sharon. I have been named, along with President Bush 41, Dick Cheney and Schwarzkopf. And also, even before anything has happened, they have named 43, President Bush and [Secretary of Defense] Don Rumsfeld. We have cautioned our

Belgian colleagues that they need to be very careful about this kind of effort, this kind of legislation, because it makes it hard for us to go places that put you at such easy risk. And I know it's a matter of concern at NATO headquarters, now, and international headquarters sitting there in Belgium where not just U.S. officials but officials from anywhere, where officials of Mr. Sharon can be subject to this kind of litigation and if you show up, next thing you know who knows ?" Powell's hint was as big as the NATO headquarters in Brussels. The Belgians realized they had gone too far, and in April 2003 they passed new legislation limiting the scope of their "universal" justice system, and quietly gave up their prosecutions.

———

Israel's diplomatic ties with Russia were complex, all the more so in my father's case. For dozens of years my father had fought in uniform against an enemy armed with the finest Soviet weaponry—tanks, artillery, planes, missiles, rockets, heavy machine guns, and rifles.

Russian history is also rife with pogroms, and the slaughter and scapegoating of Jews. For hundreds of years, since the Middle Ages, Jews have been persecuted in Russian lands. Attempts were made to convert, expel, and kill Jews, not necessarily in that order.

During the Communist revolution the White Army murdered an unknown number of Jews, somewhere between tens and hundreds of thousands, this time because of their perceived ties to the Bolsheviks. After the revolution they were persecuted for being active in the Zionist movement because the Communists disliked movements with different aspirations and goals than theirs.

During the ethnic cleansings of the 1930s many Jews were killed. After World War II, the murders and the bloodletting continued, and in August 1952 the members of the Jewish Anti-Fascist League, which was founded by the Communist Party, were murdered in what became known as the Night of the Murdered Poets.

In November 1952, the Prague Trial began. Communist Jews were accused of being agents in the service of Zionism and American imperialism, among them two Israelis visiting Czechoslovakia. One of the two, kibbutz member Mordechai Oren, was one of the leaders of Mapam (the United Workers' Party), an Israeli left-wing organization that supported the Soviet Union and idealized Stalin. Oren, who supported the Soviet side in the Cold War and was merely visiting the much-admired Communist Bloc at the behest of the party, found himself accused of a crime and tortured before serving a four-year sentence in prison. His incarceration caused a rift within Mapam, with some opposing his arrest and some supporting it; yes, some Jews are just that naïve.

With the swift summation of the Prague Trials came the Doctors' Trials, which began in January 1953. Over the course of the trials, nine Jewish doctors were tried and found guilty of belonging to a Jewish group that allegedly planned to poison and murder Soviet leaders.

The Soviet Union sealed its citizens behind the Iron Curtain. The Jews who were denied the ability to immigrate to Israel, referred to as Refuseniks, spent years in prison and in Siberian exile.

As far back as the 1950s, the Soviet Union began arming the Arab states to the hilt. After the Six-Day War, the Soviet Union severed diplomatic relations with Israel and became uniformly pro-Arab and anti-Israel, forming ties with Palestinian terror organizations. To this day Russian-made weapons continue to threaten and harm Israel and Israelis.

On the flip side there are emotional ties to Russian culture. Back in the 1970s, when I was still a child, during the heart of the Cold War, I spoke to my grandmother Vera and expressed my dislike of Russia. "I love Russia," she responded. Only years later did I understand what it was that she loved—the harmony of the language, the great poets, the classical writers, the pride, the endurance, the landscape, the vast open spaces, the trees, the forests, the snow, the rivers, the deep green grass. Smoked fish, vodka, conversations with friends around the table, cabbage and beets, borscht, boiled potatoes and sea salt, caviar, sweet tea, rolled blini stuffed with cheese and raisins (one of my father's finest childhood memories). Russian culture.

I remember how, in her small kitchen in Kfar Malal, my grandmother would take her vodka with a single, sharp jerk of the head. One weekend when she came to visit us on the farm, my mother, who knew what Vera liked, served her a small tin of black caviar. "Would be a pity to spoil the taste," she said in her Russian-coated Hebrew, refusing sour cream or a cracker, "only with a spoon." My father absorbed the mixed feelings toward Mother Russia—a stepmother but nonetheless a mother. He understands Russian entirely and can speak it passably. When talking over the phone or meeting with Russian leaders, he would often correct the translators.

During the 1990s, once diplomatic ties between Israel and the deteriorating Soviet Union had been restored, my father would tell his Russian peers that their stance on Israel over the years had been a mistake. After all, the vast majority of Zionist leaders and Israeli heads of state traced their roots back to Russian lands, a fact that should surely have contributed to a natural and firm relationship between Israel and Russia.

"Ex-Russians ruled this land until they died of old age"—that's what my father used to say. He always kept in mind that it was Russians who had liberated the Jews from the death camps in Poland, the Russians who had voted for the formation of the Jewish state, and the Russians who had been one of the first countries to recognize its establishment in 1948.

My father knew what national pride meant to the Russians, and he appreciated that. He was aware that the Russians see themselves as a superpower, an empire, and that having international influence is important to them. He knew the history of the country, the national ethic of perseverance, the fact that neither Napoleon in 1812 nor Hitler in World War II was able to bring the country to its knees.

The fact that he bested Arab armies trained and armed by the Soviets, who at times even assisted on the battlefield, only added to the great respect he received in Moscow. His understanding of what was important to the Russians and what motivated them helped matters considerably, even if there were disagreements, particularly regarding the sale of weapons and technology to Israel's enemies.

I have no doubt that my father's grasp of the Russian language and his roots through the country made the Russians proud and enhanced his excellent personal relationship with Vladimir Putin and the upper echelons of Russia's leadership.

During meals in Russia there's never any doubt about drinking—you know, based on how many people sit around the table, what the minimum number of shots is. Everyone present at the meal gets to their feet in turn, according to the hierarchy, says a few words of greeting, and then all present throw back the contents of their glass in a single shot. The only question that remains is the number of rounds. And there were rounds aplenty. During official visits to Russia as minister of infrastructure and foreign minister, my father emerged from those meals victorious, beating the Russians at their own game on their home court. Vera's genetic line probably helped.

In 1997, while serving as minister of national infrastructure, my father visited the Tamanskaya Motor Rifle Division, near Moscow, and lectured at the military academy there. It was a special occasion for him and for his hosts. Their military order and discipline and the way in which they greeted him was extraordinary. When the Russians like someone, they know how to show it.

During that same trip, my father, as he usually did, asked to visit an agricultural farm. He was taken to a *sovkhoz*, a state farm, not far from Moscow. The cows looked famished, and the manager of the *sovkhoz* introduced Comrade Gallina, the farm's most outstanding milker. "Last year," said the manager, "Comrade Gallina milked such and such number of cows in such and such amount of time, while this year she milked even more cows in less time."

Comrade Gallina, a local babushka, proudly stood up straight. Her turn had come. She stepped forward, holding herself erect, and announced, "Our *sovkhoz* manager has been able to reduce the amount of feed given to the cows from such and such amount last year to such and such this year and has at the same time raised milk production from such and such last year to such and such this year."

Someone whispered to my father, "A hundred or two hundred miles from the *sovkhoz*, they don't even know that the Communist Revolution has ended."

Once, when the Russian ambassador was visiting our home on the farm, he said that during a visit to Japan, Boris Yeltsin had asked the Japanese for a $3 billion loan to fund the Siberian gas pipeline. "You can have the money as a grant if you give us back the South Kuril Islands," was the response. "How could I go home and say I gave away holy Russian land?" Yeltsin had said.

"How can you ask Israel, a very small country, to give away some of its territory while you, an enormous state, the biggest in the world, refuse to give back the South Kurils, which you captured?" a guest asked. The ambassador, a heavy man with a red mustache, his face ruddy from the Israeli heat, held a lamb shank in one hand and a glass of vodka in the other. Laughing, he said, "That's why we have a big country and you have a small one."

On July 9, 2001, after a Russian airliner crash in Siberia, my father phoned Putin to offer condolences.

"I'd like to express my condolences over the death of the 145 people in the air accident last week."

"Thank you, and I'd like to express my condolences for the victims of Palestinian terror."

"We thank Russia for your willingness to help. As I'm sure you're aware, I've met the envoy Vidoven, and a few days ago I spoke with Foreign Minister Ivanov."

"Yes, they reported back to me about your talks."

"I hope to visit Russia this coming September."

"You have an open invitation, and I look forward to seeing you soon."

"As you are aware, Israel has accepted Mitchell and Tenet. We have unilaterally stopped the fire. We are continuing to make things easier on the population. But the Palestinian terror continues."

"I received a detailed report about the events. But I would be happy to hear what is happening directly from you, because we also hear support for Tenet and Mitchell from the Palestinian side."

My father detailed various deadly attacks that had happened in the past few days:

"Today a car full of explosives blew up in Gaza. Even after we warned the Palestinians and passed precise intelligence information over to them in advance, they did not act or try to prevent the incident . . . Terror is a strategic choice of Arafat . . . Concessions only breed more violence. International pressure must be brought to bear on Arafat. He only responds when he's isolated . . .

"I'd like to underscore again that Israel will not conduct negotiations under pressure and while facing terror . . . We will conduct negotiations only when there is quiet, and we mean complete quiet. No terror and no incitement to violence. Russia would act in the same manner . . ."

Putin said, "I completely agree with your attitude regarding terror, and you are absolutely right that any political ties should only be forged after the cessation of all terror activities. We have to work on reducing violence until it is completely stopped.

"I very much look forward to meeting with you [in Moscow] in September, and of course, until then I'll be happy to speak with you over the phone. I wish you much wisdom and courage, and we identify with Israel's struggle against terror."

———————

Putin's warmth was decidedly not mirrored in Jacques Chirac's attitude toward Israel. This was the case despite the fact that because Chirac, in his younger years, had been an officer in the French Army in Algeria, he had a deep personal respect for my father.

The ties between Israel and France were strong and well tended to by Shimon Peres, especially during the 1950s and '60s, when Peres served as director general of the Ministry of Defense and as deputy defense minister. The French had coordinated their actions with Israel in the 1956 Suez War. While preparing for the mission, French military officers had visited my father at the paratroopers base in Tel Nof. Their concerns about the ability and readiness of the troops to fulfill their role in the joint French, British, and Israeli mission dissipated when they observed the high caliber of his unit and the amount of arms they had captured during their three years of countless cross-border missions. Personal ties were forged, particularly between my father and Colonel Jean Simon, who went on to become a general and one of the most highly decorated French officers.

France had assisted with the assembly of the nuclear plant in Dimona, and the Israeli Air Force was comprised of mostly French aircraft. But after the Six-Day War, France, like a beautiful but unfaithful lover, radically changed its position on Israel, exhibiting a one-sided approach and strong self-interest. They have further managed to maintain this position for over forty years, adhering to policies that are overwhelmingly anti-Israel and pro-Arab, their theory being that ties with the Arab world serve their interests better than ties with Israel. After the Six-Day

War, the French placed a military embargo on Israel, refusing even to supply the ships and planes Israel had already purchased. This stance forced Israel to smuggle five warships from the port of Cherbourg, a move that embarrassed the French but fortified our navy. It was only the election of Nicolas Sarkozy that brought about a refreshing change in France's approach toward Israel.

With President Jacques Chirac at the Élysée Palace.
(Courtesy of the Government of France)

Jacques Chirac was one of the primary proponents of the anti-Israel movement, and the driving force behind France's decision to sell a nuclear reactor to Iraq. The Osirak—sometimes referred to as Ochirac—reactor was assembled during the latter half of the 1970s. Chirac himself once addressed Saddam Hussein: "You are my personal friend. Let me assure you of my esteem, consideration and bond." Chirac also bestowed on Arafat the adoration that France reserves for only the most esteemed leaders. The information, the very proof, that Arafat was involved in terror did not change their position in the least: in Paris, and particularly with Chirac, Arafat remained the beloved of the French Republic. His crimes were overlooked, as was his bankrolling of those crimes with aid money earmarked for Palestinians. According to the November 9, 2003, edition of *60 Minutes*, the PA itself hired Jim Prince and his team of American accountants to conduct an audit. They found close to $1 billion in a secret portfolio belonging to Arafat. Although the money for the portfolio came from international aid funds, virtually none of it was used for the Palestinian people. And, Prince said, none of these dealings were made public.

As frosty as France's overarching position toward Israel was, it was particularly harsh toward my father. That attitude was fanned by the French press and by administration officials who voiced opinions and leaked statements that were scathingly critical of Israel.

What apparently especially disturbed Chirac was that he was not as influential in Israel as he thought he deserved to be. Of course my father would have preferred a less chilly wind from the French capital, but he wasn't willing to pay for it. The French, and particularly Chirac, with his superpower pretensions, had a hard time coming to terms with the fact that Israel had a prime minister who disagreed with him, who ignored French efforts to pressure him. He was, on the matter of terrorism and security for Israeli citizens, immovable.

Instead, as world opinion shifted, Chirac found himself, over time, almost alone in supporting Arafat until the latter expired in a French hospital in 2004.

On July 11, 2001, after a short visit with Chirac in Paris, my father said. "Thank you for your hospitality in Paris and for the dinner and the conversations. I hope to host you in Israel in the future."

"Thank you. I wanted to ask how the situation is today in light of the terrible situation in Rafah."

"In Rafah we try to prevent weapons smuggling to the Palestinian Authority. We control a thin strip of territory there, and our forces are under constant fire and attacks. In one day fifty-eight grenades were thrown at one of our positions near the Egyptian border. I passed a message on to Arafat long ago, and I've spoken to the United States about this as well. The smuggling has to stop. We can't allow weapons smuggling to the Palestinian Authority, and so we are exposing the tunnels, doing so often under constant fire. Israel has accepted the Mitchell and Tenet reports, and we are restraining ourselves and offering easements to the civilian population. The Palestinians repay us with terror, violence, and incitement. The Palestinian Authority must fulfill its obligations.

"We have had eleven deaths [since Tenet.] Two IEDs were discovered this morning. A car blew up in Gaza on Monday—and although we supplied intelligence information in advance to the Palestinian Authority, no counterterror measures were taken. I do hope this comes to an end because I would like to proceed according to the Mitchell Plan. The problem is that the terrorism has not come to an end. There are no arrests. I believe pressure must be brought to bear on Arafat to cease the terror. Terrorism is a strategic choice for Arafat and he employs it by way of Force 17, the Tanzim, the Fatah. All three cooperate with the Hamas, Islamic Jihad, and Hezbollah. Concessions will only breed further violence. International pressure on Arafat is critical; he only responds when he's isolated."

Chirac replied, "Thank you. I've listened to your comments about the situation with great interest. My understanding of the situation has not changed since we last met. We're asking Arafat to govern and are trying our best, but both in the United States and in France we disagree with your estimation that Arafat rules over everything. It is true to say that there's no peace without security, but total security is unattainable without peace. We are in contact with the Americans and others and are always happy to do everything."

Chirac, for some reason, spoke for the Americans as well, perhaps thinking that it would add weight to his appeasing attitude toward Arafat's terror. My father did not take the bait.

"Thank you. I would just like to emphasize that insofar as Arafat's control is concerned—I believe that he controls far more than we all believe, and at any rate he has not ordered the arrest of the seventy terrorists that stand behind all the current activity. And regarding what you said about the obliteration of terror—we're so far from that, I do not even think we see steps being taken to reduce the activity. I believe that pressure brought to bear on Arafat is what is necessary so that he realizes that he will not reap any political achievements as a result of terror.

"The demand must be the complete cessation of terror and incitement. If we do that, I believe we can reduce the terror to a minimum, but if we do not, the terror will never stop, and we will not be able to progress with the peace process that we all seek. That's why the demand must be for a complete cessation. And in so far as the United States is concerned—I am glad that France is cooperating with the United States and I am grateful for your important efforts, but up to this point, the American position is that we still cannot begin the seven days that are to proceed the six-week cooling-off period."

Several days later, my father discussed the same subjects with Prime Minister Lionel Jospin of France. He also brought up Iran's shipments of hundreds of tons of weaponry to Hezbollah, a fact well known to every interested party. "The fact that Hezbollah is deployed along the border and not the Lebanese army is a time bomb," he told Jospin. (Five years later this time bomb went off and became the Second Lebanon War.)

Jospin replied "innocently," "I will check this with our intelligence service. This is the first I've heard of this."

The French ignored the information about Arafat and his involvement in terror. They were deaf to such charges. Unburdened by facts, they supported Arafat, period.

Despite the depth of Israel's current relations with Germany, the Holocaust—the annihilation of a third of the Jewish people at the hands of the Germans and their varied European collaborators—will never be forgotten or forgiven.

After the war Germany felt remorse, which of course did nothing to help the millions of slaughtered but was at least more honest, more real, and less hypocritical than the behavior of many other European nations. Austria, Belgium, Poland, Ukraine, Hungary, Croatia, Holland, Lithuania, Latvia, and of course France all admitted late or only halfheartedly what their citizens, either personally or in an organized manner, had done to the Jews during, and even after, the war. In Poland, for example, even after the war had ended, many Poles continued killing Jews, some of whom had returned there, looking for family members. Germany remains friendly toward Israel. The two countries have developed mutual trade relations and close security ties. The Germans have also played a key mediating role in hostage negotiations and in dealings with enemy states such as Iran and Syria. It was in this light that my father saw Israel's relationship with Germany: accepting the importance of the relations, in terms of strengthening the state of Israel and bolstering its international status, but also abiding by the vow never to forget.

On July 14 my father and Joschka Fischer spoke again: "We are worried about the developments in your region," said Fischer. "The last few days have not been hopeful."

My father gave a tally of those developments: "Since Tenet there have been some four hundred and fifty terror incidents."

"I spoke with Arafat this morning," reported Fischer. "And I told him that the killing of settlers and army personnel was unacceptable."

"I am a big proponent of the security meetings," my father said, "and we will continue having them, but absolutely nothing happens afterward. There are no results, and the terror continues. Arafat cannot distance himself from the terror because the Fatah and the Tanzim are under his command. In terms of arrests, we passed seventy names on to Arafat. He was told that he needed to arrest [those men] because they are planning mass terror attacks . . . We asked that he arrest them, otherwise we'll have no choice but to intercept them on the way to the attack . . . He knows everything but doesn't arrest them. He has chosen the strategy of terror, making the assumption that it is possible to pressure Israel. We are cooperating with everyone in the international community. Our interests and yours are the same—stability. But we can not live from funeral to funeral."

"Why not use the Security Commission with American monitors in the trouble spots?" suggested Fischer. "You trust the Americans, and so do the Palestinians."

"We have no problem with American monitors," my father replied. "But we must agree—so long as the terror continues, there cannot be negotiations. Terror cannot be concealed, and it is not a matter of reducing the numbers—are two fatalities instead of four a reduction? I am not going to enter a situation where there are daily negotiations over what a reduction in terror means."

"But on the other hand," said Fischer, "the circle of violence and the closures must be broken."

Even someone as friendly to Israel as Fischer couldn't let go of the "circle of violence" propaganda. Israel does not harm civilians deliberately. At times civilians do get hurt in military actions but that is only because the terrorists deliberately hide among civilians, putting them at risk. On the other hand, the Palestinians deliberately target civilians and even children. On March 26, 2001, ten-month-old Shalhevet Pass was shot by a Palestinian sniper. He saw her through his rifle's sight and pulled the trigger. There is no circle here.

Fischer, with his European mentality, which places great stock in the sanctity of one's word, certainly in the context of a formal agreement, simply could not understand the Palestinian mentality. Apparently he had not done much shopping in Oriental bazaars. My father continued to explain Israel's position with great patience.

———

On July 18, 2001, a few days after his conversation with Joschka Fischer, my father discussed this subject with Russian foreign minister Igor Ivanov, with whom he was in frequent contact.

"Yesterday there was another Palestinian escalation," my father said. "They've fired mortars on Jerusalem. On Monday a disaster was avoided at the Maccabiah games [an international Jewish sports competition taking place in Israel]. An attack in Binyamina, in the heart of the country, claimed two lives and twelve others were injured . . . The Palestinian Authority does not in any way use the detailed information we provide them. Every day brings gunfire, grenades, mortar fire, explosives. We intercept those on the way to attack and exercise self-defense . . . Arafat could avoid this if he made arrests, but he does not.

"Most of the attacks are perpetrated by organizations he controls—the Tanzim and Fatah . . . we're dealing with a pathological liar and murderer who has ordered the murder of children, women, and babies. And yet he continues to whine about the way Israel treats him."

"We speak with Arafat every day," Ivanov answered. "He takes measures, but insufficient ones. I understand your feelings, but understand us—we do not want a tragedy or disaster in the Middle East, but do not know what else can be done. If you suggest something, we'll do it together . . ."

"You asked what might be done to combat the terror. You have experience in Chechnya, but here the situation is more delicate," my father pointed out. "Israel is not a superpower like Russia, and we must take measures of self-defense. But Arafat must be made to feel pressure. Only then will he move forward. We will not move forward without quiet."

My father did not hesitate to bring up Chechnya. Ivanov knew that if Israel were to fight Palestinian terror the way Russia fought terror in Chechnya, there would be much less terror—and a lot less Palestinians.

"Thank you," said Ivanov. "I will call Arafat."

"Tell him," my father answered. "Tell him that the only issue is that he stop the terror, and then we can progress according to Mitchell."

TWENTY-FOUR

A UNIFIL observation post along the Lebanese-Israeli border. (Courtesy of AP Photo/Alexander Zemlianichenko)

November 29, 1947, is a significant date in Israel, for it was on that day that the UN voted to terminate the British Mandate in Palestine, accepting the partition plan, which divided the land west of the Jordan River into two states, Arab and Jewish. The live transmission of the General Assembly's vote, the stern roll call, and the alphabetical response of each nation—yes, no, or abstaining—is familiar to everyone in Israel. At the time, Jews in the land of Israel listened nervously to the vote, keeping score as it progressed and bursting into spontaneous celebrations of joy when the plan was authorized. Israel accepted the plan, but the Arabs rejected it, abiding by what has become a Palestinian tradition of missing every chance at a normal life. The rest is history.

It seems, though, that November 29, 1947, was the high point of Israel-UN relations. Since then the UN, at the hands of the Arab nations, the nonaligned countries, and, in the past, the Communist countries, has become an arena in which to launch attacks against Israel, condemning Israel incessantly and largely ignoring all Arab assaults. Relations have had several low points, including Resolution 3379 of November 10, 1975, which called for the "elimination of colonialism, neocolonialism, foreign occupation, Zionism, apartheid, and racial discrimination." Later in the same resolution came the following: "The racist regime in occupied Palestine and the racist regime in Zimbabwe and South Africa have a common imperialist origin, forming a whole and having the same racist structure and being organically linked in their policy aimed at repression of the dignity and integrity of the human being." Zionism was also labeled a threat to world peace, and all countries were called upon to resist this racist imperialist ideology.

For some reason, the assertion that Zionism is racism is what most captured the public's eye; the far graver call for the elimination of Zionism (yes, worded just that simply) and its presentation as a threat to world peace and security received less attention.

Sixteen years later, on December 16, 1991, the UN General Assembly, in one of its shorter and less explicit decisions, nullified that statement. "The General Assembly decides to revoke the determination contained in its Resolution 3379 of 10 November 1975." Nonetheless, ten years after that, in September 2001, the UN's World Conference against Racism, Racial Discrimination, Xenophobia and Related Intolerance was held in Durban, South

Africa, where Arab nations and other Third World countries used the forum to bash Israel. The often rude and anti-Semitic statements followed one major theme—that Zionism was in fact racism. On September 1, 2001, the NGO Forum against Racism came out with a declaration that "Israel is a racist apartheid state" whose actions "amount to ethnic cleansing and are evidence that genocide is being committed against the Palestinian people." On the other hand, the NGO Forum expressed its support for the Palestinian Authority just as the PA was busy financing and committing terror attacks. The representatives from the United States and Israel walked out so as not to participate in this ugly show.

In fact, simply looking at a list of the states that have held positions on the UN Security Council is enough to make the point: that body, charged primarily with tending to the peace and security of the world, has included Algeria, Iraq, Lebanon, Libya, Sudan, and Syria, states that have not only supported terror but engaged in it. This speaks volumes about the United Nations.

I mmediately after the founding of Israel in 1948, the Arab states violated the UN's decision regarding the partition plan. They attacked the fledgling state with a single, overt goal—its destruction. Israel did not provoke this reaction—there was no border dispute, no provocation beyond Israel's existence—and yet the UN, a supposedly neutral body, did not react to it. The organization further proved its bias when in 1951 the Security Council failed to enforce its own decree regarding the freedom of every vessel to sail through the Suez Canal, a ruling that only reinforced the 1888 Convention of Constantinople, which guaranteed the rights of all vessels, regardless of their flag, to sail through the canal, in war and peace alike. But when the canal was closed to Israeli vessels, the UN took no measures. Before the Sinai War of 1956, the Egyptians closed the Straits of Tiran, an overt violation of Israel's rights that was resoundingly ignored by the UN.

Throughout the 1950s the UN ignored the terror attacks launched against Israel from Egyptian, Jordanian, and Syrian soil. The same cannot be said of the Israeli responses to these attacks, which were roundly condemned, as was Israel's presence in the Sinai Peninsula to open the Straits of Tiran at the end of the Sinai War. The UN's call for the immediate withdrawal of all Israeli troops from the Sinai did not seem to take into account the reasons for the outbreak of war.

On the eve of the Six-Day War in 1967, UN troops responded with inexplicable haste and without any signs of protest to President Gamal Abdel Nasser of Egypt's demand to evacuate the border region, an area they had monitored since 1956 to maintain the peace. UN forces fled the border region and Sharm el-Sheikh alongside the straits. When the Egyptians amassed troops in the Sinai and once again shut the straits, an international shipping channel, to Israeli vessels, effectively closing Israel's southern port in Eilat, the UN once again did nothing beyond consent, withdrawing the very troops that were to have barred an attack on Israel and ensured compliance with an international treaty.

The United Nations Interim Force in Lebanon (UNIFIL) was stationed in the south of the country in 1978. Its job was to restore peace and security to the area—in essence, to guard Israel from terror attacks. In actuality, the force restricted Israel's defensive actions and did not prevent terror organizations in the area from firing at Israel. They further allowed terrorists to remain in their positions near UN posts so that the IDF would hold its retaliatory fire out of fear of striking UN soldiers. Major General Moshe Kaplan (Kaplinsky), who was my father's military adviser and later became deputy chief of staff, told me in an interview on May 17, 2011, that "the terrorists in Lebanon systematically placed their positions—both the overt and covert—near the UN positions because they knew that we do not fire towards UN positions." It is common knowledge in Israel, and I remember it myself from my service in the army.

Once Israel withdrew entirely from Lebanon in 2000, the UN acknowledged that Israel had officially redeployed to the internationally recognized border, thereby complying with UN Security Council Resolution 425 of 1978; nonetheless, when the Hezbollah terror attacks continued unabated, the UN did not interfere.

The absolute pinnacle of the organization's hypocrisy was reached in 2000. Hezbollah terrorists ambushed an IDF vehicle on the Israeli side of the fence, setting off explosive charges that first ripped through the vehicles and later tore away the international border gate. They then crossed the border and seized three downed soldiers, whom they dragged onto Lebanese soil. All this transpired next to a UN post. According to Kaplan, the point-to-point distance between the UN position there and the site of the kidnapping is only 500–600 yards, with a clear view of the site.

On top of the fact that the UN force, which was charged with stopping attacks of exactly this nature, witnessed the events and did nothing to prevent them or warn Israel of their development, they also failed to reveal that they had the soldiers' belongings in their possession as well as tapes filmed by the UN forces— important pieces of evidence that could help determine the condition of the seized soldiers. As usual, those taken captive by Arab terrorists were denied the right to see the Red Cross or any other international organization.

Israel asked if the UN possessed footage of the attack, but the UN continued to deny its existence for many long months. Even after admitting that UNIFIL personnel had witnessed the attack and the abduction and that they possessed video footage of it, along with some of the clothes and gear of the abducted soldiers, the UN continued to bicker with the Israelis about what would be handed over to Israel and whether the video footage would be censored or not.

That collaboration between the UN and Hezbollah, whether intentional or not, was one of the subjects of my father's conversation with Secretary-General Kofi Annan on August 5, 2001:

"I've called to be updated about the situation," said Annan.

"The situation is complex," my father told him. "Two hours ago a woman was killed near Alfei-Menashe, and four people were wounded seriously. This comes after this morning's attack in Tel Aviv, where ten people were injured by gunfire from a Palestinian terrorist near the gate of the General Staff's base. Since the Dolphinarium, the Palestinians have killed nineteen and wounded over one hundred.

"When an attack is in the works, we will not wait for an explosion. We will not allow the situation to escalate—we seek stability, but we will not pay for it with our lives. Arafat must take the requisite minimal steps. I'd like to take this opportunity to speak about the UN investigation into the captive soldiers in Lebanon. But as the prime minister of all prime ministers, perhaps you'd like to speak first," my father said.

"No, please go ahead."

"We are very concerned and disappointed with UNIFIL's behavior and blunders in the handling of the abduction. We are also concerned about the UN's activity: it seems, and I hope this was unintentional, that the UN delayed and prevented information from reaching Israel. The UN investigation mentions three tapes. We need all of them in order to fully probe the fate of our boys. This humanitarian tragedy is also the responsibility of the UN and UNIFIL because of the gross violation of UN Security Council Resolution 425. From what I understand you have agreed to hand over to us seven out of a total of fifty-three articles of evidence found during the UN investigation. We expect to receive everything, from Nakura [a UNIFIL base in south Lebanon] and from New York."

"I understand that the team you have sent will be arriving in New York tomorrow."

"Correct. I hope for a swift end to this affair so that the UN's credibility can be rehabilitated along with the belief in it and its institutions. I deeply regret this, and I very much appreciate you and your work, but when it comes to Israel, the UN's behavior looks very bad indeed."

"We sent a good person to conduct a good investigation, and I sent a copy of it to you. We will cooperate with the team you sent. We've looked into it and verified that there was no malicious intent on the part of the UN but rather mere carelessness. I hope we will be able to put this behind us and move on."

"We have a very bad feeling about the whole affair, the delays and the attempts at concealment. I am sure this was done without your knowledge, but the picture is very bleak."

"It should not have happened," Annan admitted.

"There should be UN pressure applied to allow a visit and see what has happened to our abducted soldiers."

"We have in fact applied pressure and have tried unsuccessfully, as has the Red Cross."

"We received a response from the Red Cross saying it is a political matter. There is nothing political about it; it is solely a humanitarian matter."

"You are right. I will speak with the head of the Red Cross. They should address this. It is not a political matter."

"Someone has to be sent to see what has befallen them."

"You are right," Annan said.

"Furthermore, the Lebanese army needs to deploy along the border."

"We've pressured them to do so, including in the Security Council, but it has not worked. They say they will only redeploy when there is peace with Israel."

"It has nothing to do with peace. We would like to have peace with Lebanon. They are being pressured by Syria, which wants to join the Security Council and is a whole different matter, a state that hosts and supports extremist terror organizations. Everything must be done to prevent this."

"We agree, and this is what I have reported to the Security Council, but up to this point we have been unsuccessful."

"Pressure must be applied on Syria, which wants entry into the Security Council. It's unclear how that is reconcilable with hosting terror organizations, helping Iran deliver thousands of Katyusha rockets and other arms to Lebanon and aiding Hezbollah. Syria should be pressured on this matter and on the matter of deploying the Lebanese army."

"We have been speaking with many parties, urging restraint."

"We are concerned by the anti-Israel and anti-Semitic trends."

"We are encouraging everyone, including Europe and the United States, to show restraint and to moderate."

"It's important. The Arab League and the Muslim states vote against us, and the same goes for the nonaligned states, so we are in a minority there. Iran, Iraq, Libya, UN members who call for the destruction of Israel, are honored members—we live in a cynical world."

O n August 9, at 1:30 p.m., a terrorist carrying a guitar case packed with explosives walked casually into a Sbarro pizza restaurant in downtown Jerusalem; the place was overflowing with a lunchtime crowd. It was during the summer, which meant that there were more than the usual number of children in the restaurant and on the street. A moment later a huge explosion tore the restaurant apart. Fifteen people who were there to eat Italian food lay dead, many others had massive injuries—the bomb had been packed with screws, bolts, and nails. Over 130 were injured, some critically. The dead included children as young as two. Five members of the Schijveschuurder family were killed, including the parents, a two-year-old son, and two daughters, aged four and eight. The bomber's handler turned out to be on the terrorist list that had been given to Arafat, but which he had, as usual, done nothing about.

My father visiting Zaire in 1981. *(© IDF and Defense Establishment Archive)*

My father giving President George W. Bush a photo taken on a visit to the Western Wall while he was still governor of Texas. *(Courtesy of the White House)*

Me in July 2008 presenting President Bush with a photo taken during my father's visit to the Bush ranch in Crawford, Texas. *(Courtesy of the White House)*

President Bush serving cookies on his ranch, April 2005. *(© Avi Ohayon, Israeli Government Press Office)*

In a military helicopter, November 2005. (© *Ziv Koren*)

FROM LEFT: Mahmoud Abbas, George W. Bush, my father, and King Abdullah at the Red Sea Summit in Aqaba, Jordan, in June 2003. (© *Avi Ohayon, Israeli Government Press Office*)

Tossing flowers on the Mahatma Gandhi memorial in India, September 2003. (© *Avi Ohayon, Israeli Government Press Office*)

My mother in the garden on our farm, 1999. (© *Shaul Golan, Yedioth Aharonoth*)

My father on the farm with our horse Rasta, January 2000. (© *Michael Kremer, Yedioth Aharonoth*)

With my oldest son, Rotem, and me by our house on the farm, 1999. (© *Shaul Golan, Yedioth Aharonoth*)

Rotem with his grandfather, riding matching ATVs. My mother took this photo in 1999.

With our twin sons, Yoav and Uri, in August 2002. (*© Moshe Milner, Israeli Government Press Office*)

With my brother Omri and his oldest daughter, Dania. (© *David Rubinger, Yedioth Aharonoth*)

With his grandchildren in December 2005. FROM LEFT TO RIGHT: Aya, Dania, Uri, Yoav, and Rotem holding Jack. Abigail is not in the photo and Talya was not yet born.

Our four children: Uri, Talya, Rotem, and Yoav, July 2011. *(Courtesy of Anat Levy)*

With my wife, Inbal, 2003. *(Courtesy of Efrat Yoffe)*

The day after the Sbarro attack, the cabinet passed a resolution to shut down the Orient House in Jerusalem. The building, built in the late nineteenth century by the Husseini family, had been serving, in violation of Israeli-Palestinian agreements in Oslo, as a Palestinian governmental institution in Jerusalem. When the Orient House was shut down, a stolen Israeli Uzi was found within. Documents found in the building showed that the Palestinian security services had been operating in Jerusalem, an activity prohibited by the agreements. Among other things, lists were found of Arab residents of Jerusalem who had apparently been apprehended by the Palestinian security services. That type of arrest could include interrogations, beatings, and threats. These activities were all part of Palestinian attempts to claim sovereignty over parts of Jerusalem that were not included in the agreements.

The PA's activities in Orient House had been a continuing sore point—they were in violation of Israeli law and of the agreements with the Palestinians signed on September 28, 1995, in Washington, D.C.

Other Palestinian institutions, such as the Governor's House and the offices of Arafat's Force 17 on the outskirts of Jerusalem, were shut down during the same operation. Weapons explosives and documents were found there, linking these institutions' activities to the Palestinian Authority.

On August 11, my father spoke with Chancellor Gerhard Schröder of Germany. Although he is a friend of Israel, he too began his conversation with the familiar "I am shocked by the circle of violence."

My father replied and reiterated:

"Circle of violence—there is no such thing. There is Palestinian terror and Israeli countermeasures."

"I agree with what you've said. The restraint you've showed in the face of such a fierce and shocking attack is admirable. The question is whether taking over the Orient House is the right course of action within the framework of this restraint."

"Taking over the Orient House is not a political escalation. Israel has decided to enforce the law. That is my duty. They are involved in terror. Those groups sit there and violate Israeli law, are involved in terror in Jerusalem. The steps we've taken are intended to ensure the peace of Jerusalem and its residents. If I have to decide what's preferable, a political or a military escalation, then I prefer the political, even though I would rather have no escalation at all."

TWENTY-FIVE

*My father with Prime Minister Tony Blair at 10 Downing
Street, July 10, 2003. (© Amos Ben Gershom,
Israeli Government Press Office)*

The Anglo-Israeli relationship has had its ups and downs.
The small Jewish community in the land of Israel rejoiced when the British occupied the area in 1917, ousting
the Turks and ending four hundred years of oppressive rule.

In Israel the story of the NILI spy ring—a Jewish network that
provided the Egypt-based British with intelligence regarding the
Ottomans—is well-known. The spies' cover was blown in the fall
of 1917, and those caught by the Turks were subjected to torture
and summarily executed. My father, born more than ten years
later, was raised on this heroic story.

The British presence was seen as a step on the path toward a national home for Jews in the land of Israel, as articulated in the Balfour Declaration, signed in November 1917 by Arthur James Balfour, the British foreign secretary.

Foreign Office,
November 2nd, 1917

Dear Lord Rothschild,
I have much pleasure in conveying to you, on behalf of His Majesty's Government, the following declaration of sympathy with Jewish Zionist aspirations which has been submitted to, and approved by, the cabinet:
"His Majesty's Government view with favour the establishment in Palestine of a national home for the Jewish people, and will use their best endeavours to facilitate the achievement of this object, it being clearly understood that nothing shall be done which may prejudice the civil and religious rights of existing non-Jewish communities in Palestine, or the rights and political status enjoyed by Jews in any other country."
I should be grateful if you would bring this declaration to the knowledge of the Zionist Federation.

Yours sincerely,
Arthur James Balfour

That concise letter was seen as extremely significant, since it was the first time an empire such as Britain had accepted the Zionist vision, an almost two-thousand-year-old dream for generations of Jews. At this time, when the British controlled the land of Israel, the possibility of establishing a national home for the Jews in Israel seemed more than ever possible.

On July 24, 1922, the League of Nations mandated British sovereignty over the land of Israel and charged the country with implementing the Balfour Declaration—founding a national home for the Jewish people in the land of Israel. During their rule, though, terror attacks and the importance they placed on ties to the Arab world eventually caused the British to retreat from that mandate.

That same year, in 1922, the British divided the Mandatory land of Israel into two parts: The land on the east side of the Jordan River, which is much larger than the west side, was designated for the Arabs. The Balfour Declaration applied only to the west side. Thus the Arab insistence on getting all the land according to the pre-'67 borders is ignoring the fact that the entire land had already been divided and most of it (the east side) was apportioned to the Arabs. In addition, they rejected the proposals brought forth in 1937 and 1947 that would have given them more land on the west side of the Jordan.

In the aftermath of three Arab waves of murder and rioting—in 1921, 1929, and 1936–39—the British appointed committees of inquiry, each of which produced a white paper. All told, these three policy papers gradually limited Jewish immigration to the land of Israel and the freedom to settle and purchase land. The third paper, published in 1939, was a total reversal of the mandate the British had received and the promise of the Balfour Declaration. That paper ruled that a single state would be founded in the land of Israel, jointly governed by Arabs and Jews—a complete violation of the vow to form a national Jewish home in Israel.

The recommendations of the third white paper, passed into law in 1940, limited the number of Jewish immigrants, despite the situation in Europe at the time, and placed severe restrictions on Jewish land acquisition, making it nearly impossible for Jews to buy land. Palestinian terror and violence had paid off. The British, in their zeal to appease the Arabs, reneged on their commitments and did not remain true to the mandate they had been given.

A classic example of British cynicism was the story my father used to tell of the way they handled Elis Kadoorie's will. A Jew of Iraqi heritage, he had bequeathed money for the founding of an agricultural school for his people in the land of Israel. The British used the money to found two schools, one at the foot of Mount Tabor for Jews, and one in Tulkarm, for Arabs. That of course was not the Jewish philanthropist's intention, but the British ignored his wishes.

British cynicism extended through World War II. The Jewish population in the land of Israel committed itself to the British side in the war; some 30,000 Israelis volunteered for the British armed forces, out of a community of 470,000. By contrast, the most visible Palestinian leader, Amin al-Husseini, supported the Nazis, was involved in the Rashid Ali rebellion in Iraq against the British and the anti-Jewish riots there, and met with Hitler during the war in 1941. He also established in Croatia the 13th Division, Handschar, of the SS, which at the time included Bosnian Muslims and Albanian Muslim volunteers from Kosovo. Due to its success, another two divisions were formed, the 21st Division, Skanderbeg, and later the 23rd Division, Kama, from among Albanians of Kosovo. Nonetheless, throughout the war the British restricted the number of Jews allowed to enter Israel, calculating that while Jewish support was a given, accommodation with the Arabs could be made through appeasement. A great many could have been spared death during the war, but the British barred Jews from entering Mandatory Palestine, steadfast in their policy of appeasement.

Refugees seeking illegal entry to Palestine were deported. Some of the ships sunk at sea. The story of the *Struma* is especially horrific: this dilapidated vessel, built to sail on the Danube in 1830, was loaded with more than 760 Romanian Jewish immigrants fleeing Europe when its engines failed. The ship was then towed to Istanbul, but the British refused to allow the refugees to enter the land of Israel, even if they came within the allowed annual quota of Jewish immigrants. Instead, they urged the Turks to return the ship to Romania.

In the end, the Turkish navy towed the ship out to sea and abandoned it there, without a working engine. The following day, February 24, 1942, it was torpedoed by a Russian submarine that mistook it, so they claimed, for an enemy craft. The British, and more specifically Sir Harold MacMichael, the high commissioner for Palestine, were widely seen as being responsible for the horrific and unnecessary deaths of the hundreds of Jews on board.

After the war, the British adhered to the 1939 white paper policies. Skeletal survivors who had managed to endure the horrors of the Holocaust were barred from entering the land of Israel, a policy brought to the fore by the *Exodus*, a ship loaded with 4,500 Holocaust survivors who were turned away and sent back to German soil, where they were kept in an internment camp for an entire year.

Although the British largely ignored the crimes and murders the Arabs committed against Jews, they didn't allow ordinary Jews to bear arms. As a child in the village, my father was charged with hiding the old Mauser rifle, kept as a defense against marauding Arab gangs, in a wooden box in the garbage pit in the dairy; and as a youth in the Haganah he, like the rest of the volunteers, was well trained in concealing weapons and disguising military drills. His squad leader course was held on Kibbutz Ruhama. It was chosen as the locale for the course because it was distant and desolate enough that if a British force approached, the dust clouds would be visible from afar, providing ample time to hide the weapons and slip into one of the wadis surrounding the kibbutz.

My father did not attend the high commissioner's cocktail parties, although those affairs, attended by the British High Commission and high-ranking officers, with the wealthiest people in the country, were not far from Kfar Malal. They were, however, far removed from his life in the village. His exposure to British officers was through the window of his Tel Aviv school on Geulah

Street, where he saw them frequenting a hotel that housed the city's prostitutes.

That was one of the ways my father experienced the British Mandate.

———————

On November 29, 1947, when the UN General Assembly voted on the partition plan, Britain abstained. Even when it had been decided that they would be leaving the area within six months, they still restricted Jewish immigration, hindered Jewish residents from using their defensive forces, froze the Jewish Agency's assets in Britain, and turned a blind eye toward Arab military activity.

On April 13, 1948, Arab forces ambushed a convoy of doctors, nurses, and patients as they made their way to Hadassah Hospital on Jerusalem's Mount Scopus, killing a total of seventy-nine people. Not only did the British, who were responsible for the safety of the convoy, witness the attack and not help, but they actually prevented Haganah reinforcements from reaching the scene. Earlier, in the 1920s, the British formed the Arab Legion, which eventually was to evolve into Jordan's regular army. Known to be the best of the Arab fighting forces during the War of Independence, it was commanded by British officers. During the War of Independence the legionnaires fought against the state, and it was they who blocked the road to Jerusalem from Latrun. On May 13, 1948, one day before the Declaration of Independence, the legionnaires, along with local bands, took the Etzion Bloc (Kfar Etzion, Ein Tzurim, Mesuot Yitzhak, and Revadim), and close to 130 defenders were massacred after they had laid down their arms and surrendered. Once the legionnaires had taken the Old City, they expelled the Jews from the area, they detonated synagogues, and they destroyed the Jewish Quarter.

While attending the British army's staff college at Camberley in 1957–58, my father met British officers who had commanded troops in the Arab Legion. They were interested in the details of battles their troops had participated in, and proud of the units they had led.

Personally, my father had a wonderful time in England. He found his year at Camberley, southwest of London, to be fascinating. Arriving in the fall of 1957, he spent his weekdays on campus, soaking up British culture, customs, and mentality, and meeting other officers from a host of foreign armies. Those were the guidelines he'd been given before leaving by Chief of Staff Moshe Dayan. Margalit and Gur lived in London and saw him on the weekends.

He also spent time with his close friend Cyril Kern, who at the age of seventeen came to Israel and volunteered to fight in the Alexandroni Brigade's reconnaissance unit, and has been a close friend ever since. They would go to the theater, museums, and restaurants and take long hikes outside London. Everything about life abroad was different from his childhood in the village and his time in uniform. On campus, a batman was assigned to him who had held the same position for twenty-seven years, and was still considered young and inexperienced. The oldest batman was eighty-five, and he remembered well the officers he had served during the Second Boer War (1899–1902). My father's batman was curious and quite surprised to find that my father always responded in the affirmative to his daily question: Would he be interested in a bath tonight?

I remember well my father telling us about the Thursday blacktie dinners and the British gentlemen in their tuxedos, standing in the lounge with their backs to the fireplace in order to warm their backsides. He also shared his somewhat embarrassing interaction with the queen. Her Majesty had come to celebrate the venerable college's hundredth anniversary, and despite a considerable amount of preparation, my father strayed from protocol. The expected response to her "How do you do?" was "Thank you, ma'am." My father, shocking the officers around him and perhaps the queen herself, tacked on an additional "How do you do, ma'am?"

During his time there he wrote a letter to Basil Liddell Hart, the famous British military historian and strategist, who agreed to meet him to discuss my father's thesis topic, "Command Interference in Tactical Battlefield Decisions: British and German Approaches." Throughout his life, the subject interested my father a great deal. At one of the festive Thursday dinners the matter of Liddell Hart's ethnicity was raised, and when my father said Liddell Hart was not Jewish, the officers asked, "Well, how else would you get to see him?" His explanations about the letter he sent and the invitation he received were greeted with polite British incredulity; it seemed far more reasonable to them that the invitation was due to a global Jewish network. My father and Liddell Hart remained friends.

⸻

"If you're invited to dinner with the queen, you'd better know your table manners," our parents would say. I also use the good old queen to try to improve my children's behavior. "What would happen if you ate like this next to the queen?" I say. "She would faint," or "Imagine waving your fork like that with the queen sitting next to you—you'd pull the gold and diamond earrings right out of her ear." And so we have found ourselves for the second generation (if not more) employing the queen to shape the manners of our children.

My father held British steadfastness in great esteem, along with the nation's courage and composure, as witnessed by the brave determination of British civilians under German bombardment and the evacuation of Dunkirk early in World War II. He appreciated the way the British never relinquished their fighting spirit or their tenacious grasp on their heritage. But insofar as we are concerned, British negativism has never subsided. More than sixty years after the British Mandate, academic groups and others try to exclude Israelis from their ranks and have boycotted Israeli universities in the name of liberalism and the free pursuit of knowledge. Iran and Syria, meanwhile, totalitarian states where human rights are trampled without any hope of judicial recourse, states that actively support terror, are never called to task. The murder of Israeli citizens is smiled upon by students and faculty alike—a shrine to terrorism was erected in September 2001 at An-Najah National University in Nablus, where, without any protest from the Palestinian Authority, one could blithely reexperience the Sbarro bombing in Jerusalem, with re-created slices of pizza and strips of flesh on display.

One positive note in Israel's relations with Britain was the prime minister at the time, Tony Blair. Although he did not always agree with my father, his attitude was always practical, and his understanding of Israel's terror-plagued predicament only deepened after 9/11 and the subsequent attacks in London.

On March 8, 2010, in Jerusalem, Blair talked to me about my father:

"He was, as you know, very straight-talking. Very strong—great presence. At first, to be honest, I found him quite difficult because he is so, you know, he is almost fierce; he didn't get into the diplomacy much. He was just straight to the point—and nailed it. But over time I came to have a huge respect for him and liking for him, and I thought he was very courageous and actually quite a visionary leader.

". . . I was saying that despite the Intifada, despite all the pressure, we have to get to a point where some people can see some prospect for peace. And he was saying to me, Look, as long as terrorism is still happening there is no chance for peace, and I was trying to say to him that we have to offer a way out of this impasse we're in. And I think that at that time it was hard for people from the outside to understand what it was like in Israel at that moment. So one thing that I understand now far better than I did then was the impact the Second Intifada has had on the whole psychology of the Israeli people toward the peace process.

". . . He articulated the position that I think is still the basic, the sort of center ground, of Israeli politics today—which is what I call the tough road to peace. In other words, he understood the need for peace; he understood that it had to be based on a fair solution between Palestinians and Israelis, but he felt that after the Second Intifada the only way this peace could be achieved was through a very tough position from Israel. You couldn't just rely on the good faith of your partner—that good faith had to be explored and proved. I think that was how I defined his position.

". . . He had a very simple lifestyle, I think, but he was a very, very determined man, and he did actually have a vision for peace that can still lead to peace."

My father spoke with the prime minister on August 17, 2001, when Blair called to inquire about the closing of the Palestinian institutions in Jerusalem:

"Hello, Ariel."

"Are you in the Caribbean?" my father asked.

"I've come back here."

"I am jealous of you. Everyone is on vacation, and I am involved in commando operations against terror."

"I don't understand, Ariel. What is your intention with the Orient House?"

"Previous Israeli governments have neglected this matter. The Palestinian Authority broke Israeli law. They are not allowed to do what they are doing. Not in Jerusalem, and not in Abu Dis, where Palestinian organizations work under the control of Arafat. They are the most active perpetrators of terror. The Fatah and the Tanzim took action against Palestinians living in East Jerusalem. They threatened and tortured them. They started to control part of Jerusalem. Arafat increased the volume of terror in Jerusalem. We won't allow them to violate Israeli law and the Oslo Accords. They cannot maintain military and police forces there. I had to decide what to do after the terror attack in Jerusalem. It was an awful incident. A family was killed—a mother, a father, and three children, leaving four children behind. And a mother and her daughter were killed. The terror continues, and Arafat has not taken a single measure. Arafat could have prevented this. We gave him the name of the person behind this operation, but he did not arrest him. We told him they planned to carry out terror attacks. Only once we ousted them from the Orient House and from Abu Dis did he take measures and make arrests. I sent Arafat clear messages to stop the terror. I spoke with President Bush, and we discussed this. I want to avoid taking measures, but I am committed to the security of the citizens and I am doing just what any other country would do."

"I understand the nature of the terror attacks. I am not surprised by the public's reaction. People are agitated and angry," said Blair.

"I went to visit the four children. On Sunday I will go visit the girl in the hospital. It is a terrible tragedy."

"I totally understand it."

"In addition to terror, in a few years there will be a new threat to regional stability, and that is weapons of mass destruction. In the meanwhile Arab terror represents a threat to the Middle East. The terror stretches from Afghanistan to Lebanon, which has become a local and regional center for terror. Terror cannot be rewarded, and it must be combated. We appreciate your understanding."

"I agree with you."

"Tony, are you still on vacation?"

"Yes, I am in a rural area."

"I also want a few days of vacation. I'll go to the farm. I miss the horns of the rams."

Sbarro was an awful terror attack, but far worse was yet to come.

*With New York mayor Rudy Giuliani at
Ground Zero, November 2001. (© Amos Ben
Gershom, Israeli Government Press Office)*

TWENTY-SIX

The events of September 11, 2001, altered global priorities and, for some, sharpened the distinction between good and evil. A sudden understanding of Israel's predicament descended on the United States and Europe. What had seemed like a local problem—an issue linked to a conflict being played out in some far-off corner of the Middle East—landed on the doorsteps of those who had demanded restraint, proportionality, and other ideas about how to handle terror when it occurs elsewhere.

For Israel, insofar as its public diplomacy was concerned, life became easier: suddenly everyone in the West grasped the meaning of terror and the fact that it could strike them and their families. Moreover, those who had carried out these attacks were not linked to the conflict with Israel. They were not refugees; the usual "justifications" did not apply, and so Israel could not be blamed. The perpetrators of these attacks were mostly linked to Saudi Arabia and the global jihad, the war against those who renounce Allah. "Live as we say, or die"—that, in essence, is what the terrorists say to each and every person in the Western world.

Throughout the Palestinian Authority's realm, on 9/11 the public rejoiced. The celebrations accurately charted the mood of the Palestinians. They were elated by the murder of thousands of civilians, people who had never done them harm; on the contrary, the victims' tax dollars had been funneled into the PA's and aid organizations' coffers. The footage of these spontaneous expressions of joy embarrassed Arafat and his men, and they tried to keep them off the air, to disown them, but they could not. Israel grieved as though the attack had taken place on its own soil.

Palestinians celebrating on 9/11.
(BBC News, September 12, 2001, http://news.bbc.co.uk)

One of my father's foremost characteristics is his ability to grasp new developments, to take in the full picture amid the turbulence of complex situations. He did this on the field of battle as well as in the world of diplomacy. This new world order, he recognized, presented some difficulties for Israel. He believed that it gave rise to the notion that terror attacks upon Israel were a thing apart from attacks in the rest of the world.

To rebut this notion, he took pains to tell every politician and journalist he spoke with that there was no distinction between terror directed at Israel, which was deemed partially legitimate, and the terror visited on the United States, which was widely seen as illegitimate. Time after time he underscored the fact that terror is terror, and murder is murder.

The second matter that concerned him was the expectation that the Americans, in order to build a coalition against terror, would want to enlist the cooperation of the Arab states. He knew that the Western powers would want the coalition to be universal, representing not the Christian West against the Muslim world but a diverse band of states against only those engaged in the terror trade. That Arab cooperation, he knew, would come at a price.

My father wasn't willing to allow Israel to pay that price. He made that point in his dealings with world leaders and also expressed it in what became known as the Czechoslovakia speech. In response to a reporter's question after the events of 9/11, he said that Israel would not be Czechoslovakia, implying that Israel would not allow itself to be sacrificed by the West to appease the Arab world, as the West had sacrificed Czechoslovakia to appease Germany in 1938. The parallel offended the Americans. That was certainly not my father's intent, but his message was clear.

Even in the aftermath of the September 11 attacks, Israel's actions to defend itself were being questioned. On September 24 my father met with Jean Vardin, the French foreign minister, in Jerusalem. Vardin expressed his concerns about the "measures" Israel used in its war against terror. "If we prove that Arafat is talking from both sides of his mouth," my father asked him, "what will your position be? Will you then support our stance on the need for military measures?"

The coming months would reveal France's position.

––––––––––––

My father's worry about the price Israel would be asked to pay came to the surface during a conversation with Prime Minister Tony Blair on September 25:

"Hello, Ariel."

"Hello, Tony, how are you?"

"I'm fine—much like you, very busy."

"We admire your stance on terror and your support of the United States. Please accept our condolences for the British nationals killed in New York. We value our relations with Britain and with you personally. We have been facing terror for years. It is the most dangerous and the most immediate problem facing the free world and stability—alongside the threat of weapons of mass destruction."

"I completely agree, and I thank you for the condolences," Blair said. "We are most explicitly with the United States for the long haul, and we would also like to work with you to make progress on this matter so that something good will come of the evil done."

"We've offered all possible assistance: intelligence, antiterror, airports, and more. You and the United States surely have the best antiterror units, but we also have many years of experience, and we will help in whatever is needed, without any conditions. The dangers of terror, which we have faced for years, are clear to us.

"We were disappointed to find that Hezbollah, Hamas, and Islamic Jihad were not included in the [EU] list of terror organizations. All of them have ties to the most extreme Islamic movement. Hezbollah and Iran were involved in the terror attacks in Argentina, against our embassy and the Jewish community, attacks that claimed hundreds of lives. There is no ranking of terror . . . There is no good or bad terror, and all those who believe in democracy and liberal values hope that all the terror organizations are included in the list.

"I ordered the cease-fire activities but since then there have been one hundred and twelve terror incidents, with nine wounded, some critically, and two dead, one of whom was a mother of three with a month-old infant who is used to nursing and refuses to take milk. The man responsible for the murder is free. Even though Peres asked Arafat to arrest him, he was detained for a few hours and then released. They are broadcasting on the radio that they don't want to kill civilians, only soldiers. We will not accept that as a cease-fire. The terror continues, and the pressure on him must continue. Peres will meet him tomorrow and maybe convince him . . .

"Again," my father concluded, "keep in mind the support and assistance we're offering you and the United States."

Blair seemed to be in agreement. "One of the messages [British foreign minister Jack] Straw passed on to Tehran and one we will be following is Hezbollah's unacceptable activity. States must decide, are they part of the campaign or do they continue to support terror? I'm aware of the massive internal pressure that is brought to bear when terror is inflicted on us, and that is just a fraction of what you face. I understand you fully, and the tragedy of murdered civilians is awful. Our aim is to ensure that via this tragedy we construct a process that combats terror in a way that guarantees all of our futures. I know this will be very difficult for you, and my advice to you is, if it's possible to have the Peres-Arafat meeting, have it."

"They will be meeting tomorrow at nine thirty in the morning," my father said, "and I will brief Peres beforehand. I'll meet with him at the farm before he departs for Gaza."

"Thank you," Blair said, "for agreeing to meet with Straw. I am aware of the problem, and I want you to know that I will always be a good friend to Israel, as will Straw."

"We know and appreciate that," my father said. "There is outrage and disappointment over Straw's statements."

Earlier that week, while visiting Tehran, Straw had said, "One of the factors that helps breed terror is the anger that many people feel in the region at events over the years in the Palestinian Territories." He certainly remained loyal to the traditional British cynicism by blaming Israel for the fact that Saudis who may have never seen an Israeli in their lives had decided to murder thousands of Americans. He made that statement while he was visiting Iran, a country that generates terror attacks the world over. Upon learning of his remarks, my father had canceled his scheduled meeting with Straw. But after a personal request from Blair, he had relented.

"It was a terrible mistake to link Israel to Bin Laden as though it was America's support for Israel that caused the terror attacks," my father continued. "We should not have to foot the bill for that. The Arab states will demand that Israel pay with concessions. As I told you when we met, we will make painful concessions for peace, but we will not compromise the security of Israel and its citizens. I hope that when the Arabs come up after the meeting and demand concessions as a condition of support for the coalition, you will refuse; otherwise it will put us in a very difficult situation."

"Of course."

<hr />

Although Shimon Peres and my father were not always in agreement, my father considered Peres an important member of his government. For years, dating back to the 1980s, my father had been behind the establishment of unity governments. He always believed in the unity of the people. There were enough haters around us and enough problems that required "a broad national consensus," a phrase he used often. But although he felt it was important to have Peres in the government, maintaining good relations with him required a special touch. If I had to summarize my father's approach toward Peres as a minister in his cabinet, I would say it was one of stroking and neutralizing.

Peres is a congenial man. You can drink vodka with him and enjoy his company, as we did on the farm. He knows how to appreciate a woman's beauty and a good meal. Long years of visits in Paris strengthened these abilities. I remember one of his visits to the farm, late one night when he was a minister in my father's government. Toward the end of their meeting, I pulled a bottle out of the freezer and took some smoked fish from the refrigerator. The atmosphere was good until the discussion turned toward his plans for peace with the Palestinians—those steps he'd already had the chance to enact and those, thank God, he did not. Although Peres was logical and efficient and a person one could work with, when things came to his contacts with the Palestinians and his peace plans, he simply lost all perspective.

I contended that the whole intent of Oslo had been to solve our differences peacefully, in conversation, and what we received in return was terror. The problem, I said, was that our differences were not territorial, not of a nature where the sticking points could be smoothed over and the conflict resolved— rather, they were rooted in the Palestinians' fundamental unwillingness to accept our basic rights to exist. We'd both had a few drinks, and as the conversation became heated, it is possible that my tone of voice went beyond what was called for. Peres's face had grown red with drink and the strain of conversation. At a certain point my father stepped in to defend Peres. He scolded me for my manner of speaking, but I could see in his face and hear in his voice that he was rather pleased with what I had said.

Personally, my father and Peres had been on good terms for years, and Peres's late wife, Sonia, had always liked my father. Moreover, Peres's international standing certainly helped Israel explain itself to the rest of the world. But along with those virtues came certain difficulties. Peres displayed an unstoppable zeal for meeting with Arafat and other Palestinians and devising plans and agreements. Even when it had become clear to one and all, including Peres, that Arafat was involved in terror, he did not let up. It was as though he had to prove that he had not been mistaken in Oslo. His attempts to conduct political meetings at a time when the government had decided that there would be no such negotiations under terror hamstrung efforts to bring international pressure to bear on the Palestinians. This was where the attempts to neutralize his actions came in. But each time Peres was prevented from meeting with Arafat, his face would fall, like that of a child who'd had his favorite toy taken away. That long, sad face would grow even sadder.

My father had neither the patience nor the desire to engage in mollifying Peres; it was my brother Omri who would urge my father to invite Peres over to the farm for breakfast or a late-night meeting. He'd come in bristling and leave soothed. My father knew such invitations were important and would admit as much afterward, and with his charm he was always able to make things right again—until the next time.

In any case, Peres remained an active and often helpful participant in the government—hence Blair's eagerness to have him keep up the dialogue with Arafat.

With Shimon Peres, minister of foreign affairs, during my father's first term as prime minister, October 2001. (© Amos Ben Gershom, Israeli Government Press Office)

My father was able to find moments of satisfaction, despite the daily struggle with terror. Although tourism suffered, there were those who remained loyal to Israel. Among these were the Christian pilgrims who continued to visit. My father spoke to such a group—the International Christian Embassy of Jerusalem, who each year come to Israel to celebrate the "Feast of Tabernacles." In a speech on October 2, 2001, he said, "The land of Israel and Jerusalem are holy for the Jews, the Christians, and the Muslims. But it is promised only to the Jews. It is true that we speak of Judea and Samaria from a security point of view, but [the real significance] is the historical right of the land. Judea and Samaria are the cradle of the Jewish people. Israel was willing to give up parts of its historical homeland for peace although it won all its wars. There is no other country like that in the world."

With Rehavam "Gandhi" Ze'evi, during the Sinai War, October 31, 1956. (© Avraham Vered, "BaMahane," courtesy of IDF and Defense Establishment Archive)

On the morning of October 17, 2001, Rehavam "Gandhi" Ze'evi, the minister of tourism in my father's government, was murdered in the Hyatt hotel in Jerusalem. He and my father had known each other for about fifty years, since their days in uniform together. The murderers fled to Ramallah and hid in Arafat's headquarters.

For my father, that was the final straw: he was done with Arafat. All he needed was to make sure that the world also understood with what and with whom they were dealing.

During that period most of my father's conversations with dignitaries traced a similar arc: they began with the condolences often expressed after a brutal terror attack, and progressed to suggestions of steps we should take so that the Palestinians would stop killing our citizens. Following that pattern, the day after Ze'evi's murder the calls started to pour in from all over the world.

One of those calls came from Kofi Annan:

"Allow me to begin by expressing my sympathy and condolences for the death of Minister Gandhi," he said. "We were all shocked by this. We've sent a message to Arafat via Terje Rød-Larsen [the UN secretary-general's personal envoy to the Palestinian Authority] to arrest those responsible. I understand some of them have been arrested. We hope the violence will stop . . ."

"Thank you for the condolences," my father said. "An hour ago a man was killed and a woman was gravely injured."

"I haven't heard about that," said Annan. "Just three days ago it seemed like things were heading in the right direction."

"On the ground nothing has happened," my father said. "Arafat took some steps, but not enough. What's happened, and I regret this, is that Terje Rød-Larsen has become Arafat's spokesman. It has been published that Arafat regrets and condemns the attack, but no one has heard it from him, in his voice. Instead it's been the voice of Terje Rød-Larsen. Rød-Larsen has become Arafat's spokesman. I don't think that is his job."

"Terje went to him to explain to him how serious the situation is and that he must take action."

"He went to explain to him," my father acknowledged, "but in actuality he delivered his message."

"Arafat himself did not speak?"

"No, he did not speak. Arafat has sixty thousand armed men under his control. The Jihad are just a few, and in terms of the Hamas we're dealing with two to three hundred men. He can [make the arrests]. In the past I have advised him that if he cannot do it in one sweep, then he could do it in stages, and we would assist him.

"We opened the road to Ramallah three days ago, and immediately Minister Ze'evi was murdered. It happened because the road was open and we didn't conduct inspections . . . Arafat has to be put under immense pressure."

———

C hancellor Schröder called on October 19:
"I understand that Mr. Ze'evi was a personal friend of yours," he said. "Please pass my condolences on to the family."

"I will do that," my father said. "The terror has not stopped. There are terror-related activities every day. The problem is that Arafat has never arrested them seriously. He has only arrested them in what he called a *jail*. They were kept in apartments, where they could continue to plot and plan terror attacks. They told those who were arrested that they were arresting them to protect them.

"The headquarters of the Popular Front, the organization that murdered the minister, are located right opposite Arafat's offices in Ramallah. We knew that an attack was on the way, but they took no measures . . . There were things that he could have done.

"Where he wants quiet, there is quiet. I have ordered the IDF not to act in Hebron because it has been quiet. Up until yesterday there was also quiet in Jerusalem in the neighborhood of Gilo. Yesterday suddenly gunfire and mortars were directed at Gilo. It's a civilian neighborhood. I had to go in [to Bethlehem Beit Jallah]. I said that if they started the fire again, we would be forced to go in again. We don't want to be there, but we have to protect our citizens. We will protect ourselves just as Germany would. There will be no escalation—I am exercising caution, but we will take all measures necessary to protect our citizens. Yesterday was the minister's funeral. Today a boy who was shot while hiking will be buried.

"I know that the world is busy and we wish the war on terror success," my father said. "We also have terror. Arafat hosts terror organizations, allows them to train, and therefore his status is that of a terrorist. We support the coalition and the campaign against terror, but we will not foot the bill of the coalition."

"I completely understand," said Schröder. "We hope that you carry on with the policy of restraint as much as possible. An agreement must be reached. I am willing to clearly present the matter to Arafat. Minister Fischer will meet with Arafat on October 25 and will discuss the matter with him. There must be an end to the violence."

M y father viewed the world's involvement in Israel's affairs with a hint of irony. He knew well that some of these world leaders were well-meaning, but others wanted to improve and strengthen their country's status using the Israeli-Palestinian situation. And some wanted to improve their own status and appear to be global peacemakers. Often it was a combination of both. Thus, my father had to deal with many leaders offering him advice.

It is hard to fathom a greater divide in mentality than the one between Scandinavians and Middle Easterners. In the culture of the former, it is accepted that a person means what he says. In the latter, as my father had to explain to the prime minister of Norway, Kjell Magne Bondevik, just because somebody says he will do something, it does not mean that it will ever get done—particularly in Arafat's case.

On October 20, Bondevik called my father:

"Prime Minister Bondevik here. How are you?"

"Thank you, how are you?"

"Are you aware of the change of government in Norway?"

"We follow the developments over there just as you follow the developments over here," my father answered.

The irony was evidently lost on the Norwegian. "I've established a government consisting of three parties," Bondevik continued. "I stand at the head of the Christian Democratic Party."

Even though my father may not have been an avid follower of Norwegian politics, he always came prepared to meetings and phone conversations. "You were prime minister from 1997 to 2000. I am glad you are back in office. We appreciate your friendship and the friendship of Norway. We met in September 2000."

"I remember the meeting. We met at your party headquarters. We were both heads of the opposition, and now we are both prime ministers," Bondevik observed.

"Better to be prime ministers," my father said, "than heads of the opposition."

"Easier, though, to be head of the opposition," admitted Bondevik.

"Easier, but many would like to be prime minister," countered my father.

"You are going through a tough time with the Palestinians," Bondevik said. "We were shocked by the murder of Minister Ze'evi. Norway condemns terror and the murder in the most forceful terms . . . I spoke with Arafat two days ago and demanded that he act to put an end to the violence and the attacks against Israel. I asked him to arrest those behind the murder . . . He told me he has arrested seventy-three people."

My father laughed. "He has not arrested a single person involved in the killings."

That same day, my father spoke to Silvio Berlusconi, the prime minister of Italy, and updated him on events in Israel. My father liked his straightforwardness regarding foreign affairs. Berlusconi says what he thinks. He did not close his eyes to Palestinian terror and did not ignore the Iranian threat. They had a good relationship and once he even invited my father to his summer house.

"Come on, Dad, go," I said. "What can be bad? I'm sure he has a fabulous home in Sardinia and serves wonderful Italian food." We didn't know back then about the parties that were thrown there. I am sure he would have been happy to go but the endless war against terror did not give him a moment of rest.

On October 27, 2001, Norway's prime minister called again. At this point, Arafat was still a world traveler meeting foreign leaders. He had met with Bondevik in Oslo. Bondevik updated my father on the substance of their meeting:

"Good evening. I am happy to hear your voice again and to continue our conversation from last week," Bondevik said. During their previous conversation, Bondevik had called for Israel to withdraw from Area A, and for restraint. "I understand you had a very busy week."

"Yes, a very busy week," my father said. "In terms of the issues you raised: the evacuation of Area A—we've withdrawn from Hebron, Beit Jallah, Bethlehem, and Jericho. I would like to withdraw our troops from the remainder of the territory, but [the Palestinians] must take measures that they have yet to take: responsibility for the preservation of the quiet. In the areas where they take action, we will leave, and I do not want to go back in."

"I tried to pressure him," Bondevik said. ". . . I strongly urged him to bring Ze'evi's killers to trial, to arrest known terrorists, and to renew his commitment to the cease-fire that has been declared."

"Declared but never implemented!"

Arafat was frightened about the fallout from 9/11, and thus on September 15, he published an open letter to the citizens of Israel in which he wrote, "I order the Palestinians to obey my previous order to cease-fire. I gave strict orders with a full commitment to a cease-fire." But it was worthless.

"He reiterated his commitment to a one hundred percent effort to combat terror and said that he is making a one hundred percent effort. I know that that is doubtful—but that is what he said."

"Thank you for your efforts and your assistance towards peace," my father said. "We look forward to peace negotiations, but only once the incitement, the terror, and the violence has stopped. Our demands of Arafat are not for one hundred percent effort but for results. Effort is difficult to measure—results are known. It does not matter what he says, but what he does."

"I understand entirely," said Bondevik. ". . . I understand that Peres and Arafat met today in Majorca at the international aid forum. What was the result?"

"As far as I know, they met for lunch. I have no further information at the moment."

"He has a mandate for talks with Arafat?"

"No. In my opinion that would be a big mistake and would encourage [Arafat] to [commit more] terror. Every meeting with him legitimizes him and gives him the option not to take measures. We are interested in taking part in international aid forums such as Majorca, and they can, as a matter of etiquette, sit around the same table, but there will be no talks . . . First there must be absolute quiet."

But instead of calm, a string of horrific terror attacks occurred. On November 29, three people were killed and nine wounded when a suicide bomber blew himself up on a bus traveling from Nazareth to Tel Aviv. That Saturday night, two terrorists blew themselves up on Jerusalem's teeming Ben Yehuda Promenade, an event that was followed by a third, coordinated blast from a booby-trapped car on the nearby Rav Kook Street. Ten youngsters were killed, and over 170 were injured. At the time my father was in the United States. He moved up the date of his meeting with President Bush and hurried back to Israel.

The day after he returned, my father convened the government, which declared at the close of their meeting that the PA was a "terror-supporting entity." The Tanzim and Force 17, two forces controlled by Arafat, were both declared terror organizations. The Ministerial Committee for Defense was given greater leeway in the fight against terror. Shimon Peres, along with ministers from the Labor Party, threatened to leave the government. It was an empty threat.

On the third of December, a terrorist blew himself up on the Number 16 bus on Giborim Street in Haifa, killing fifteen people and injuring forty.

W e are concerned about what is happening in Israel," remarked Prime Minister Bülent Ecevit of Turkey in a phone call on December 4.

At that time, relations between Israel and Turkey were good. There was close cooperation between the two countries in many spheres, including the military.

When he was young, my father had traveled through Turkey. He had visited his uncle Solomon, Vera's Istanbul-based brother, and toured the region on his own. As he was getting off a bus to go for a hike in the mountains, the other passengers— locals who were worried about the tourist—tried to stop him from setting off alone. Undaunted, he continued on his journey. The experience he had in the countryside—especially when he happened upon a grove of fig trees in eastern Turkey with unforgettably ripe fruit—was so powerful that even as prime minister, every time he met a Turkish representative, he had to recount the story. And in fact, the dealings between my father and the leaders of Turkey were characterized by warmth and excellent relations.

Turkey is eager to be involved in negotiations between Israel and Palestinians and the Syrians. As part of its perception of itself as a regional power, it strives for influence.

My father was never willing to accept disrespectful behavior toward himself or Israel, greeting such behavior with a sharp and swift reaction. I have no idea what Prime Minister Ecevit was used to or what he expected, but I am sure that my father's reaction to his discourteous manner of speaking surprised him:

"We share your pain at the violent events," Ecevit said that day, "but we are concerned about your calls for war and the attack on Arafat's offices."

"Thank you for calling," said my father. "We have many who have been hit by terror, by the coalition of terror that Arafat has formed. Fifty percent of the attacks are executed by Arafat's Presidential Guard, his security personnel, the Fatah, and the Tanzim."

"He arrested one hundred people."

"You know exactly what it means to arrest someone. They are not under arrest. He carried out very few arrests, and the majority of those arrested have nothing to do with the terror attacks. In addition, we did not declare war against Arafat."

"We listened to your press conference, and you used the word *war* several times." The tone of the conversation was surprising, especially coming from Ecevit, who had been prime minister during the Turkish invasion of Cyprus in 1974, and in light of Turkish actions against the Kurds who seek independence in the eastern part of the country. And of course there was the slaughter of the Armenians in the early twentieth century.

"I spoke in Hebrew. I don't know what type of reports you received. We declared that the Palestinian Authority is harboring and providing shelter for terror organizations and allowing them to act."

"The headquarters of the Palestinian Authority were bombed."

At this point in the conversation, my father's patience ran dry. "That is just the beginning, and if Arafat does not do something, it will continue. We are very familiar with your history, and we know what it is you did when you were attacked. We are attacked every day. We have had terror here for over one hundred years."

"We understand your position," said Ecevit.

". . . Allow me to suggest something, as I did with the United States and some of the European states," my father said. "I sent the head of the security apparatus to them. If you have the time, I'll send him to you, and he can update you with the details."

"You do not think the dialogue should be started anew?" suggested Ecevit. "There was supposed to be a meeting in Norway."

"I didn't hear about it," my father said.

"I spoke with Arafat yesterday, and he told me that on the eighth and ninth of the month there would be a meeting in Oslo."

"I haven't heard of it. Arafat is a pathological liar. He looks you in the eye and lies . . ."

"You do not think that reinitiating the dialogue would be useful? I am willing to put Ankara or Istanbul at your disposal for the dialogue," said Ecevit.

"I'd like to do that in Turkey, but only when the time is right."

"Would you like us to start coordinating the meeting?"

"We will not negotiate under terror . . . When the terror stops, we will conduct negotiations," my father said. "I am acting just as you would act. Just as you acted in Cyprus to protect your civilians, as you acted against the Kurds. We have the full right to protect our civilians."

". . . I hope that your people want peace," said Ecevit. "It is hard to achieve full peace. Negotiations must start immediately."

"The dialogue will begin only when there is absolute quiet."

"That is a very harsh stance."

"Why?" said my father.

"Because it will be impossible to implement," said Ecevit. "If there is absolute quiet, there will be no need for dialogue."

"The steps we have taken are necessary. We have not declared war. There will be dialogue for a cease-fire, nothing more," my father said. "When there's quiet, we will start political negotiations. I have my doubts about Arafat's intentions."

On December 5, my father spoke yet again with Bondevik. In Norway apparently all was tranquil. The irrepressible Bondevik continued to express his interest in what was happening in Israel.

This conversation again began with an expression of sorrow for the loss of life due to terror and swiftly switched to the offering of advice. "I spoke with Arafat over the phone," said Bondevik, "and I told him that we condemn terror, that it is not acceptable anywhere in the world."

After thanking him for his words, my father explained that Arafat was not taking measures against terror. He then detailed the facts of a suicide bombing earlier that day and stressed a point:

"He has to feel pressure. He has to feel that he is isolated."

"I feel that he is under pressure. He told me he is serious about the war on terror. He has arrested 120 people, and his forces are fighting terror . . . He told us that you attacked a position seventy yards from his headquarters. He made clear that he would like to do more, but that he cannot because of your attacks and his isolation."

". . . The problem," my father said, "is that we who live in democratic regimes find it hard to digest [the fact] that someone looks you in the eye and lies. He lies. We and the Americans know that true arrests have not been made . . ."

"It is important to avoid violence and an escalation. Arafat must be tested," replied Bondevik.

"I am willing to test him, but not under his fire. There is something I told you in the past: for true peace I am willing to make painful concessions. There will be no concessions on the lives of our civilians or on something that could put the state in danger."

I admit that on more than one occasion I lost patience when I heard my father explaining the same things over and over to leaders across the world. I am filled with admiration but also with sadness when I think of the endless explaining he had to do.

Once more my father repeated the demands he had put before Arafat as conditions for political negotiations:

"There are many people who have suffered from Arab and Pal-estinian terror over the last one hundred years. I am capable of looking those people in the eyes and telling them that they must make painful concessions," he said. "I don't see anyone else capable of doing that. Arafat's missing [an opportunity], and maybe we will wait for someone else, [someone] pragmatic."

"Who is the alternative?" Bondevik asked. "Arafat understands that the hour of truth has come."

"It has to be in deeds and not in words," my father said.

"He told me he has arrested the leadership of the Islamic Jihad," said Bondevik, "but that it is difficult for him to convene his cabinet because of the attacks."

". . . All of his officers are with him in Ramallah," my father said, "he has no problem . . . There has been quiet; we haven't done anything for thirty-six hours. We haven't been taking action because of the rough weather. It is raining hard, and I am pleased about it. As a farmer, I mean."

"You have not attacked Palestinian positions in the last thirty-six hours?"

"The last time was yesterday at eleven."

"I hope that continues over the coming days."

"If they carry on, we will respond," my father said. "We will protect ourselves."

They continued the conversation two days later, on December 7:

"Mr. Prime Minister, how are you?" my father asked politely.

"Okay. I understand that you are in a tough situation," said Bondevik.

"We have been in a tough situation now for one hundred years."

"This is true. Arafat says there are Israeli tanks positioned seventy yards from his headquarters. He is scared."

"If he is scared, maybe that will help him carry out preventative measures. He is an obstacle to peace. But we will not interfere and choose the Palestinian leadership," my father said. "There have been developments since our last conversation. Mortars, gunshots, a suicide bomber in Jerusalem. So long as this goes on, we will take the necessary steps, but we will not harm Arafat."

"I'm glad. I spoke with him yesterday," Bondevik reported, "and made clear that he must arrest those behind the terror. He has to take action, to keep his promises."

"In terms of his efforts," my father responded, "he pretends that he has arrested thirteen people, but we have no evidence that even one of them has been jailed. One of the main mortar and bomb factories," he continued, "was situated under the headquarters of the Palestinian chief of police, Ghazi Jabali, who had been appointed by Arafat to start the factory. We had to attack there. We did, and we ruined the factory."

"You attacked a police station," Bondevik said, shocked.

"Underneath which was a mortar factory. The chief of police was instructed by Arafat to manufacture mortars. That is where they worked. We used special devices that enabled us to strike the underground factory . . ."

"There are difficulties on the Palestinian side, protests . . . The balance must be found. Arafat needs his police force in order to conduct arrests."

"He does not need them to manufacture mortars," my father said.

"He must take measures," my father repeated, "and he knows exactly what must be done . . . Have you returned to Norway?"

"I am in Oslo."

"Is everything white? Covered with snow?"

"Enough snow for skiing, but I have no time," said Bondevik.

"I want to thank you that in spite of all of your own troubles you find time to deal with ours. I value our friendship."

"I will be in touch with you. We will continue to pressure Arafat and will be happy to be of assistance. I told President Bush that American involvement was important."

No doubt the president had taken note of Bondevik's suggestion.

As a body, the European Union has always been pro-Arab, less balanced in its stance toward the Israeli-Palestinian conflict than the United States. This attitude was reflected in a conversation my father had on December 12 with Javier Solana, the EU's representative for foreign affairs:

"Dr. Solana, how are you?" my father asked.

"Thank you, how are you?"

"Where are you?" My father was always interested in the people he was in contact with. He might well have been wondering how, before settling down to dinner at one of the finer restaurants in town, Solana had found the time to immerse himself in our affairs. We certainly do not call the Spanish every Monday and Thursday to ask them whether they would like our help in their dealings with ETA, the Basque separatist organization.

"In Paris," Solana responded. "I've come for talks with the heads of state. I wanted to express my condolences both personally and officially for the tragedies that have befallen you in the last several days . . . As a personal friend of yours, I know you are going through a tough time. I wanted to know if you could share with me what we might be able to do."

"You are familiar with the situation," my father said. "The terror continues . . . This morning there was a suicide bomber near the King David and Hilton hotels. We were lucky, and he blew himself up early. Seven people have been lightly injured. Islamic Jihad claimed responsibility for the attack. Arafat has arrested none of them. This evening we apprehended another suicide bomber . . . The situation is such that Arafat must take several measures: he must seriously arrest terrorists, their senders, the planners of the missions, and the heads of the organizations; dismantle the terror organizations; disavow unlawful weapons and hand them over to the Americans, as he agreed to do in the past; and foil terror attacks before they happen and cease the incitement. Those are our primary demands. We will not harm him directly, I have told you that in the past. In addition, we will not attack prisoners."

"That applies for the police headquarters, too?" Solana asked.

"No, no, I didn't say that," my father replied. "The measures we are taking against them are a result of the fact that they have played an active role in the terror attacks. Last week, one of the suicide bombers in the Afula region was a policeman, and the other was a member of the Preventative Security Service. We have no intention of dismantling the PA or of hurting Arafat."

B y the next week, the situation had changed. There had been forty-five deaths and hundreds of wounded due to terror attacks on civilians over the preceding two weeks. On December 15, my father spoke with Joschka Fischer, who was in Brussels:

"We have decided to no longer trust Arafat, only ourselves," my father told Fischer. "We will take action against terror. I am sorry that this is the situation."

"What are the political prospects?" Fischer asked. "We were all shocked by the attacks, and we understand what has happened to your public opinion but what will the political prospects be if there is no partner?"

"There is no partner," my father replied. "For now, that is the situation."

"Is your intention to rule directly over the territories?"

"You asked if we intend to return to the territories? The answer to that is no. We come and go in order to make arrests. We have arrested many in the past weeks. We have no intention of keeping Gaza, Jenin, or Ramallah in our hands. Three days ago an act of terror was committed. We passed the names of the perpetrators to Arafat, also via General Zinni, so he [Arafat] would take measures. He did nothing. We will not trust them. We want to negotiate when it will be possible. We will have no such discussions with Arafat, not now and not in the future.

". . . I hope there will be someone else with whom we can negotiate; if not, we will continue to live as we have lived up to now, with Arab, Palestinian terror. Whoever continues to speak

to Arafat delays the rise of someone else, [someone] more pragmatic. It will encourage Arafat to continue on with the terror . . ."

"I understand," said Fischer. "We are concerned that the collapse of the PA will lead to chaos, violence, and terror. How can
you be sure that in the end there won't be the same level of terror
as there is now?"

"We are not taking action against the PA," my father explained,
"not because they are helping in the war against terror but because it is possible that someone with whom we can negotiate will
emerge from their midst . . . In addition, we will not take any
physical action against Arafat. He is in Ramallah—he will not
leave the area unless he wants to leave the country . . .

"One thing there is a consensus about: we cannot trust the PA
with our security. That was the mistake of Oslo . . . If you come to
the area we will talk."

"The problem is the humanitarian issue," said Fischer. "We
hear that there is a shortage of food supplies, and that causes
concern."

"That is not the situation . . . I recommend that one of your
people meet with the coordinator of government activities in the
territories. I checked the situation with him. We will make every
effort to mitigate the suffering. The problem is that as soon as we
open the roads, there are attacks and fatalities. Arafat is the only
one who is responsible for the situation. He doesn't care. We are
in a war . . . I imagine that with the steps we will take, we will be
able to lessen the terror significantly . . . That is my duty, and that
is what I will do."

TWENTY-SEVEN

Weapons seized from the Karine-A, *January 2002.*
(© Moshe Milner, Israeli Government Press Office)

The Israeli message that my father conveyed, on the matter of Arafat's involvement in the terror trade, was beginning to sink in.

In mid-December 2001 my father's diplomatic adviser reported to my father about two meetings: one between Chancellor Schröder and Secretary of State Powell and another between Schröder and President Putin. According to Schröder's political adviser, the Germans had left the meetings with the impression that the Americans were fed up with Arafat and that Putin was expressing a positive attitude toward Israel and its prime minister.

In contrast to their earlier position, which supported Arafat, now the Russians seemed to be critical of Arafat and the Palestinians.

Israel's diplomatic campaign, aided by my father's hard work, was finally on track.

The American position and the change in the Russian stance had great influence over the positions of other countries.

———————

It never occurred to Arafat that he would have to pay a price for terror, and certainly not a personal price. Ever since he had entered the territories in 1994, after the signing of the Oslo Accords a year earlier, Arafat had made a point of attending Christmas Mass in Bethlehem, a celebration that was broadcast to millions of Christians around the world.

During the month of December the PA sent a routine request to coordinate Arafat's helicopter ride to Bethlehem so he could take his honorary seat in the first pew during the midnight mass.

My father had other plans for him.

I was standing next to him in his study on the farm after he made the decision. I expressly recall the conversation my father had that evening with Uzi Landau, the minister of internal security. Notwithstanding his reputation as a staunch opponent of the PA and of all concessions to the Palestinians, Landau, fearing massive international repercussions, opposed barring Arafat from the mass.

I expressed my opinion that under no circumstances should Arafat be allowed to attend the religious service; but my father did not need convincing. Despite the fears of some ministers, the cabinet reached a firm decision: there would be no reward for terror.

Arafat was never again allowed to take part in the Christmas celebrations in Bethlehem. There was no diplomatic fallout; on the contrary, it was understood that Israel would not live with terror was still an everyday event, and that those responsible for it would suffer the consequences.

Ever since the city had been turned over to the Palestinian Authority, many of the Christians of Bethlehem had suffered harsh persecution, and many had fled the city. Those who remained were secretly joyous at the decision to exclude Arafat from the Christmas mass—I've heard this sentiment from Christian residents of Bethlehem, although, out of fear, few would admit it publicly. But then, it seems obvious that a religious ceremony in a holy place for millions is best conducted without a murderer as the guest of honor.

One never knows what the breaking point could be, that point until which one can keep going in a certain direction but after it everything changes. In retrospect, we can say that the *Karine-A* was the straw that broke the camel's back and revealed Arafat as he truly was: a major part of the problem and not the solution.

On January 3, 2002, in a combined air and naval operation, the freighter *Karine-A* was caught in the open waters of the Red Sea. Some fifty tons of weaponry and ammunition had been loaded onto the ship in Iran. From there, the ship, acquired by the PA, was to sail through the Suez Canal, after which the weapons would be off-loaded near the coast and delivered to the PA in Gaza.

Arafat denied any ties to the weapons ship, but soon enough his claims were contradicted by irrefutable information. Even after it was established that the weapons had been destined for Arafat, he granted an interview in which he contended that the entire affair was a Mossad conspiracy.

Days later he sent a letter to President Bush, acknowledging that the Palestinian Authority had been involved with the ship but contending that he himself had known nothing about it. In shifting the blame to others, he betrayed one of the men closest to him, Fuad Shubaki.

The attempt to smuggle in weapons and his connections with the Iranians concerned the Americans. But what really got to them was the fact that Arafat lied to their faces. The nerve to lie to the president in a personal letter was too much. As far as the Bush administration was concerned, Arafat was in fact out of the picture. On January 12, my father spoke with President Putin. They discussed the manner in which Iran was arming itself, its support for Hezbollah, and its involvement with the Palestinians and Israeli Arabs. The question of the post-Arafat era came up, too. It was already on the table.

O nce my father had decided that Arafat was out of the picture, he started to discuss in internal meetings the need to create an alternative among the Palestinians.

This move had two purposes. The first, for which seemed to be a slim chance, was to try to reach a cease-fire, and in that eventuality begin negotiations with an alternative leadership; the second was to ensure Arafat's isolation. Since he had not yet been rendered wholly irrelevant and illegitimate—the shift was only in its nascent stage, and there was not yet a consensus on the matter in Israel, let alone Europe—it was important to offer the possibility of an alternative leadership so people could see that there was life—a better life—after Arafat. At the same time, it served to mitigate the worldwide pressure that was sure to come if Israel were to refuse to speak with any Palestinian official.

In late January and early February 2002, my father met with three senior Palestinian officials—Mahmoud Abbas, Abu-Alaa, and Muhammad Rashid—in Jerusalem.

This development loomed large in my father's conversation with Joschka Fischer, who was in Israel at the time, on February 15:

"I intend to meet with the Palestinians," my father told Fischer. "I met with them before I left, and I will meet with them again. I want to move things forward. I appreciate Germany and Britain's stance, coordinated with the United States, that new initiatives for the region will not be adopted. We have to continue to strive for quiet. Last week we caught eight suicide bombers; we found explosives on six of them, and two committed suicide when stopped at a police roadblock. The subject I am discussing with the Palestinians is what can be done to fight terror.

"You have helped in the past [to] reach a cease-fire," my father said, "even if it didn't last for a long time, and in Bethlehem your measures brought quiet. There is quiet there now, even though there are some small incidents. There is quiet in Hebron and Jericho, and those cities are open. We have to check how we can move forward and reach a cease-fire. Maybe we will do it in stages, city by city."

"I will meet with Arafat and then I am off to Kabul." Despite his support for Israel, Fischer still hadn't given up on Arafat.

"I know you are active over there," my father said. "What is going on?"

"We have forces there," Fischer said. "We seek stability. The situation is fragile."

"I would like to hear about it. We have a fragile situation in Lebanon. We will be careful not to escalate. The Iranians have provided Hezbollah with arms—thousands of rockets and Katyushas—whose range is the entire north of Israel, including Haifa, Nazareth, Tiberias. The arms came through Syria. The Iranian actions are very dangerous, they have direct ties to Chairman Arafat. He can deny or send letters but the facts are known. The ties started well before the weapons ship was seized."

On February 18, a suicide bomber blew himself up in a car outside Jerusalem, killing a police officer. Two hours later, two soldiers and a woman—a mother of two children—were murdered. Several hours later the indefatigable Norwegian prime minister was back on the phone.

"Hello, my good friend," Bondevik said.

"Mr. Prime Minister, how are you?"

"I know that you are in a tough situation."

". . . You are concerned, and rightfully so, but we have funerals every day, and the public does not accept that. They know that we have the power to send them to hell, and that I am refraining from doing so."

"I understand. [Regarding] the conversations with Mahmoud Abbas and Abu-Alaa—Arafat said you would call once you returned from Washington, and you did not do so."

"Arafat said you would call, and you didn't call"—holy mother of God, what patience the man had to employ! I'm sure he would have been happy to send him off to fish for some wild salmon or groom some reindeer or do something else useful, but my father, like a good soldier, did what had to be done.

"I do not speak with Arafat, but directly to them. I returned from Washington with the flu and had a high fever. I was laid up in bed till yesterday. I stayed on the farm, and I heard the fire of the mortars. The farm is near Gaza, and at night you can hear the fire. You feel like you're back on the front lines again. That is why I could not have another meeting. I had said I would meet them again."

"Will there be another meeting?" asked Bondevik.

"Yes. Yes," he said replied.

"What is your plan if there is a cease-fire?" Bondevik asked. "What are the goals?"

"For there to be an agreement, then an accord and then peace."

"Do you see a solution with two states?"

"In the end, as part of the accord. A completely demilitarized state," he emphasized. ". . . If it is not demilitarized, they will sign an agreement with Iran and we will find Iranian arms boats in Gaza. What will we do then? Attack an Iranian ship?"

A meeting of the League of Arab States was scheduled for March 2002 in Beirut. The proposal to be presented at that meeting—which would later become known as the Saudi peace initiative—was already being discussed informally among world leaders.

The Saudi initiative called for peace and normalization between Israel and its neighbors in return for a full Israeli withdrawal
to the June 4, 1967, lines, and the untangling of the Palestinian
refugee problem in accordance with UN Resolution 194.

On the surface, the proposal looked appealing, with its provi-
sion that the Arab states welcome peace with Israel—something
they had been unwilling to do since the state's inception. But the
details made the offer unacceptable.

Withdrawal to 1967 borders was impossible for Israel. At some
points, those borders reduced Israel's width to approximately ten
miles. After the War of Independence in 1948 and the armistice
of 1949, Hamat Gader and the lands northeast of the Sea of Galilee, parts of the Hula Valley, and the Banias were deemed to be
part of Israel. However, soon after, the Syrians forcibly occupied
these areas and constantly sniped at Israelis tilling their land. According to the Saudi initiatives, even those areas that Syria had
taken forcibly from Israel before 1967 would be returned to Syria.

The issue of the Palestinian refugees was problematic as well.
Hundreds of thousands of Jewish refugees had been forced out
of their homes in Arab countries; they arrived in Israel penniless
and stripped of their property. Arabs left the state of Israel during
the course of the War of Independence. But instead of granting
them citizenship where they settled, the Arab states exacerbated
the problem by purposely creating refugee camps, where they
have languished for over sixty years.

The Saudi initiative proposed a situation in accordance with
UN Resolution 194, Clause 11, which calls for, among other
things, the resettlement of the refugees in their homes in Israel
proper, an action that Israel cannot accept. Israel cannot afford to
allow a Palestinian population to settle in its midst.

It's also interesting to note that no one—not the UN, not the
EU, and certainly not the Arab world—has acknowledged that all
that the Arabs demanded today had once been in their hands but
it was not enough for them. In 1947, before the war, there were
no Palestinian refugees, and up until 1967 the Gaza Strip, Judea,

and Samaria were in Arab hands, yet it wasn't enough for them. The aim of the Palestinians has always been the destruction of the state of Israel and not, as is often portrayed today, a territorial dispute over borders.

M y father broached the issue of a "Marshall Plan" for the Palestinians with Fischer.

"There must be quiet and stability, otherwise there will be no investments," said Fischer.

"We want stability, but there will be none so long as the terror continues."

"I have a final comment," Fischer said. "I was shocked by the murder of Daniel Pearl. I was saddened by the final question he was asked: 'Are you a Jew?' He said that he is a Jew and that his father is a Jew, and then he was slaughtered. That is unacceptable. I am shocked that in our day this is even possible. He was an American citizen, but this type of anti-Semitic act is unacceptable."

D espite the desire for quiet, the terror continued unabated. Until late February 2002 hardly two days went by without the murder of Jews, but all that was a mere prelude to what transpired during the following month.

On Saturday night, March 2, a Palestinian terrorist belonging to Arafat's Fatah organization blew himself up in the Beit Yisrael neighborhood of Jerusalem, killing eleven people, including two babies and four children. Eight members of the Nechmad family, who had traveled from Rishon LeZion to celebrate a bar mitzvah, were killed in the attack.

On March 5, a terrorist sent by Marwan Barghouti, a member of the Tanzim and of the Palestinian Legislative Council, using an M-16 automatic rifle, opened fire on the patrons of a restaurant called Seafood Market. Three people were killed in the attack.

(Barghouti was convicted for five murders on June 6, 2004, in a Tel Aviv district court.) Among those slain were Eliyahu Dahan from Lod, the owner of a café in the city who had been known for his welcoming attitude to Jews and Arabs alike. Yosef Habi of Tel Aviv was killed while shielding his wife, and police officer Saleem Barakat, a Druze from the village of Yarka, was killed, too—but he managed to kill the terrorist before he died.

On Saturday night, March 9, a Palestinian terrorist blew himself up inside a popular Jerusalem café called Moment, only a few yards from the prime minister's official residence. At the time, my father was out at a meeting, where he was given a note with the initial report about the attack. The blast had killed eleven young revelers, and dozens more were injured.

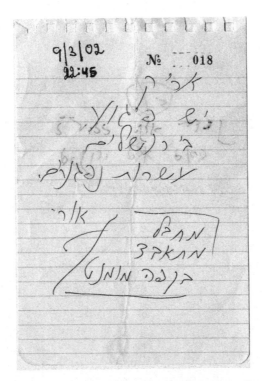

While at a meeting, my father received a note about the suicide bombing at Moment Café, near the prime minister's residence in Jerusalem, March 9, 2002.

On the afternoon of March 21, a terrorist blew himself up in the middle of downtown Jerusalem, on King George Street. The blast killed three people and wounded seventy. The terrorist, from the Nablus area, was a Fatah man. In fact, because of Israeli pressure, he had been arrested several weeks earlier by the Palestinians. PA personnel had been authorized to move the man from Tulkarm, where he had been arrested, to a jail cell in Ramallah, traveling through the IDF roadblocks that had been put in place around the Palestinian cities in an attempt to prevent terror attacks. Once they reached Ramallah, the PA had promptly released the terrorist, and he set out again to kill Israelis.

The fact that the man belonged to his organization, of course, did not deter Arafat from publicly condemning the attack. "We will spare no measures" in trying to catch those responsible and bringing them to justice, he said. All he would have had to do was put himself behind bars.

Two of those murdered in the attack were Gadi and Tzipi Shemesh. Tzipi was five months pregnant. The couple had just come out of a medical facility, where an ultrasound exam had revealed that Tzipi was carrying twins. The parents left behind two girls.

Yitzhak Cohen was killed, too. He was on his way home from work and left a wife and six children.

Three days later, on the morning of March 24, kindergarten teacher Esther Kleiman was murdered when a terrorist opened fire on a bus she was taking from Neve Tzuf to Jerusalem. She was twenty-three years old.

TWENTY-EIGHT

*My mother took this photo of my father
setting the table for Passover.*

For our family, the holidays are special occasions. The holiday meals are traditional in custom, atmosphere, and, of course, the menu. The Seder—the ritual commemorating the exodus from Egypt and the transformation of the Jews from a people of slavery to a free people—is no exception. Our Seder has always been both festive and intimate, celebrated with family and a few close friends.

My father always set the table. That was his part in the preparation, and he was never willing to give it up. Leaning over the embroidered white tablecloth, he would lay out the fine silverware with pride, precision, and care—exactly according to protocol, in order of usage, from the outside in. He set a soup spoon, a fork, and a knife for the fish on either side of the plate, and finally a fork and knife for the main course—my mother's traditional lamb, the kind she had been making for decades. Up above the plates he laid a small fork facing right and a small spoon facing left, for dessert. To the right of the plate was a wineglass, and to the right of the wineglass a water glass. Each person had a plate with small dishes of the traditional Seder foods—horseradish, the sweet apple- and pecan-based *charoset*, parsley, egg, and salt water. The main Seder plate had all of the above, plus a shank bone with bitter herb. The matzoh had its own plate. Great care was also taken with napkins, which he placed to the left of each plate, and with salt and pepper, placed on either side of the table.

Each person had a Haggadah placed on his plate, including English versions for those who spoke English, like our close friend Cyril Kern and his family; Maya, my father's cousin, and her daughters, Cory and Maria. We read the entire Haggadah, including the part to be recited after the meal, and we hid the *afikoman* from the children and reached an arrangement with them for its return. The songs were rendered according to the traditions of my mother's and my father's homes, with the knowledge that the Jewish people have celebrated Passover for some 3,500 years—a record of the celebration of a religious event perhaps unmatched in world history.

On March 27, 2002, our Passover Seder started according to custom. During the Seder there was a telephone call. It's always a call or a note. There were reports of an attack at the Park Hotel in Netanya.

Every year the Park Hotel holds a Seder. That night, on the ground floor of the hotel, two hundred and fifty people were in the middle of the Seder when a Palestinian terrorist disguised as a woman entered the hall and blew himself up. Married couples, a father and daughter, and others were murdered, among them the elderly—some of whom had survived Auschwitz and Bergen-Belsen, only to be slaughtered on Israeli soil. Thirty people were killed; one hundred and forty were wounded. The youngest of the dead was twenty; the oldest was ninety.

It says in the Haggadah, "For not just one alone has risen against us to destroy us, but in every generation they rise against us to destroy us, and the Holy One, blessed be He, saves us from their hand." On this day, too, Palestinians had chosen to murder Jews.

That night, a helicopter landed on the farm and took my father to his office in Tel Aviv for a meeting with the heads of the IDF and the Shabak, Israel's internal security service.

My father was livid, and he let that be known in his conversations with the army officers and other high-ranking security personnel. As someone who saw himself as—and, as prime minister, was by definition—the defender of the Jews, the harming of Jews anywhere in the world drove him mad, but the anger was only exacerbated when it occurred in Israel, on the night of the Seder. For him it was over; it had gone too far.

At our kitchen table. (© Ziv Koren)

Near dawn of the next day, my father landed back on the farm and came into the kitchen.

In our house we regularly stop at the kitchen before leaving the house and upon returning to it. The kitchen is built of wood and red brick; the room is decorated with copper pots and pans and the ceramics we made with my mother when we were children. The countertop is always adorned with jars of olives and some type of homemade dish. My father never failed to stop in the kitchen for a cup of tea, seating himself at his place at the head of the table and relaxing after a long day at work.

Before dawn, only Roni Schaiak, who was visiting from Jerusalem, awaited him. He had been unable to sleep. As my father drank his cup of tea in his usual spot, Roni approached him with trepidation.

"Commander"—Roni insists on calling him that—"why did you have to go in to the office tonight?"

"Because," my father answered, "I wanted to make sure that they understood exactly what I want of them."

———

On the evening after the attack, with the close of the first day of the Passover holiday, the government convened and authorized the prime minister and defense minister's proposal to embark on "Operation Defensive Shield." Two ministers, Foreign Minister Peres and Minister of Science, Culture, and Sport Matan Vilnai, voted against the proposal. These two were so disconnected from reality that they did not understand we were literally fighting for our lives.

On that same evening, a Palestinian terrorist sneaked into the settlement of Elon Moreh not far from Nablus and murdered four members of the same family—the eighty-one-year-old grandfather, Yitzhak, his daughter, Rachel, her husband, David Gavish, and their son, Avraham, who was survived by a pregnant wife and a daughter.

———

The operation began on Friday, March 29. Its objectives were to debilitate the infrastructure of Palestinian terror and to stop the wave of terror attacks. To this end, the IDF entered Jenin,

Nablus, Ramallah, Qalqilya, Tulkarm, and Bethlehem, along with many other cities and towns.

In the early-morning hours of that same day, during the morning prayer service, a Palestinian terrorist burst into a synagogue in the settlement of Netzarim and stabbed seventy-nine-year-old Tuvia Wiezner; the terrorist later stabbed seventy-year-old Michael Orlinski, who had been on his way to the synagogue.

That same afternoon, while many Israelis were doing their pre-Sabbath shopping, a female Palestinian terrorist blew herself up at the entrance to a supermarket in Jerusalem. The blast killed the security guard, Haim Smadar, who had barred the terrorist from entering the store, preventing an even greater tragedy, and Rachel Levy, seventeen, who was chatting with a friend at the time of the attack. Thirty people were wounded. Eyewitnesses say that the terrorist gave warning to two Arab women who were selling vegetables outside the market before carrying out the attack. (Two weeks later, the Saudi Arabian ambassador to England wrote a "poem" praising the suicidal terrorists and paid special homage to this female bomber.)

Rachel and Ben-Zion Charhi went out to a Tel Aviv café on the evening of March 30. Between a cappuccino and an espresso a terrorist from Arafat's Fatah organization entered the café and blew himself up, killing Rachel and wounding her husband, along with thirty other people. Rachel Charhi left behind three children.

On Sunday afternoon, March 31, a terrorist blew himself up in the Matza restaurant in Haifa, killing fifteen people and wounding over thirty. The manager of the restaurant was an Arab-Israeli; Jews and Arabs alike were killed in the attack. Among the dead were fathers and their children, later buried side by side. Families were ripped apart—and this in an area known for peaceful coexistence between Arab and Jew.

These were only some of the terror attacks in that month, perhaps the worst of the Palestinian terror onslaught: over 130 Israelis were murdered and hundreds injured during that month, which became known in Israel as "Black March."

In the meantime the IDF was operating at full force. The emergency call-up of the reserves was widely answered. There was a feeling among the public that this campaign was a fight for existence. Even those who counted themselves as stalwart members of the Israeli left understood that there were at this point only two options: terror would prevail over us, or we would prevail over it. If allowed to continue, the terror would annihilate some and break the spirit of others, debilitating the Israeli economy, slashing foreign investments, and irreversibly damaging our ability to deter future attacks. Many Israelis, including friends of mine, people who considered themselves centrists and to the moderate left, people who supported the Oslo Accords and believed in the peace process with the Palestinians, said, "That's it—we've had enough."

Insofar as the IDF was concerned, this operation was a revolution, a complete turnaround from the period during which Ehud Barak was prime minister, a period when wounded soldier Madhat Yusuf was allowed to bleed to death near Joseph's Tomb in Nablus. It would not be a simple transition from that situation to a full-fledged offensive, seeking out the terror strongholds in the Casbah in Nablus, the alleys of Jenin, and other terror centers.

Since the Oslo Accords and the arrival of Arafat and his followers to Judea, Samaria, and the Gaza Strip, the army had grown accustomed to seeing Area A, the region under Palestinian control, as inviolable. Even if we had intelligence that showed that a terror operation was being plotted against us, running along unfettered, the army had been instructed to rely on the Palestinian security forces, leaving our security in their hands. Of course, the Palestinians never played their part. But there was another, more subtle obstacle to this new undertaking.

At the point when the government decided to take Israelis' security into its own hands via Operation Defensive Shield, my father was aware that a problem had taken root in the IDF: some officers expressed the opinion that the solution to certain military problems was political in nature.

My father spoke about the problem at home and with people he was close to on many occasions, but never publicly, so as not to weaken state security and Israel's power of deterrence. It also would not have been wise to voice those concerns, as there is nearly universal adoration and appreciation of the IDF among the Israelis. (Only after the IDF's failures during the Second Lebanon War were the public's eyes opened to the force's limitations.)

Nevertheless, the attitude of these officers angered my father. He had always believed that it was the army's job to continuously provide the political leaders an array of military options, allowing the government to decide which if any to enact. Those in the military, he felt, had no business justifying their blunders by citing a political solution that should have been employed. A military man should offer a military solution. If he cannot he can leave his position and leave the task to an officer who can. An officer who believes that the solution to a problem he is faced with is political is welcome to leave the army and go into politics. That is where they deal with political solutions. In my father's view, the military initiates, and the political leadership—the branch with the wider perspective—tempers. This had been the rule when he was the twenty-five-year-old commander of Unit 101, and later of the paratroopers. He had presented many operational plans, some of which had been authorized and some of which had not. Now, as prime minister, he found himself in the awkward position of having to spur the army to action. He proceeded to do this in meetings with officers and in lectures to the IDF Staff and Command College, but the main way he dealt with it was by relentless pressure on the IDF to take action.

Apart from this internal problem, many people, including military officers, doubted the IDF's ability to attain the goals of Operation Defensive Shield, and some warned that the death toll would be much higher than we actually experienced.

O peration Defensive Shield began with the incursion into Ramallah. Arafat's compound was placed under siege. Along
with Arafat were his CFO of terror, Shubaki, and the murderers
of Minister Ze'evi.

Nablus and Jenin were two of the hotbeds of Palestinian terror.
At the time both cities were populated by thousands of armed
men in addition to the PA's security forces. The IDF seized control of both cities, including the narrow, booby-trapped alleys in
the Casbah of Nablus and the labyrinthine streets of the refugee
camps, where the terrorists hid easily amid the civilian population
who provided cover. Despite these obstacles, the IDF was able to
inflict harm on hundreds of terrorists, arrest many more, including the most-wanted leaders, and unearth many demolitions labs
and illegal weapons factories, and to minimize the toll on the civilian population.

At a briefing near Jenin with Minister of Defense Binyamin
Ben Eliezer to his left during Operation Defensive Shield,
April 10, 2002. (Courtesy of Reuters)

International pressure on Israel was quick to arrive. Media outlets in Europe depicted Israel as using unreasonable and un-justified force, and European and American leaders attempted to induce Israel into easing its siege on the Palestinian cities, and particularly on Arafat.

During a phone conversation on April 6 with my father, Igor Ivanov, the Russian foreign minister, described the anti-Israel demonstrations in Morocco and Tunisia:

"I'm sorry you did not have the opportunity to see the funerals of the Israeli citizens," my father replied.

"We understand that you have to fight terror. We ourselves are committed to that. Along with that, the messages [you convey] and the force you use are bringing a series of condemnations against Israel in the world. Believe me, I am speaking as a friend. I am in daily contact with the world leaders."

My father responded, "Regarding our public relation problems and Israel's isolation, I say to you as a friend that if I have to choose between good PR and our casualties, it will take me a second to decide. You say that we are losing the world's sympathy; I'm ready to lose the sympathy and get bad PR to save lives. We will not remain in the Palestinian areas. We entered to rid ourselves of the terrorists, and caught hundreds of murderers, weapons, and explosives."

Ivanov replied, "We will be in touch."

"Thank you, Igor Sergeyevich," my father said in Russian.

———————

That same day, my father also spoke with President Bush, who asked that Israel leave the Palestinian cities as soon as possible.

"I am very much aware of your desire to see this operation finished as soon as possible," my father said. "I do want you to understand, however, that our forces are operating in very difficult conditions in cities and towns with huge amounts of weapons, explosives, and active terrorists . . . all around, when we are trying very hard to minimize . . . any civilian casualties, and that makes it so difficult, and takes longer."

The president inquired as to the duration.

"I will try my best . . . I hope you know, we have found materials that were given to your people that clearly connect Arafat to the recent suicide bombers and terrorist activity.

"With Arafat it is impossible. If they want a Palestinian state, that is understandable, but [Arafat] wants a Palestinian state instead of the state of Israel."

"It will be difficult to achieve peace in the current situation," the president said. It was a friendly conversation. "I support Israel, and I say this from a position of support."

TWENTY-NINE

With President George W. Bush at the White House, July 2003.
(© Moshe Milner, Israeli Government Press Office)

The same day, April 8, 2002, that the IDF started to pull out of those areas where they had been successful in routing the terrorists, the Danish foreign minister came out with a strong critique against Israel. "They say they are hunting down terrorists," he said, "but it seems they are making the entire civilian population pay the price . . . clearly Israel has crossed the line."

Continuing the UN's long-standing tradition of anti-Israel bias, Kofi Annan also criticized Israel that day, saying, "The whole world is demanding that Israel withdraw. I don't think that the whole world, including friends of the Israeli government, can be wrong . . . the longer this goes on, the more it erodes the moral and political position of Israel in the world." He did call on the Palestinians to "honor the demands" the UN resolutions placed on them, including an absolute cessation of violence.

The day after the speech in the Knesset, Condoleezza Rice, then national security adviser to President Bush, called my father's office. His adviser noted the following points of their conversation:

> *She thanks us for leaving Tulkarm and Qalqilya and for responding quickly to her request.*

> *It is very important to them that the momentum be continued and that we leave additional areas in the next twelve hours.*

> *She understands that Jenin is a serious problem and that we found many terrorists and weapons. They will raise this matter with the Arab leaders, but in order for them to demand that the Arab leaders pressure the Palestinians, they need for the withdrawal to proceed.*

> *The national security adviser noted that the continuation of the withdrawal would lessen the pressure on them and enable us to withdraw in a more orderly fashion from the places that are truly important to us.*

> *Ramallah—source of great concern. They ask that we ease the siege around Arafat's offices so that Colin Powell does not have to enter under the barrels of our weapons. It was suggested that they meet with Arafat outside Ramallah—she stressed that that was not a realistic option.*

> *She was told that we cannot ease the siege because of Ze'evi's killers— she asked that we weigh the option of their removal from the premises by British prison guards.*

On April 12, my father met with Secretary of State Colin Powell. Powell laid out the problems of the Middle East as he saw them: growing turmoil on account of the Israeli military actions, mass demonstrations, the attacks on the U.S. Embassy, UN pressure, and most important from his perspective, a concern for the United States' standing in the region.

He asked how long the operation would last.

"Till we finish it," answered my father. ". . . In the areas that we finished with, the troops left the cities. They will now encircle it—and if it stays quiet, then they will redeploy to the security areas, guarding the roads and villages. If the [terror] activity starts again, then we will have to go back in and operate again."

Powell was shown photographs depicting the horror of terror attacks. He brought up the issue of especially complicated situations like the stalemate at the Church of the Nativity in Bethlehem, where terrorists had taken monks hostage, knowing that Israel would not storm a place held holy by millions of people.

The matter of the siege on Arafat's headquarters was raised; my father explained to Powell that as long as Ze'evi's killers and Fuad Shubaki were not arrested, there would be no capitulation.

"I have no faith in Arafat," Powell said.

Powell was concerned about Jenin. The Palestinians had rushed to proclaim that a massacre had occurred there. Terje Rød-Larsen, the UN secretary-general's special envoy, was quick to condemn Israel, and through that strengthening the Palestinians' case. Terje Rød-Larsen was awarded, together with his wife, the Norwegian ambassador to Israel, $100,000 from the Peres Center for Peace—to which the Norwegian government was a contributor, in an illustration of the circularity of the peace industry. Apparently, there is no business like the peace business; according to the media, the couple did not report this side income.

My father mentioned the pressure coming from Europe. The rumors of a massacre and the wave of anti-Semitism in Europe were likely linked, he suggested. As always, he would not let others ignore the reason behind his country's actions. "Israel is the only democracy in the world that has to guard every kindergarten and every school from terrorists," he reminded Powell.

Despite Powell's derogatory remarks about Arafat, two days after his meeting with my father on April 14, he met with the Palestinian leader—much to my father's displeasure. The meeting had been given little credence by at least one Palestinian minister, Hassan Asfour, who commented the week before, "Powell is a neo-Nazi agent who can be expected to bring with him some of the new poison which President Bush has started to spread in his speeches. We call upon the Arab leaders who intend to receive the neo-Nazi agent to beware."

After his own meeting with Powell, my father was interviewed by Dan Rather, former longtime CBS News anchor. Early in the interview, Rather said that governments in the world, aside from Israel, believed that in order to reach a political agreement my father would have to talk to Arafat.

"Less than two years ago," my father responded, "Mr. Arafat was offered everything that one could have expected by my predecessor Mr. Barak and the efforts of President Clinton . . . But Chairman Arafat didn't accept it and instead adopted a strategy of terror, and therefore I don't think that one can get into peace with Chairman Arafat.

When Rather raised the subject of Jenin, my father rejected the claims that a massacre had taken place there. "There was very hard fighting going on in a built-up area where any other country in the world, including Americans, would not have endangered their soldiers. They would have brought bombers, and they would have avoided all those casualties."

(There were those who were disturbed by the manner in which the IDF operated in Jenin, taking special care not to harm civilians, a practice that put our soldiers at greater risk. In general I agree with this line of reasoning: in a conflict of this type, those who don't wish to take part in the battle should remove themselves from the scene of combat. In other places, the IDF managed to save Palestinian civilian lives and paid with fewer casualties than the fourteen who were killed in the difficult battle in Jenin. However, taking into account the international press, the international pressure, the UN and its resolutions and investigative reports, along with the overall hostility toward Israel, it's possible that this approach was the only option in this case.

During that same CBS interview my father said, "As a soldier that participated in all the wars, I know that it happens that there are civilian casualties, and every casualty is a tragedy. But the difference between us and the Palestinians is, we are looking for terrorists. When they choose targets, they are looking for civilian targets—celebration places, weddings, bar mitzvahs, people coming out from synagogues—and that is the most terrible thing.

"I know the world is worried . . . I got many phone calls from heads of state complaining that Mr. Arafat has only a candle for light there. But I have not gotten a call about a family—father, daughter, son—that was killed, and the mother is in very bad condition in the hospital. Nobody said a word about it. I didn't hear anything about a funeral for two children of a family, and the parents do not know [of their deaths] because both of them are in the hospital in critical condition. Nobody is worried about it. Nobody has said a word about it, and it brings me back to the point that Israel is the only place in the world where the Jews have the right and the capability to defend themselves by themselves. This nation is strong. Much stronger than our enemies believe . . . Nobody will enforce upon us any decisions or resolutions that might affect our future. Israel is a wonderful country, a courageous country, and we should be left to defend ourselves in this historic place. And I look forward with great optimism."

Meanwhile, reports came from the field that hundreds of guns, rifles (some stolen from the IDF), machine guns, shotguns, pistols, silencers, mortars, and Hebrew prayer books (to enable the terrorists to masquerade as religious Jews) had been found in the Beitunia headquarters of Palestinian Preventative Security chief Jibril Rajoub. Rajoub himself left the site before the battle but urged his remaining men to "fight to the death."

The searches continued.

As the antiterror operations began to prove successful, international pressure on Israel gathered steam. Joschka Fischer, in a conversation with my father, made a plea to meet with Arafat.

"We cannot allow that," my father said. "We are in the middle of an operation—we need time to uproot terror activity."

"What can the Palestinians do?" Fischer asked. "It must be made clear to them so that they understand that Arafat is responsible, they need to understand that terror has a price and that it is their leadership leading them into this chaos."

My father asked Fischer for a balanced approach. "If you are unwilling to declare Arafat a terrorist, at the very least don't ask to meet with him . . . Only the successful completion of the current campaign will enable us to progress with the political matter. We are taking precautions regarding civilians . . . It takes time because we call out on loudspeakers to avoid harming civilians."

Prime Minister José María Aznar of Spain sent a letter to my father calling on Israel to lift the siege on Arafat's Ramallah headquarters, withdraw from Ramallah entirely, and immediately implement a cease-fire. He further requested that the EU be allowed to reestablish ties with the Palestinians via Miguel Moratinos, their special envoy to the Middle East.

After sending the letter, Aznar called my father. The conversation was difficult.

"You said no to everything and even asked for several things," Aznar said. "The European public is growing more and more hostile; the European states will not be able to buy you more time. The states have a problem with public opinion."

"We have a problem with fatalities . . . ," my father said.

Aznar spoke of public opinion but was silent on the natural right to self-defense. He did not mention the freedom to ride a public bus without fear that one of the passengers might blow himself up, the freedom to send your children to school without wondering how susceptible those buildings were to attack. "Arafat is humiliated by the siege," he said, as though the lost honor of that infamous terrorist should be a consideration at the heart of Israel's attempts to stop the spiraling terror. ". . . If the operation continues—the situation will worsen."

"It cannot get any worse," my father said.

"Israel has the right to defend itself, but what is the political plan?" Aznar said, adding, "I am soon going to visit Arafat."

"We would be happy for your visit," my father said, "but at this point such a visit to Arafat is problematic."

Aznar said that while "the Spanish presidency pledges its condemnation of terror," he also warned that "there are many states that oppose Israel's policies—including Spain. We oppose terror, but we oppose your policies. There is a Palestinian Authority, and they are responsible for the area, and one cannot refuse to speak with their leaders."

Aznar pointed out that as president of the European Union, he had prevented many anti-Israel actions by member states. Again he mentioned a visit to Arafat, but my father spoke firmly. "Not right now."

"Yes, right now," said Aznar, "for the representatives of the European Union,"

My father was not a graduate of the Foreign Ministry's diplomacy course, but his straightforwardness had won him much respect and admiration, among the Arabs as well. His yes meant yes, and his no meant no. He kept things simple: "I'll bring it before the cabinet—and I'll recommend not to allow it."

———

During this time some rather unbelievable discussions were being held in Oslo. For background, recall that Shimon Peres, along with then prime minister Yitzhak Rabin and Yasser Arafat, had won the Nobel Peace Prize in 1994. Yes, it is hard to imagine, but in 1994 the threesome won the prestigious prize. Giving Arafat a Nobel Peace Prize is a lot like choosing former Italian porn star Cicciolina to replace Mother Teresa as head of the Missionaries of Charity. The bestowal of that year's prize—unusual in that it was given for a peace that might have happened and did not, rather than one that had in fact transpired—remains a historical fact

Yitzhak Rabin and King Hussein had been worthy of the prize for the peace deal between Jordan and Israel; but that had happened too late for their eligibility. Shimon Peres, whose powers of manipulation were formidable, pushed for the prize for the Oslo Accords, and he pushed especially hard for his own inclusion as the third party to the prize.

Shortly after the prize was awarded, it became evident that the Oslo Accords had brought not peace but bloodshed. And Arafat, a terrorist responsible for the taking of so many innocent lives, certainly was unworthy. In this case too the prize was more tarnished by Arafat's acceptance than he himself was honored by its bestowal.

Now, with uncanny irony, members of the Nobel Foundation expressed dismay at their inability to rescind Peres's prize, condemning him for being part of the Sharon government. In fact they went so far as to say that Peres's participation in that government might taint the prize. They said nothing about Arafat, whose involvement in terror was common knowledge.

On April 27 my father received a phone call from Kofi Annan:
"Mr. Prime Minister, how are you?"

"How are you, Mr. Secretary-General?"

"Good to hear your voice. I called to thank you for sending the [Israeli] delegation to meet with my people. The meetings were constructive. We met their demands, and I hope that the commission does not widen its jurisdiction and deals only with Jenin, as the UN decision requesting the Israeli cabinet allow the UN delegate to investigate Jenin indicates."

"Today we had a tough day. I will say this in a polite way: the UN is less interested in operations carried out against Israel."

"We condemn terrorist activity," said Annan.

"Today two terrorists entered a small village and killed a woman and her four-year-old daughter in their beds. Both of them were asleep. Two men were shot to death as they left synagogue. Seven were wounded, one critically. I hope the others stay alive. I must say, Mr. Secretary-General, that when I see the terror attacks, the murders, the assassinations, the hundreds of dead and thousands of wounded—what the world is dealing with is a Palestinian-made blood libel, and Israel, rather than joining ranks with everyone else in attacking this blood libel, is being brought to trial by the world. As a citizen of this world, I have grave thoughts. I am sorry I say this but it is what I think."

"You always speak candidly," Annan said. "On the matter of suicide bombers both the UN and myself personally have condemned them, and we will continue to do so. We know that anti-Semitism is on the rise, and I condemned that before the Human Rights Council . . . I know that there have been reports in the European press and in other places, but my approach and that of the commission will not be one-sided or put Israel on trial. Your delegation has voiced candid and legitimate concerns."

"The murders were condemned, but no one thought to send a delegation [to investigate them]. There is all the proof of Arafat's involvement and direction of terror and other actions. These days the matter is well-known. Before we had intelligence pointing in this direction, but now it is clear. No one thought to appoint a commission to check these facts. There was a hard battle there. I do not know of another country in the world that would not have taken the necessary steps. With all due respect to you and the organization you head, when you say balanced approach, what could possibly be balanced between victims and murderers? We sacrificed soldiers so as to avoid having greater [civilian] casualties . . ."

On May 2 the siege was lifted, and Arafat was once again allowed to receive international visitors, an opportunity seized upon by France, Germany, Turkey, Japan, Russia, South Africa, and the Vatican, which was primarily interested in the crisis at the Church of the Nativity (which was finally resolved on May 10). In his first speech after the siege was lifted, Arafat proclaimed, as he often did, "We're marching towards Jerusalem with one million *shahidim* [martyrs]." This was his declaration of his plans for peace—to pave the path to Jerusalem with a million dead Palestinian terrorists. One of the absurdities of his pronouncement lay in the fact that at no point has Jerusalem served as a capital for the Arabs, let alone for the Palestinians, who have never been a nation. By contrast, for three thousand years the Jewish people have never had a capital other than Jerusalem; nor has Jerusalem served as the capital of any other nation.

At that point, Arafat was given the opportunity to travel around Judea and Samaria. However, with a resurgence of terror, he was quickly quarantined again.

On May 5, my father flew to the United States for a meeting with President Bush, his fifth in his fourteen months in office. My father met with the American president frequently. He believed in the importance of trust and the need for a personal bond, and he also wanted to ensure that his messages were not misunderstood. Moreover, he recognized the importance of an American Middle East policy that allowed Israel to defend itself freely and did not try to impose a solution to the conflict but rather strove toward an understanding with Israel and the Palestinians. Such an understanding would also curb pressure from the UN and Europe.

The primary goal of the trip in early May was to convince the Americans to accept his position—first the cessation of terror, violence, and incitement, and massive reform in the PA, including the replacement of Arafat. Once that was accomplished, long-term interim agreements could be reached, during which the actions and the intentions of the Palestinians would be assessed. Only then could a final, permanent settlement be reached.

Operation Defensive Shield had already yielded documents that served as proof of Saudi Arabia's funding of terror, via money sent directly to Hamas and to the families of suicide bombers. My father went to his meeting with Bush equipped with a stack of dossiers relating to Arafat's involvement in terror.

My father met the president on May 7. President Bush said he was "disappointed with Arafat." The two leaders agreed on the need for fundamental reform in the PA, which should be started, as Bush said, "as soon as possible." Bush announced that CIA chief George Tenet would come to the region shortly in order to establish a single Palestinian security force—a development Bush regarded as "very important."

Toward the end of the meeting, my father received word about a terror attack in a Rishon LeZion billiard hall. A terrorist had come in at night and blown himself up, killing fifteen people and wounding fifty. My father immediately decided to cut short his visit and return to Israel.

Before he returned home, my father held a press conference. "This attack is proof of the Palestinian Authority's true intentions," my father told the reporters. "We cannot continue the peace process with a terrorist and corrupt entity. He who rises to kill us, we will rise earlier and kill him. It is completely clear to me that we will have to act forcefully. This situation cannot go on."

O n May 9, the day after his return to Israel, my father spoke with Kofi Annan:

"Prime Minister, you had a difficult return home. I am calling to express my deep condolences."

"I was scheduled to be in New York for several hours and had intended to call you to say hello. I thought that we might show you documents we found proving that Arafat was involved in terror acts. The Saudis, I'm sorry to say, because I liked the peace vision of the Saudi prince, are financing Hamas and the families of the suicide bombers."

A few days later, my father had a conversation with Prime Minister Aznar of Spain. This conversation, on May 11, was far friendlier than the last one—perhaps because Aznar's perception of Israel's position had changed. Or perhaps Israel's clear and defiant stand on principle had brought it new respect:

"I'm in the middle of work; are you also working on the weekends?" asked Aznar.

"We have no other possibility. We have many problems here."

"Are you working because of the circumstances, or is it the high salary you're getting as a prime minister?" inquired Aznar.

My father laughed and said, "I don't know how it is in Spain, but here the salary is not high. I work because the situation here is complicated."

"If it was because of the salary, I would be working only one morning a week," said Aznar.

"I want to thank Prime Minister Aznar for the invitation to a meeting, but because of the circumstances I had to return home."

"I condemn that terror attack," replied Aznar.

"We heard about the terror attack in Spain prior to the football match. Terror has no limits. There is no difference between terror and terror, and we need to uproot it anywhere in the world."

They agreed that my father would send the head of the Mossad to update the Spanish prime minister. "We'll work together. We both could ask for a raise in our salaries," ended Aznar.

―――――――――

While it did not eliminate the attacks completely, Operation Defensive Shield was a turning point in the war on terror. At its close, the gunfire from Bethlehem and Beit Jallah toward Jerusalem's Gilo neighborhood came to a complete stop.

On May 10, after a thirty-eight-day siege, the thirty-nine terrorists in the Church of the Nativity were stripped of their arms. The monks and civilians were freed, and the terrorists deported— twenty-six of them to Gaza and the remaining thirteen, the more dangerous of the bunch, those that had committed acts of terror, to Cyprus and from there to other countries. Among the freed hostages were eleven European activists who had joined the terrorists out of a sense of solidarity.

The terrorists had been armed with rifles and explosive charges. They shamelessly booby-trapped the holy place for millions. With the confrontation over, the IDF neutralized the explosive charges, the church was cleaned of its filth, and they allowed the shrine to be reopened for prayer for the first time on May 12.

Since the terrorists had been exiled rather than arrested, the manner in which the standoff was resolved was not ideal. But considering the sensitivity of the location and the grave concern of the Vatican and all of Christendom, the solution seemed moderate and reasonable.

Another outcome of Operation Defensive Shield was an agreement whereby the killers of Minister Ze'evi were taken to Jericho, along with Fuad Shubaki, the CFO of the *Karine-A* affair, and jailed there by British prison guards. (Israel did not trust the Palestinian guards, and so the British were put in charge.)

Meanwhile, the UN fact-finding committee had proceeded with its investigation, pursuant to UN General Assembly's Resolution ES 10/10 of May 7, 2002, into the battle of Jenin. They found that no massacre had occurred in Jenin. The committee's findings were backed up by Amnesty International and Human Rights Watch reports, which found that according to their opinion, Israel had used excessive force, but there was no massacre.

Operation Defensive Shield had taken roughly forty days, during which twenty-nine soldiers lost their lives and over one hundred were wounded—a steep and painful price. However, the IDF had managed to seize control of the Palestinian cities, eliminate hundreds of terrorists, and arrest thousands of terror operatives. Labs used for producing bombs had been discovered and destroyed; vast amounts of explosives and weapons had been confiscated.

The most important turnaround was psychological. The IDF and the Israeli public understood that there was now a viable response to terror. The mood had changed. No longer would we wait for the tens of thousands of armed Palestinian security officers to fight and abolish terror. The rules had changed. From this point on we would defend ourselves. There was no longer anywhere in Judea and Samaria that the IDF could not enter to prevent terror. The IDF had been given license to operate—a status quo that endures today.

The military success of Defensive Shield and subsequent operations and the progress in building the separation fence that helps prevent Palestinian terrorists from entering the country brought a sharp drop in the number of terror attacks in the years after 2002. We still endured attacks, but the tide had turned. We have not had another month like Black March.

Ever since its inception Arafat had made sure that the PA would have somewhere between ten and fifteen different security organizations, each with overlapping authority. The services included the National Security, the Palestinian Police, Preventative Security, General Intelligence, Military Intelligence, Naval Police, Force 17, the Presidential Guard, and others. By fragmenting the services, he had ensured that no single body amassed too much power. As one would expect, the different organizations clashed with one another.

Now my father was leading a move to transform the PA from a group of armed gangs into a body with which one could cooperate.

On May 31, he spoke with an aide to President Mubarak, Osama el-Baz. "The situation is dangerous for us all," said el-Baz.

"It's dangerous for others," my father said. "[But] here, there are murders taking place." In order to fundamentally change the situation, my father told el-Baz, terror had to stop, and the PA had to be reformed.

El-Baz acknowledged that Arafat had made mistakes.

"Not mistakes but a strategy of terror," my father told him.

Contradicting el-Baz's contention that there was "no military solution," my father reminded him that there had been "a drastic reduction in terror" since Operation Defensive Shield. My father then suggested splitting the region into eight different security areas in order to help the Palestinians fight terror. He also made another thing clear. "Negotiations," he said, "will be handled by me."

———

The fact that Arafat's power had been severely diminished was reflected in Egyptian president Hosni Mubarak's statement in an interview with the *New York Times* on June 5, 2002. On the eve of a trip to Washington, Mubarak addressed Arafat's failure as a leader and called for his retirement the following year, at which point he would assume a more symbolic role.

It is hard to know how sincere Mubarak was in condemning Arafat, given Egypt's longtime attitude toward Israel. But then, the Egyptians have always been two-faced. On the one hand they allow for the massive smuggling of arms from Egypt into Gaza, along a narrow, isolated strip of land that they could easily control if they truly wanted to stanch the flow of weapons. On the other hand they present themselves as opposed to terror. While they show great hostility toward Israel in all international forums, they try to assume the role of mediators, portraying themselves as a people in pursuit of peace when, in actuality, this is done in order to preserve their status and to ensure the coninuance of American aid.

The Egyptians harbor considerable hatred for the Palestinians, and the way they speak about the Palestinians behind closed doors is quite severe. ("Sign already, you dog," Mubarak hissed at Arafat on May 4, 1994, when Arafat refused to sign on the maps in the Cairo Agreements [part of the Oslo Accords] at the festive signing ceremony. The insult was caught on a live television broadcast.)

The Egyptians don't mind if the Palestinians shed our blood—as evidenced by their willingness to allow Hamas terrorists to travel freely in their country; on the other hand, they don't want Palestinian terror to be too successful, fearing that the connections between Palestinian terror groups and their own could provoke civil unrest or the assassination of the president, as in 1981, when Anwar Sadat was killed by a branch of the Muslim Brotherhood. (Although one cannot attribute to the Muslim Brotherhood the instigating of the demonstration that led to Mubarak's downfall, nevertheless they were the most organized and focused.)

Mubarak had a clear incentive to portray himself as the regional peacemaker—on a par, at the very least, with the Saudis. And certainly before his trip to the United States. Nonetheless, for the first time, the message began to resonate in the Arab world: Arafat was not a partner.

Nine years later, Mubarak himself was deposed by the masses and became irrelevant.

THIRTY

At the site of a bus bombing in Jerusalem, June 2002.
(© Avi Ohayon, Israeli Government Press Office)

On June 5, 2002, a Palestinian terrorist got into a car packed with dozens of pounds of explosives. He pulled the car, stolen by Palestinians in Israel, alongside the 830 bus at the Meggido junction and detonated the charge. The bus flipped over and burst into flames. Seventeen people were killed, and over forty wounded. The terrorist was from Islamic Jihad, an organization whose headquarters are in Damascus, Syria. (Ironically, during the month of June, Syria presided over the UN Security Council.)

In response to the attack the IDF swung into action in several villages in Judea and Samaria. Tanks surrounded Arafat's headquarters in Ramallah, but by now such maneuvers were greeted with relative silence by the international community. Fourteen months earlier, such an act would have caused international condemnations against Israel.

———

Toward the end of my father's visit to Washington in May, Condoleezza Rice had told Dov "Dubi" Weissglass, my father's chief of staff, that there was a lot of pressure on the administration to issue a statement of some kind on the Israeli-Palestinian conflict. She said it was not clear yet what President Bush would say, but it was evident that he would have to make himself heard on the matter.

In June, understanding the importance of the American position and wanting to influence it as much as possible, my father set out for Washington again to meet with President Bush. It was his sixth such visit, and the second in the course of a month. The trip was delayed by a single day because of a terror attack at Meggido Junction.

During the hour-and-a-half-long meeting on June 10, it was clear to my father that the Americans were leaning in a direction favorable to Israel. Afterward, the president held a press conference. The success of the peace process, he told reporters, was contingent on reform in the Palestinian Authority, and Israel had the right to defend itself. In response to a question about Arafat, President Bush said that Arafat was not the issue, and that he had already voiced his disappointment with his leadership.

———

The terror attacks continued in June. Israelis were murdered in Ofra, Karmei Tsur, Alei Sinai, Itamar, and Herzliya, and in two gruesome terror attacks in Jerusalem. In those terror attacks, dozens were killed, among them children, and more than one hundred people were injured. Javier Solana, in response to the attacks, said, "I am horrified by the level of violence. Innocent civilians continue to be the main victims of this conflict." Still, he did not speak out against Palestinian terror. The Spanish foreign minister, Josep Piqué, said that "the attack was carried out by the enemies of peace and the enemies of the Palestinian people." Again and again, such vague responses avoided the clear and stark condemnation of Palestinian terror.

The attacks in Jerusalem delayed the unveiling of President Bush's Middle East peace initiative. "Today it is hard for people to focus on peace while there are those suffering from terror," was the official White House message. The president refused to condemn Israel for entering the Palestinian areas, emphasizing Israel's right to self-defense.

President Bush's official declaration came on June 24. He expressed his support for the establishment of a Palestinian state but with certain conditions. The president's main points were:

Two states for two peoples.

Combating terror.

New and different Palestinian leadership, the election of Palestinian officials unlinked to terror.

The United States supports the establishment of a Palestinian state whose borders, and certain aspects of its sovereignty, are temporary, until a final Middle East peace deal is reached.

The Palestinian state will be established on the basis of reform—the creation of new political and financial institutions that are rooted in the ideals of democracy and the free market, and decisive action against terror.

The establishment of a Palestinian state demands the founding of an independent and efficient judiciary branch.

The United States will not assist in the establishment of a Palestinian state until its leaders take an active and continuous role in combating terror and dismantling its infrastructure.

The Palestinian state can be swiftly established and can quickly reach agreements with Israel, Jordan, and Egypt on matters such as security. Final territorial borders and the capital will be determined during the final settlement. The Arab states have offered their assistance.

Israel must, with progress on the security front, withdraw to the pre– September 28, 2000 lines, stop settlement, restore freedom of travel, and release the Palestinian funds.

The Israeli occupation begun in 1967 must be resolved in a settlement based on UN Resolutions 338 and 242, with an Israeli withdrawal to safe and recognized borders.

The conflict over Jerusalem and the Palestinian refugees must be resolved, along with a peace agreement with Lebanon and Syria that will support peace and the fight against terror.

The president expects Israel to respond and to work towards an overall peace deal within three years.

Regarding Israel, the president said that the occupation threatens Israel's character and democracy; Israel's actions in the territories must stop.

The president expects the Arab states to tighten their ties with Israel, both diplomatically and economically, normalizing relations between Israel and the Arab world.

By now the picture was clear, and it was a positive one. President Bush had accepted my father's and Israel's position.

Recognition that the Palestinians had to halt their terror before the establishment of a state or the beginning of final peace negotiations was a welcome departure from the previous attitude of world leaders. Bush was the first to accept the Israeli position— that negotiations would not be held under fire. The murders could no longer be used as a negotiating tool.

Kofi Annan initially resisted Bush's call to replace Arafat. Javier Solana, the EU foreign policy chief, concurred. The French foreign minister assailed the Bush speech and expressed dismay that the United States would become the sole actor in the Middle East. "Europe must play an active role in the peace process," he said. As for the Palestinian Authority, the reaction was one of rage; judging from the appeals and protests from the PA and the Arab states, the speech was quite an achievement for Israel.

Eventually the UN, the EU, and Russia all accepted Bush's position. In Europe, though, they still weren't ready to give up on Arafat.

The president did not let up. Two days after his speech he said, "I have faith in the Palestinians. They understand what we're saying and will make the right choice. I can guarantee you that we will not transfer money to a corrupt and nontransparent society, and I believe that other states won't either." With that, he called for the replacement of Arafat and threatened cutting off financial aid to the PA. Discussing the notion of exerting military pressure on the PA, he was even more forceful. "All options are open to us," he said.

Prime Minister Tony Blair of Britain, whom my father visited in London on his way back from Washington, expressed his own disappointment with Arafat. "There must be Palestinian leadership with which one can conduct negotiations," he said, "the kind that wants peace and shuns terror absolutely." At the G8 conference held in Kananaskis, in Alberta, Canada, at the end of June, the Italian prime minister, Silvio Berlusconi, expressed his wholehearted support of Bush's message and called on Arafat to step aside.

The U.S. State Department traditionally shows less empathy for Israel than the White House does and tries to please the Arab world, placing more emphasis on American interests as they understand them and less on its values. But this time the State Department also fell in line with the president. Referring to Arafat, the State Department spokesman said, "We were disappointed again and again." He noted that the United States was not of the opinion that Arafat was capable of instituting the necessary reforms and that the United States was in touch with Palestinian officials regarding alternative leadership. Arafat, Secretary of State Powell said, had "rightfully earned" the president's reproach: "There is a price for not fighting terror and for not instituting reform." And despite his direct talks with Arafat, he acknowledged, "we have not seen any improvement." Powell confirmed that the information they had received regarding the Arafat-approved transfer of money to a terror organization had resulted in the hardening of the American stance. He also noted that President Bush's policy had widespread international support.

In Sudan the representatives of Muslim states issued a statement supporting "the Palestinian people, under the courageous leadership of Yasser Arafat." The Saudis released their own statement criticizing the demand to replace Arafat.

Another person displeased with Arafat's predicament was our own foreign minister, Shimon Peres. In all matters concerning the Palestinians, he still seemed detached from reality. If Arafat would make the reforms, he insisted, he could work with him. In private meetings he criticized President Bush's speech. But despite its critics, the speech represented a significant diplomatic victory, proving the importance of the trust that had evolved between the Israeli prime minister and the American president.

On July 5, my father spoke with Anders Rasmussen, the prime minister of Denmark and the newly appointed president of the European Union:

"I'd like to wish you the best of luck as the president of the European Union."

"Thank you, Mr. Prime Minister."

"I will make every effort so we can tighten our ties during this period," my father said. "I hope the EU position will be balanced. It has not always been so, and you know it."

My father raised the issue of the spread of anti-Semitism in Europe: "The EU states must take the necessary measures to prevent this. It is a dangerous development."

"We will fight anti-Semitism wherever it rears its ugly head," Rasmussen replied.

"I've written down your comments, and we will weigh our steps. In terms of the crisis in the Middle East and the peace process, the Bush speech is a good base.

"I know that you are skeptical about Europe."

"That is based on our experience."

———

Acceptance of the tenets of Bush's speech had yet to be translated into action when my father spoke with Kofi Annan on July 15.

———

On July 16, Palestinian terrorists wearing IDF uniforms set off an explosive charge near an Israeli bus not far from the town of Emmanuel in Samaria. As soon as the charge went off, they opened fire, killing a baby, her father, her grandmother, and five other people. A premature infant born to one of the injured died, as well. Sixteen people were hospitalized. Two days later, two terrorists blew themselves up in Tel Aviv, killing three people and wounding forty.

"This just shows that so long as Arafat is in control there the terror will not stop," my father said to Kofi Annan in response to his now-regular condolence call. "There are no reforms. Anyone can make announcements and declarations, but in actuality nothing is happening. I wrote to you about this. When I asked to speak with you, I did not know there would be an attack. I wanted to call you and tell you that we're interested in taking widespread humanitarian measures. I thought there was an opportunity. I wanted to suggest that. To help those who are not involved in terror. I am speaking of an international mission. I'd like for you to raise this before the Quartet [the United Nations, the United States, the European Union, and Russia]."

"Thank you, Prime Minister, the humanitarian aspect is very important. I am happy that you are personally engaged in this. I will raise the matter and get back to you after the meeting."

And two days later, on July 18, Kofi Annan called again. "Mr. Prime Minister, how are you? As promised, I have called back after the Quartet's meeting. Before that I would like to express my deepest condolences. It is awful that every time we talk, we speak of condolences." Well, it seems he finally noticed.

"There were two suicide bombers today," my father said. "Two days ago seven were murdered. Another baby died today, one day old."

"What a disgrace," said Annan.

"Members of the Quartet were encouraged by your proposal and are interested in getting it implemented."

"Thank you for raising the issue. We are committed to providing humanitarian aid and an international effort in that regard will be important. The important thing is for all the states to exert pressure now and to move as quickly as possible towards the reorganization of the terror organizations."

This was in the hopes that they would shift from terror to security organizations.

"We discussed this yesterday," Annan said, referring to the Quartet meeting.

"I'm glad. The other important thing is how to grant the minister of finance the authority over the budget and the financial side of things He is a nice man. He sits in his room with his secretary but the money goes straight to Arafat. [With] our goodwill all of ours, yours, the Europeans', the Americans'—nothing will move forward as long as the terror goes on."

It's important to remember that Arafat was saddled with some serious expenses; after all he had to support his wife, Suha, who was living in Paris. And everyone knows that the good life in Paris does not come cheap.

While the Europeans said they accepted Bush's stance on terror, it still did not stop them from meeting with Hamas.

On July 28, Alastair Crooke, a representative of the EU, reportedly met with Ahmed Yassin, the head of Hamas, an organization whose openly declared goal is the destruction of Israel. The EU confirmed the contacts with Hamas.

Jacques Chirac, in a conversation with Shimon Peres on July 29, opposed including Hezbollah in the EU's list of terror organizations, despite all of the organization's terror attacks, not only in Lebanon and Israel but also in Argentina, at the Jewish Community Building and the Israeli Embassy, and against the American and French embassies in Kuwait.

In contrast to the Europeans, the U.S. representative to the United Nations, John Negroponte, had informed the Security Council on July 27 that from that point forward the United States would support calls for peace negotiations between Israel and Palestinians only if UN resolutions expressly condemned terror. This was new; previously, the United States had wielded its veto power against anti-Israel proposals in the UN Security Council, but now it made known that it would not so much as review proposals that did not include this element.

On July 31, a Palestinian terrorist planted a backpack full of explosives in the Hebrew University's Frank Sinatra Building in Jerusalem, killing seven people and wounding seventy. Five of the dead were American citizens, mostly students, and the American officials announced that the FBI would take part in investigating the crime.

The following Sunday morning, August 4, a suicide bomber blew himself up on a bus near the grave site of Rabbi Shimon Bar Yochai, not far from Safed in the north. Nine passengers were killed; forty were wounded.

With U.S. secretary of defense Donald Rumsfeld
reviewing troops at the Pentagon, March 2001.
(© Yaacov Saar, Israeli Government Press Office)

Two days later, August 6, Donald Rumsfeld spoke at the Pentagon.

Those problems have been going on since the country was estab-
lished in the late '40s. It is a complicated set of issues. And it has
been—it has tended over time to have been dominated by a couple
of facts. Several. One is periodic warfare. Second is the fact that the
surrounding areas from Israel have preferred that Israel not be there.

And third is that the people that Israel has been trying to in-
teract with and find as interlocutors have, for whatever reason,
not been effective interlocutors. That is to say, they have not had
a structure and an accountability that would enable them to make
a deal or keep a deal. And Barak made a proposal that was as
forthcoming as anyone in the world could ever imagine, and Arafat
turned it down.

If you have a country that's a sliver and you can see three sides of
it from a high hotel building, you've got to be careful what you give
away and to whom you give it. If you're giving it to an entity that

has some track record, that has a degree of accountability, that has the ability to enforce security that's promised in whatever arrangements are made, it seems to me that's one thing. If you're making a deal and yielding territory to an entity that cannot or will not do that—and there is no question but that the Palestinian Authority have been involved with terrorist activities, so that makes it a difficult interlocutor.

My feeling about the so-called occupied territories are that there was a war, Israel urged neighboring countries not to get involved in it once it started, they all jumped in, and they lost a lot of real estate to Israel because Israel prevailed in that conflict. In the intervening period, they've made some settlements in various parts of the so-called occupied area, which was the result of a war, which they won.

They have offered up—successive prime ministers have offered up various portions of that so-called occupied territory, the West Bank, and at no point has it been agreed upon by the other side. I suspect, even in my lifetime, that there will be some sort of an entity that will be established. Maybe it will take some Palestinian expatriates coming back into the region. It may be that the neighboring countries, Egypt and Jordan and Saudi Arabia and others, will have to assist in providing a degree of accountability.

The settlement issues—it's hard to know whether they're settlements in portions of the real estate that will end up with the entity that you make an arrangement with or Israel. So it seems to me focusing on settlements at the present time misses the point. The real point is to get an effective interlocutor. The real point is to get a condition so that you can have a peace agreement. And those are exactly the things that President Bush and Secretary Powell have been working on, and indeed, working particularly with Egypt and Jordan and Saudi Arabia.

Rumsfeld's statement was yet another testament to the strong connection between the United States and Israel during my father's term.

The Europeans had yet to part ways with Arafat, but even their allegiance was beginning to weaken. In August, EU Commissioner for External Relations Chris Patten of the United Kingdom sent Arafat a letter demanding, for the first time, that the PA stop its incitement against Israel. "It is incumbent upon you to take measures to stop the incitement in the PA schools, where incitement to hate and violence is being taught." In addition the EU sent letters to the minister of finance and the minister of planning and international cooperation, demanding that a budgetary law be passed and that a detailed monthly report regarding the use of European funds be sent to EU financial authorities. They also insisted that the funds set aside for investment be included in the general budget so they could monitor the monies and stop them from being stolen. It seemed as if things were finally moving in the right direction.

THIRTY-ONE

With Hosni Mubarak at the Sharm el-Sheikh Summit, February 8, 2005. (© Avi Ohayon, Israeli Government Press Office)

I remember a trip I took with my parents to Egypt when my father was defense minister. As we crossed the desert by car, we saw Egyptian soldiers on the side of the road, not too far apart, their uniforms crisp, their boots polished, standing ramrod straight at attention, for mile upon mile in the open desert. When the Egyptians want to show respect, they most certainly know how to do so, but it came after many long and bitter battles.

Israel's ties with Egypt began with the Egyptian invasion of the young state of Israel, which had just declared its independence. The Egyptian army, along with armies of the other Arab states, ignored UN Resolution 181 and invaded Israel with the express aim of eradicating it. The Egyptians advanced in two columns: one reached Lachish, a stream near Ashdod, some twenty miles south of Tel Aviv. The other reached Kibbutz Ramat Rachel, on the southern edge of Jerusalem, where they were brought to a halt in a heroic battle. As the war continued on, the young Israel Defense Forces managed, amid severe clashes, to surround some of the Egyptian forces—including a young officer by the name of Gamal Abdel Nasser—and break their offensive entirely. The Egyptians also facilitated Palestinian terror from the Gaza Strip, which continued to be under their control until the Six-Day War in 1967.

With my parents at the Suez Canal, January 1982. (Moshe Milner)

Egypt was hostile to Israel even during the years when they controlled the Sinai Peninsula and the Gaza Strip, and Judea and Samaria were in the hands of the Hashemite Kingdom of Jordan. The occupation, as they like to call it, had not yet been born. What's more, neither the Egyptians nor the Jordanians ever considered granting the Palestinians in Judea, Samaria, and Gaza any form of independence in those places. Nor did the Palestinians seek it. Egyptian hostility, then, can only be explained as the manifestation of the nation's hatred for Israel and its desire to eradicate it. "To destroy the Zionist entity" was the accepted phrase then, and it is still heard today, particularly in Iran and Hezbollah in Lebanon. Arab unity in general has also never found a more effective, more unifying cause than enmity for Israel, and it was that single shared sentiment that Nasser employed in his bid for hegemony in the Arab world.

In his meetings with senior Egyptian officials after the signing of the peace accords, my father always spoke highly of the Egyptian soldiers he had faced in battle. The jokes in Israel about the Egyptian soldiers who kicked off their boots and fled the scenes of battle for the desert during the Six-Day War infuriated him.

I particularly remember a meeting in our house with the Egyptian minister of defense, Kamal Hassan Ali, and several Egyptian generals, in the early 1980s when my father was minister of defense. The Egyptian minister looked around and said to my father, "This house is like you—big, strong, and stands alone."

Everyone was seated in the living room, greeting each other. My father knew the other generals in the room only from the IDF intelligence dossiers. But seeing them face-to-face was an emotional experience. Seated in his customary spot on the couch, he said, "I've seen you fight in all the wars and I learned that the Egyptians are brave soldiers who will fight to the death." He emphasized the final three words. It was a compliment they undoubtedly appreciated. But it was also, to the attuned ear, which I am sure they had, a sort of warning—that peace with Israel was far preferable to war—especially when it came from "General Sharon," as they always referred to the man who had vanquished them in every military encounter.

His respect for the Egyptian soldiers notwithstanding, my father realized that maintaining perspective was important. That perspective had to include the fact that Egypt had attacked Israel in the past, without provocation and with the goal to destroy it.

Other Israeli prime ministers have rushed off to Cairo to meet with former Egyptian president Hosni Mubarak, who greeted them patronizingly. My father did things differently. Mubarak's invitations were greeted with my father's stock response: "I am willing to meet in a tent that is set up on the border, half in Israel and half in Egypt."

On August 9, *Yedioth Ahronoth* published the story that became known as the "sausages incident."

Mubarak complained to Foreign Minister Peres that whenever he sent Omar Suleiman, his minister of intelligence, to see my father, his man would receive two hot dogs and a tomato.

"We were so insulted by the incident that I simply could not ignore it," Mubarak said, "and when I spoke to Sharon I protested against the diet that he imposed on my general. And how did Sharon respond: I promise you that next time he'll get three hot dogs."

Mubarak also used to complain that my father did not speak with him for a sufficiently long time: "Our conversations last a minute and a half. That's all."

I sat next to my father in his bedroom on the farm when he teased Mubarak about the hot dogs, and again, later in the conversation, when he said, "Here I am holding the receiver with one hand and my watch with the other so as to ensure that more than a minute and a half has passed." The Egyptian president received exactly the right amount of consideration, nothing more. That was not what Mubarak had grown accustomed to from Israeli leaders.

The only time my father replied positively to the many Egyptian invitations to visit was in February 2005, four years after he was elected, to a summit in Sharm el-Sheikh. That was a onetime gesture after Israeli citizen Azam Azam was freed from prison.

Azam Azam's story is a perfect example of Egyptian hostility toward Israel. On November 5, 1996, Azam, a technician employed by an Israeli-Egyptian textile company, was arrested in Cairo. Eight men from Egypt's security forces apprehended him on the way to his hotel. At first, for several days, the Egyptians denied he was in their custody, and it was unclear what had happened to him. Six months later his trial began. The Egyptians presented the evidence they had against him—two pairs of underwear and a few markers. The prosecution contended that he was planning to send messages to Israel in invisible ink written on his underwear. He was found guilty and sentenced to fifteen years of hard labor. Benjamin Netanyahu was the Israeli prime minister at the time, and all of his attempts to liberate Azam were unsuccessful.

Azam remained in prison for eight years, in a nine-square-foot cell, sleeping on a mat on the ground, with only a bucket for a toilet. The Egyptian Bar Association denounced his Egyptian lawyer, Farid al-Dib, and called for his dismissal. The same organization subsequently wanted to deny the accused the right to representation.

Azam was released on December 5, 2004. Afterward, Mubarak told my father over the phone, "I did it especially for you."

Azam Azam after his release, December 2004.
(© Avi Ohayon, Israeli Government Press Office)

While peace has endured between Israel and Egypt since 1979, regretfully the anti-Jewish and anti-Israeli sentiment in Egypt has not dissipated. The sentiment is expressed in various ways—such as the article that appeared in the state-monitored *al-Ahram* newspaper in October 2000 in which the writer claims, as in the old blood libel, that the Jews of Damascus murdered a Christian holy man and used his ground-up bones and his blood for the purpose of making Passover matzoh.

During that meeting between my father and Mubarak in Sharm el-Sheikh in February 2005, they agreed that each one of them would be interviewed simultaneously by a journalist from the leading newspaper of each country. Smadar Perry, a senior correspondent for the Israeli newspaper *Yediot Ahronoth*, was chosen from the Israeli side. She went to Cairo, where she sat waiting for quite some time because not a single journalist from the Egyptian *al-Ahram* newspaper was willing to go to Israel to interview the nation's prime minister. Although more than a quarter century had elapsed since the signing of the peace accords, the editor of *al-Ahram*, who concurrently served as the chairman of the Egyptian Journalists Union (EJU), banned all ties with Israelis. In the end they found a solution. *Al-Ahram*'s correspondent in Gaza, Ashraf Abu-Alhul, would conduct the interview. He did, but he was tried by the EJU for his offense. Only direct intervention from the president's palace saved the journalist from sanctions. For months afterward he refrained from returning to Egypt.

At times the hostility comes from those in high places. The Egyptian minister of culture, Farouk Hosny, who has been in office for over twenty years, said in May 2008 from the floor of the Egyptian parliament, "If I find any Jewish books in the bookstores of Egypt I guarantee to burn them myself." A year later, when he sought the position of chairman of the United Nations Educational, Scientific and Cultural Organization (UNESCO), the minister recanted his earlier statement. He was not elected.

In the contacts that many Israeli leaders had with former Egyptian president Mubarak, they were outrageously self-depre-cating, though there was no reason for them to behave that way. The groveling and conciliatory approach of Ehud Barak, who succeeded Netanyahu, brought him nothing but scorn from the Egyptians.

It is true that peace with Egypt is important to us. But it is no less important for the Egyptians, since it is the only way they could obtain the Sinai Peninsula, along with Western aid in the amount of billions of dollars. Still, more than thirty years after the signing of the peace accords, the signs of ill will remain. Israel still does not appear on the maps in Egyptian schools.

The depth of the Egyptian public's feeling is reflected in its reaction to an incident that occurred at the UN Building in No-vember 2008. During the course of an interfaith dialogue, Mu-hammad Sayyid Tantawi, the senior Sunni imam of the most important religious institution in Cairo, shook hands with Israeli president Shimon Peres. The handshake caused such a storm of criticism in Egypt that the imam was forced to lie and issued a statement saying, "I did not know whose hand I was shaking."

It is important to remember that agreements with dictatorships are always much more problematic than those with stabilized democracies. Inherent in a dictatorship is the lack of certainty: if and when the existing regime falls, if the new regime is not a Western-type democracy, agreements can be declared null and void. The situation in Egypt after Mubarak's departure is a good example of this uncertainty.

———

Looking at a map of Jordan makes it clear that other than the border with Israel—along the Yarmuch River, the Jordan River, the Dead Sea, the Arava, and into the Gulf of Eilat (a natural border)—all its borders, with Syria, Iraq, and Saudi Ara-bia, were drawn with a ruler, in London. The British promised the Hashemite family heaven and earth for their help in con-

quering the Ottoman Empire in 1915, promises written in letters exchanged between Hussein ibn Ali, Sharif of Mecca, and
Sir Henry McMahon, British high commissioner in Egypt. All
that remains today of these promises is the kingdom of Jordan.
After losing supremacy to the Saud family in 1924, the rulers of
Mecca and Medina, the two holiest sites in Islam, fled the Hijaz,
their ancestral home for hundreds of years. Hussein ibn Ali pro-
nounced himself king of the Arabs and a caliph, in other words a
link in the chain of caliphs that traces its lineage to Muhammad
and the birth of Islam. The many titles bestowed on, and taken
by, the Hashemites—king of Syria, king of Iraq, king of Transjordan, caliph of Amman—are reminiscent of *One Thousand and
One Nights* stories. Although Hussein ibn Ali, after joining hands
with the British and helping defeat the Turks during World War
I, lost control of Saudi Arabia and the holy sites, he received, as
compensation from the British, two kingdoms. One of his sons,
Faisal, was appointed king of Syria in 1920. He ruled for several
months before being ousted when the French were given Mandatory control over Syria, a decision that ignored British promises
to the family. One year later the British anointed this same Faisal
the king of Iraq. This was a comfortable arrangement for the
British. In this way they could perpetuate their control over Iraq
while Faisal enjoyed the pageantry and the honor, serving all the
while as a glorified contractor for the British crown, preserving
their interests.

His brother Abdullah was made king of Transjordan. Both of
these states (Iraq and Jordan) were founded by the British, and
in both instances their leaders were loyal to the British and their
interests. The borders of Jordan were drawn so as to leave a corridor under British control, a path that cut through Syria and Saudi
Arabia, providing territorial contiguity from mandate-era Israel
all the way to Iraq. A military coup in 1958 in Iraq ended the
Hashemites' rule there, leaving Jordan the only land ruled by the
transplanted Saudi family. One can talk as much as one wants

about the Jordanian people, but the truth of the matter is that there is no such thing. There is only a Saudi family crowned by the British as payment for their role in defeating the Turks and as keepers of the British national interest in the region. Alongside them reside the Bedouin tribes, who are considered loyal to the crown, and alongside these are the majority, the Palestinians. In addition, in recent years Iraqi immigrants have settled in Jordan.

King Abdullah and his brother Faisal maintained ties with the Jews in the land of Israel. In 1919 Faisal signed an agreement with the man who was to become Israel's first president, Chaim Weizmann. Their agreement put in writing the Arabs' acceptance of the Balfour Declaration. Abdullah was known for his ties with the leaders of the Jewish community in Mandatory Palestine, and yet, in the type of duality common to our region, ties were one matter and war was another.

The War of Independence ended with Jordan occupying Judea and Samaria, including East Jerusalem and the Old City. These territories were annexed by Jordan, which had changed its official name to the Hashemite Kingdom of Jordan. King Abdullah, with whom Israel maintained unofficial ties, was murdered by a Palestinian soon afterward, in 1951, on the Temple Mount. His grandson, Hussein, sixteen at the time, witnessed the slaying. One year later, at the age of seventeen, he ascended to the throne, his promotion expedited by his father Talal's mental illness. Six years after that, in 1958, Faisal II, his second cousin, the king of Iraq, was hanged in a public square in Baghdad. Then, with the British Empire crumbling, the monarchy of Jordan was left to its fate, a lone remnant of British colonial rule. The memories of his murdered grandfather and that of his hanged second cousin; his country's large Palestinian populace; the secular, presidential, anti-imperialist regimes all around, many of them seeking affiliation with the Soviet Union—all these factors explain the staunch caution that characterized King Hussein's rule until his death. The enduring fear is realistic: King Hussein survived several assassination attempts over the years, and his rule hung by a thread on several occasions, most notably in the late 1950s, when he was faced with Nasser's Pan-Arabism movement and the domestic forces that attempted to undermine him, and in 1970, during the conflict with the PLO on Jordanian soil.

By that time, his rule was in serious jeopardy. Syrian armor had already rolled into the northern part of the country, with the intent of taking down the regime. The Americans requested help from Israel's military.

My father was opposed to aiding Hussein. As a member of the IDF's General Staff, he was in the minority—a position he was familiar with—but he voiced his opinion clearly and definitively. Of course he did not want another pro-Soviet regime, such as Egypt and Syria, on that border; nor was he under the delusion that the demise of the Hashemite regime would bring peace to the eastern border. While he understood the advantages of acquiescing to American requests, he thought that what seemed like American foreign considerations were in this case matters of direct importance to Israel and its future, which should take priority over the concerns of the Americans. His position was that a country with a Palestinian majority, and some Palestinian prime ministers and Palestinian members of parliament, and with one Saudi Arabian family parachuted into the country by the British, had in essence become a Palestinian state. Making that transition official, my father thought, would have radically altered our conflict with the Palestinians. The conflict would have been transformed into a border dispute rather than a conflict over self-determination, human rights, and Israeli rule over a civilian population. The entire Palestinian-Israeli dynamic would have been changed.

His line of thinking—original and groundbreaking at the time—was based on the understanding that territorial disputes are an entirely different matter from disputes regarding Israeli rule over Palestinians in Judea and Samaria, which has caused bitter strife, domestically and internationally. In September 1970,

after the War of Attrition was over and with Israel at the height
of its powers, with the Palestinians and the Arab states still in
disbelief over Israel's overwhelming victory in 1967, my father,
whose own role in securing the victory was so great, foresaw the
problems inherent in the gain of such territory.

In the end his objections were overridden. Acting on the American request, Israel amassed forces on its northeastern border, near
Jordan. The Syrians, understanding the message, withdrew from
Jordan. Thus, it was to Israel that Hussein owed thanks for the
continuation of his rule. Henry Kissinger, the national security
adviser at the time, sent a thank-you letter to Israel.

With King Abdullah in Agaba, Jordan, June 2003.
(© Avi Ohayon, Israeli Government Press Office)

In the menagerie that is the Middle East, the Jordanian monarchy is akin to a pedigreed Siamese cat who is able to relate to all of the predatory animals. He is capable of pleasing

them when he can or lashing out when he feels trapped, all the while maintaining a most noble expression. Let there be no mistake: refined British manners and an excellent education do not change the fact that, as is common in Arab states, the Jordanian monarchy is a dictatorship. I remember speaking with Maya, my father's cousin, and saying to her, "He's nice, but he's a dictator," and her response: "Yes, but he is a nice dictator." The Jordanians' endless maneuvers to maintain power caused them to shift their loyalties over the years. King Hussein met frequently with Israelis but also allowed terrorists to operate freely from within his territory. He attacked Israel in 1967. He did not attack in 1973. He drove Arafat from the country in 1970 after the PLO's attempt to overthrow the king and murder him and then later kissed him warmly. He supported Saddam Hussein during the first Gulf War but was renowned for his ties with the CIA. He was close to Egypt but never trusted the Egyptians and feared both Egypt and Syria. In fact he trusted Israel more, but kept his ties to the Jewish state clandestine until the 1994 peace accord.

King Hussein died in 1999, and his son Abdullah inherited the crown. Relations between King Abdullah and my father were excellent. The king has been hosted at our farm. Their friendship reached the point that the king decided he would like to give my father a purebred Arabian horse. My father tried explaining: "Look, all presents to the prime minister are kept in the basement of the prime minister's office. What will they do with the horse? Tie it up? Who will feed it and take care of it down there?" The king dropped the subject for some time, but when an Israeli paper published photographs of our farm and, in the background, stabled horses, the Jordanians raised the matter again. By then my father's refusal was dangerously close to touching off a diplomatic incident.

Responding to a request from the Mossad, I met with two men from the Mossad and gave them a brief tutorial on the care of horses. "They need the veterinary authorities' authorization," I told them, "and you have to come to the bridge with a horse trailer hitched to a vehicle and then transfer the horse from the Jordanian trailer to the Israeli one." They even showed me the horse's pedigree, detailing his distinguished lineage.

In the end the horse remained in the king's stables in Jordan. Even today, when I ride my horse, Bedouins ask me if it's the king's. My horse—named Wild Spring Mustang—is good-looking, but he is not a gift from the king. (Each one of my three sons contributed a name.)

A bdullah has inherited from his father his respect for mine, as well as his deep fondness for him. Abdullah knew, as did his father, that my father could be counted on. They both wanted to be allied with him. That furthered Israel's interests but also, to no lesser extent, the kingdom's. The Jordanian regime's greatest nightmare is the establishment of a Palestinian state adjacent to Jordan without the barrier of Israel's Jordan Valley. Although they would not publicly admit it, such a development would put the kingdom in danger.

In turn, my father learned to appreciate the benefits of having a friendly neighbor to the east. It's impossible to know what will happen in Jordan in the future; but in our region every moment of quiet is a divine blessing. In addition, unlike the Egyptians, the Jordanians have taken decisive action against the terror stemming from their region, although our common border is long and hard to defend and many Palestinians live in Jordan, in close proximity to that border.

O n November 7, my father spoke with King Abdullah:
      "It's good to hear your voice. How are you?" Abdullah asked.

"Fine. There are some problems, but everything is under control . . . I want to wish you all possible blessings for the holiday [the coming month of Ramadan]."

"Thank you, that is very kind of you," Abdullah said.

"I hope that there will be quiet, and that you enjoy the time. I am pleased with the coordination between our two countries," my father continued. "It's important for both sides. It will help us move forward and strengthen the ties. I would very much like to meet with you."

"I would like that very much, too," Abdullah replied. "Perhaps our people will coordinate a meeting. I know that the lambs are ready."

"The lambs are ready?" my father said. "The whole herd is ready."

The ever-changing loyalties of the Middle East are like the tones of an Arabic song, difficult to comprehend for the Western listener: In June 2009, during a celebration of the ten-year anniversary of King Abdullah's rule to which I was invited, held in Tel Aviv, a slideshow of photos played in a continuous loop on the wall, including photos of Arafat, who had tried to have King Hussein killed on several occasions.

───────────────

By September 2002 there were already signs that the pressure on Arafat was having an impact, even among Palestinians, as criticism began to surface.

These critics did not have the power to alter the policy of terror, but they were being heard as they questioned whether Arafat and the PA, with their reliance on terror, hadn't dragged the Palestinians deeper into poverty, despair, and violence. Arafat's campaign had made things worse for them than for Israel, and it was looking more and more futile. The chaos he nearly created in Jordan in 1970 and the chaos he was able to create in Lebanon until his ouster in 1982 was not duplicated in Israel, despite his most sincere efforts. Take, for example, the words of Nabil Amr, who stepped down from a post in the Palestinian cabinet because of the corruption within the PA. In a letter to Arafat he wrote, "Did we not throw mud in the face of Bill Clinton, who dared to offer us a Palestinian state, with a few adjustments? Were we right? No. After two years of violence we now seek what we rejected. What became of the Palestinian Legislative Council? What have we done with the judiciary? What have we done with the money? What have we done with the bureaucracy? We need a new policy. We need a true and serious dialogue among ourselves, and we need real reform." That same month shots were fired at Amr's house. Two years later Arafat supporters shot him in the legs. This is the Palestinian democracy, a place where you can express an opinion but also be shot in the legs if it turns out not to be the "right" one.

On September 11, 2002, Arafat was handed his first defeat in the Palestinian Legislative Council. As a result, the government he appointed resigned out of fear that it would be ousted in a no-confidence vote, worsening the defeat. It is not that the members of the Palestinian Legislative Council had suddenly become Israel supporters or antiterror activists, but they realized that Arafat's reign of terror and corruption had led nowhere. The messages that my father and his government conveyed had resonated throughout the world and could not be ignored anymore by the Palestinians.

Muhammad Dahlan, who had just stepped down from his post as security adviser to Arafat, was quoted as saying, "The Palestinian leadership did not leave a single mistake unmade."

Dahlan was further quoted as saying, after the slaying of a Palestinian policeman by Hamas operatives, that, "we issue a determined call to arrest the twelve killers involved in the murder, even at the cost of a confrontation. If one police headquarters is burnt down, we'll burn down every one of the Hamas' centers." Five years later, in the summer of 2007, his people fled Hamas operatives and begged Israel to save them— they preferred putting their faith in Israel's hands over those of their own brothers, and they knew very well why. And, Dahlan, the supposed hero of the story, was not in Gaza when his men fled for their lives. His back was giving him trouble at the time and prevented him from commanding his troops in the field.

Even then Mahmoud Abbas sought the seat of power, but he was weak and fearful, as he continues to be. In October 2002 Shimon Peres, then foreign affairs minister, met with Abbas for three hours. In the wake of Peres's report my father jotted down the gist of what Abbas had said. His notes included the following points:

—*My [Sharon's] approach is acceptable to him.*

—*Salam Fayyad [minister of finance] is privy to this information.*

—*Abbas says of himself: I am not a clerk of Arafat's, but rather someone equal in worth to Arafat, who led the Palestinians to ruin.*

—*Israel should not be seen to be appointing [Abbas]. Stop praising him.*

—*American aid to the Palestinians [should come only] after he is elected.*

—*When he takes power, he will also take control of Hamas.*

—*Arafat is not grounded in reality. Solana supports [Abbas], as does Saudi Arabia.*

—*Arafat, according to Abbas, will not let the reforms pass but they will force his hand. He [Abbas] was very determined.*

As courageous as Mahmoud Abbas was in this conversation with an Israeli, all this could only be said in secret, for he knew that if any of it leaked he would be a dead man, as he himself said to Peres.

And the terror continued. In between the high-profile attacks the deadly trickle of terror continued, claiming two or three lives each time. While he was dealing with the Palestinians' terror attacks on Israel, my father was increasingly worried about the threat from Iran. On November 5, 2002, the London *Times* published an interview with him in which he touched on issues that he had been dealing with for a long time. "Iran is a center of world terror," he said, "and it without doubt should be included in the axis of evil. The matter is dangerous to the Middle East, to Israel, and to Europe. Once the action against Baghdad is over, the action must be continued against Tehran."

During the 2003 general election.
(© Avi Ohayon, Israeli Government Press Office)

THIRTY-TWO

O n the night of November 11, 2002, a member of Arafat's Fatah and a son of a member of the Palestinian Security Apparatus snuck into Kibbutz Metzer and murdered a woman and then a man whom he encountered on the kibbutz path. Then he broke into the apartment of thirty-four-year-old Ravital Ohayoun and killed her and her two children, Matan, five, and Noam, four.

On the evening of November 15, twelve Israelis were killed in Hebron, on the route Jews from Kiryat Arba take to and from the Tomb of the Patriarchs on the Sabbath. Four soldiers were killed in the ambush, and five border policemen, including Superintendent Samih Sawidan from the Bedouin village of Aramsheh in the Galilee, who had assumed command after Colonel Dror Weinberg fell. Three other civilians, members of Kiryat Arba's security squad, were also killed.

On November 20, my father's political adviser, Shalom Turgeman, delivered a handwritten memo to him.

Prime Minister,

I spoke yesterday with Andrey Vdovin—the Russian envoy— who returned today to Moscow . . . He met with Ahmed Qurei, speaker of the Palestinian parliament, who heaped compliments on you, saying to him that you are a leader with strategic foresight and the only one with true plans and that at this time you are the only one they feel they can work with in a practical and not just declaratory way.

He met with Mahmoud Abbas, who had similar things to say, but he is very angry about the leaks to the press regarding his meetings with Israelis.

Both of them stated that they hope that after the elections they will have the opportunity to reach a breakthrough with you, starting with a considerable improvement of the security situation.

It is possible that Mahmoud Abbas and Abu-Alaa understood that the path they had followed for the previous two years had taken them nowhere. They might well have liked the idea that Arafat had been personally invalidated and they would take his place, even if they were too weak and fearful to say so. When speaking publicly, though, they expressed the opposite sentiment. In fact, on that very day, November 20, the day Amram Mitzna, an opponent of my father's, won the post of party chairman in the Labor Party's elections, Abu-Alaa expressed his faith in Mitzna, saying, "Mitzna's plan has resuscitated the hope for the reestablishment of negotiations on the path to peace. If he is elected prime minister, a permanent accord can be reached within a year."

On the morning of November 21, a terrorist boarded a Number 20 bus in Jerusalem filled with passengers, many of them children on their way to school. Eleven were killed; fifty were injured. The victims included a mother and her daughter, a grandmother and her grandchild.

Prior to the attacks of November 15 and November 21, the IDF had retreated from Bethlehem and Hebron; authority in those two cities had been transferred to the Palestinians, with the understanding that they would prevent terror. On the day after the attack in Jerusalem, it had already been decided—during a security meeting in the prime minister's office in Tel Aviv—that these understandings with the Palestinians were now canceled.

The following day, Israeli forces entered Bethlehem, the city from which the terrorist had departed. Aside from increasing protection for civilians and arresting terrorists, they secured the Church of the Nativity to prevent the terrorists from seeking a haven there, as they had done during Operation Defensive Shield. Both attacks had been launched from areas that were supposed to be under Palestinian control.

Arafat, realizing once again that he would not be on the world stage, celebrating in Bethlehem, canceled all Christmas festivities. "*Ma-fish* Christmas," he said in Arabic: Christmas is over with. He added, "The reoccupation of Bethlehem is an international crime, and the world maintains its silence. It is hard to believe." But there were festivities in Bethlehem, and the mass, of course, took place—only without him.

———

The day the Likud Party held their primary to determine who would be their party's candidate for prime minister—November 28—began with two synchronized attacks in the vacation destination of Mombasa, Kenya.

An explosive-laden car blew up at the entrance to the Paradise Hotel, a popular destination for Israelis. The blast, orchestrated by al-Qaeda, killed fifteen people, three of them Israeli, and injured eighty. Shortly afterward, terrorists fired surface-to-air missiles at an Israeli Arkia flight as it took off from Mombasa, bound for Israel. Thankfully, the missiles failed to hit their target.

The primary proceeded; Benjamin Netanyahu ran against my father. Then the day continued with an attack perpetrated by two terrorists—both Fatah members—in Beit Shean. They threw grenades and fired their AK-47s at the Likud Branch in the city, where people had congregated to vote. They also fired at the central bus station. All told, they killed six people and injured dozens more.

The primaries ended with a sweeping victory for my father.

———

The London Conference on Palestinian Reform, scheduled for mid-January of 2003, was being organized by the British to discuss reform in the Palestinian Authority. At this time there was some tension between the Israelis and the British, since, in response to Operation Defensive Shield, Britain had decided to "punish" Israel with an embargo on weapons.

On January 5, two suicide bombers blew themselves up in unison in south Tel Aviv, near the old central bus station. Twenty-three people were killed, and over one hundred injured. The terrorists were members of Fatah; the group rushed to take credit for the attack even as the PA condemned it, in a classic case of schizophrenia.

In response to the attack, Israel barred the Palestinian delegation from attending the conference in London. In any event, my father never accepted the notion that Arafat might ever reform the PA, for had he acted rationally, there would be no need for these reforms in the first place.

The British, however, wanted the event to go forward, and so on January 14 the conference began. It was widely attended, but neither an Israeli nor a Palestinian delegation was present.

The president of the EU issued a condemnation: "The [Israeli] decision [to exclude the Palestinians] perpetuates hatred and extremism." The EU did not seem to consider the over one hundred victims of the attack nine days previously.

Following the conference, my father received a handwritten memo from Shalom Turgeman, his political adviser, summarizing the events there.

PM,

. . . The British foreign minister [Jack] Straw addressed the media and said that [the conference] was a comprehensive and effective meeting in which security issues were raised and the commitments of the Palestinians to stopping all violent actions [were] welcomed . . . He noted favorably the improvement in the economic reforms and noted that they discussed the governmental reforms re-

garding a constitution and the judiciary. He noted that the Palestin-
ians have committed to presenting a new draft of their constitution
within the next two weeks and that it will establish the parameters
of the role of prime minister and government within the framework
of a presidential regime. In terms of the security issues, he said they
had received assurances from the Palestinians regarding progress
on the security front that would enable Israel to live in peace and
tranquility.

Turgeman's second memo of the day concerned the statements of British Prime Minister Tony Blair.

PM,
*   The ambassador [Zvi Shtauber] in London has met with PM*
Blair and passed him your letter. Blair said that he is a true friend
of Israel's; that he was so in the past and will continue to be so in the
future. Blair said he understands Israel's difficulty as a small and
sole democracy surrounded by non-democratic enemies.
*   . . . Of his meeting with the Syrian president, [Blair] . . . said that*
[it] was severe and that he criticized him sternly for the fact of terrorist
headquarters being situated in Damascus and the need to disassemble
them. Blair noted of his own accord that he will work towards lifting
the arms embargo on Israel . . . as soon as possible.
*   Blair did not voice any criticism about the Israeli decision to pre-*
vent the Palestinians from participating in the conference. He noted
that he understands Israel's security needs.

No diplomatic harm had been done.

———————

Since the brand "Sharon" was very strong in Israel, while the brand "Likud" was weak. The election campaign, again run by Reuven Adler, focused on Sharon. Its slogan—"The People Want Sharon"—was on the mark; he enjoyed an unprecedented percentage of public support.

On January 28, 2003, Israel's elections shifted the balance of power in the Knesset: the Likud Party led by my father doubled its previous seats under Netanyahu to thirty-eight, while the Labor party made the worst showing in its history, by adding only nineteen seats. (In the elections held in 2006, in which my father could not participate, the Kadima Party, which he founded, won twenty-eight seats. The Likud won twelve seats. He won the election without even running. All the polls and indications showed that if he had led Kadima, the party would have won over forty seats. It was, retroactively, yet another demonstration that he carried on his back most of the Likud parliament members that were elected in 2003, many of them the same members who later made the running of the coalition difficult.)

After winning the election, my father was faced with the task of assembling his cabinet. Netanyahu, who had replaced Peres as foreign minister after Labor left the coalition in November 2002, was the obvious choice. After all, speaking is something he knows how to do. My father intended to appoint him foreign minister and Ehud Olmert finance minister, a position that had already been promised to him.

As the process of assembling the cabinet came to a close, I realized that the self-evident move is perhaps not the right one on this occasion. I spoke to my father and suggested that he offer Netanyahu the position of finance minister. Two and a half years of Palestinian terror had wreaked havoc with the country's economy and Netanyahu, I believed, with a strong prime minister above him, could make a good finance minister. In addition, my father handled matters of diplomacy on his own, in a very intensive manner.

Netanyahu was not overjoyed at the prospect: he had envisioned himself touring the world and speaking, and the offer came as a surprise. But he could not turn it down. How would he explain such a refusal? And in any event, he probably did not want to stay out of the government and go back home.

The remaining loose end was Olmert. Since my father had already promised him the position of minister of finance, Olmert played the role of injured party. By way of compensation, he received a wide-ranging portfolio that included the Ministry of Industry and Trade along with the Israeli Employment Services and the Israel Broadcasting Authority. He sought an additional appointment: deputy prime minister. He received that role, too, and that was a mistake. Speaking from firsthand knowledge, I know that my father did not for a moment believe that Olmert should replace him or that he was worthy of the role; nor did he intend to allow that to happen. My father relied too heavily on his strength and his good health, and the scenario that eventually occurred never crossed his mind. Moreover, he had no intention of bestowing that title on Olmert again after 2006, but, unfortunately, my father was not able to participate in those elections.

The most conspicuous interactions between my father and Olmert during cabinet meetings in which Olmert participated would occur when my father would fling a comment in Olmert's direction, blessing him upon his return to the homeland from yet another visit abroad. Olmert, for his part, would pass on insulted notes.

Olmert is smug as well as arrogant. He assumed the office of prime minister without the necessary awe and reverence. Olmert is a shrewd individual, but his main problem, in my opinion, is that he preferred his own good over the country's.

In the end, his flaws did not serve him well. By the time he was forced from the post of prime minister, he had no public support. As he himself testified: "I am an unpopular prime minister."

On January 29, the day after the elections, Arafat granted Israel's Channel 10 an interview. In response to a question about a possible meeting with my father, Arafat answered immediately, "Tonight! We want to reestablish the negotiations." He said this even after my father defined him as "irrelevant," isolated him, and encouraged the PA to find an alternative.

The response from my father's office was negative. The distinguishing characteristics in the few phone conversations they did have were the obsequious mumblings of Arafat, which, my father told us, disgusted him. "The chairman of the Palestinian Authority, Yasser Arafat, has continued to fund, initiate, cultivate, and launch terror and will not be a partner to negotiations. As part of Israel's efforts to progress with the political negotiations and bring quiet, and in the end peace, Israel will be willing to speak only with Palestinians who are not involved in terror of any shape or sort."

On February 28, a month after the elections and two days after his seventy-fifth birthday, my father presented his new government.

There was little opportunity for basking in the election victory, though, for the Palestinians had their own plans that day. At around two o'clock in the afternoon a Palestinian suicide bomber blew himself up while riding on Haifa's 37 bus line, killing fifteen people, most of them teenagers. Among the dead were a Druze girl, Kamar Abu Hamed, twelve, from Daliyat al-Karmel, and Abigail Leitel, fourteen, originally from New Hampshire. The shrapnel did not discriminate between Arab and Jewish victims: the driver of the bus, a Christian Arab from Shfaram, was wounded. The terrorist, for his part, carried a letter that praised the 9/11 attacks.

"The President condemns in the strongest terms today's attack on innocents in Israel," noted a statement from the White House. "The President stands strongly with the people of Israel in fighting terrorism."

Once again, the reactions from some of the Europeans and from the UN were equivocal. Foreign Secretary Jack Straw of

Britain said, "There is no justification for attacks on innocent civilians. Attacks like these will not help the Palestinian cause. Once again I urge all parties to do everything they can to prevent further bloodshed on both sides." EU foreign policy chief Javier Solana expressed his condolences to "all the family members and friends of those who lost their lives over the past few days in the Middle East."

UN secretary-general Kofi Annan condemned the attack—even labeling it "a terror attack"—but was careful to call on both sides to avoid "this dangerous and inhuman situation." These leaders spoke of "the Palestinian cause," violence on both sides, and the very general harming of civilians all across the Middle East.

For some, there still remained a failure to distinguish between the moral positions of terrorists who board a bus full of children and blow themselves up with the intent of killing as many people as possible and those who, in trying to stop them, inadvertently kill civilians, among whom the terrorists knowingly hide.

E ventually, the efforts to isolate Arafat and to find an alternative to him came to a head. The Quartet had been using its envoys to exert pressure on Arafat to appoint a prime minister, who would tend to matters of state, making Arafat's role, as president, largely symbolic. The factor that most likely pressed him into accepting the change was the coming war in Iraq. Undoubtedly he realized that after Saddam Hussein was deposed, his own name would be high on the American list. Refusal to respond to the Quartet's pressure would leave him in the hands of the Americans and the Israelis without European backing.

In the end, Arafat agreed to appoint a prime minister, Mahmoud Abbas. On March 8, 2003, in a nighttime meeting in Ramallah, the Palestinian constitution was amended and ratified.

Three days later, on March 11, the Russian envoy, Andrey Vdovin, arrived in Ramallah and spoke with Shalom Turgeman over the phone. Turgeman reported on this conversation with Vdovin in a handwritten message to my father. Mahmoud Abbas had told Vdovin that "he will need Israeli assistance, especially regarding security matters. [Mahmoud Abbas] will want to talk to you immediately after he forms a government . . . to work with you to establish a security plan." Turgeman further conveyed Vdovin's impression that "[Abbas] is serious in his intentions and that he is interested in taking decisive steps against terror." Turgeman also passed on Vdovin's observation that "after a long period without hope . . . an opportunity has been created for progress towards a war on terror that will bring quiet, and that [Mahmoud Abbas] is primed and ready to do just that."

In closed meetings Mahmoud Abbas requested Israeli help. This request is in contrast to his public rhetoric, which is due to his fear of his own people. This big difference makes achieving peace agreements with the Palestinians difficult, if not impossible, to achieve: agreements by nature require concessions from both sides, and the Palestinian leadership cannot stand in front of their people and tell them compromises are necessary. This is by definition an inherent and internal contradiction.

THIRTY-THREE

With Mahmoud Abbas and President Bush at the Aqaba Summit,
June 2003. (© Avi Ohayon, Israeli Government Press Office)

Going back to October 22, the presentation of the road
map for peace between Israel and the Palestinians drew
near, as did the U.S. war against Iraq. These two events
were not disconnected. Not only were they attempting to remove
a tyrant from an Arab country but they also were serious in
their intention to deal with an issue of great concern to the Arab
states—the Israeli-Palestinian conflict. There were those in Israel
who felt that toppling Saddam Hussein's regime and the extended
presence of American forces in Iraq would mean the end of what
is known as the danger from the "eastern front." The threat was

that Iraqi forces, among others, could go through Jordan and attack Israel from the east. My father was not one of them. In an August 2, 2003, interview with the *Herald Tribune*, he was asked if he thought the toppling of Saddam Hussein's regime would prompt change throughout the Middle East.

"The Americas will not be in Iraq forever and Israel will stay, and I have yet to see a democratic Arab state."

In mid-October 2002, during the seventh of my father's twelve visits to the United States as prime minister, he had received a first draft of the peace plan—or, as it came to be known, the Road Map.

Israel had studied the document and rejected it. As it stood at the time, it did not reflect the outline for peace given in President Bush's June 24 speech. The demands that the Palestinians cease terror and incitement were not sharp or clear enough. Moreover, the stages were not sufficiently delineated, and it was not clear that progress hinged on the completion of each stage.

My father saw no purpose in progressing toward negotiations while terror continued. Were he to begin such discussions, he knew, the terror would be used as leverage during negotiations. Nor did he feel there was any point in proceeding toward peace talks when the central tenet of the Palestinians' commitment to peace—the cessation of terror—was not being met.

In November the Americans sent an additional draft of the Road Map. Israel had a list of reservations regarding the plan. Dubi Weissglass, my father's chief of staff, was chosen to head the team that would conduct discussions with the Americans.

On April 6, 2003—two and a half weeks after the beginning of the war in Iraq—as Weissglass and the rest of the team prepared for a visit to Washington, they met with my father. He summed up his instructions to the team in a handwritten note that included the issue of Palestinian claims of return and the insistence that Israel's security needs were to be determined by Israel alone. "Once victory in Iraq is achieved," the note said, "Israel still faces national security issues—*Iran, Libya, Syria, local, regional,* and *international* terror." As to the rest of the arrangements, Israel wanted "*minimal* European involvement." One of the instructions from my father was that the delegation was to "explain what we cannot accept and *insist* upon it."

The point in which he mentions explaining to the Americans "what we cannot accept" is very typical of my father—he kept things clear, aboveboard, saying, this is what we can do, this is what we cannot do.

Israel submitted fourteen reservations about the plan prior to its official submission. These included fighting terror and its incitement, the dismantling of the terror organizations, fulfilling one stage before going to the next one, and the right of Israel to exist as a Jewish state.

O n April 30, at close to one o'clock in the morning, a Muslim British national blew himself up at the entrance to Mike's Place, a seaside pub in Tel Aviv, during a live jam session. Three people were killed—two musicians and a waitress, a new immigrant from France.

That same afternoon, as previously scheduled, the Road Map was submitted to both the Palestinians and the Israelis. My father received his copy from the American ambassador, Daniel Kurtzer. The document contained no surprises: after months of meetings and discussions, the terminology was well-known.

On May 1, 2003, a senior White House delegation, including Steve Hadley and Elliott Abrams, came to see my father and discuss the Road Map. "With all due respect to our friends," he told them, "the security of the state of Israel will always stay in Israeli hands."

It was of great importance to the Americans that the Israeli government approve the Road Map. Likewise, it was of great importance to my father that the Americans accept Israel's fourteen points regarding the plan. It was agreed that there would be "further negotiations regarding our remarks—the team [headed by Weissglass] would continue to work on it and would be in Washington in May."

W hen the Israeli team arrived in Washington later that month, the two main issues on the table were Israel's fourteen points and the freeze on expansion in the settlements.

The Americans said that they could not amend the plan because if they did, the Arabs would demand their own amendments, and the matter would never end. However, the White House did agree to issue a statement acknowledging and addressing Israel's remarks.

Israel released its own statement two days after the American one, agreeing "to accept the steps set out in the Road Map," and resolving "that all of Israel's comments, as addressed in the Administration's statement, will be implemented in full during the implementation phase of the Road Map."

The matter of settlements was a major issue for Israel. Since 1967, the U.S. administration's policy had been against Israeli expansion beyond the pre-1967 boundaries, the so-called Green Line. The Road Map's wording of that position—"Consistent with the Mitchell Report, Government of Israel freezes all settlement activity (including natural growth of settlements)"—was unacceptable to Israel. During the April 2003 discussions regarding Israel's fourteen points on the Road Map, it had been decided that the matter of settlements would be discussed in a separate forum.

D uring that same May visit, Steve Hadley and Elliott Abrams discussed the matter with my father and Weissglass's negotiation team. Their agreements were as follows:

No new settlements would be founded.

No Palestinian land would be appropriated for the purpose of settlement.

No construction in the settlements, aside from within the already constructed areas.

No budgetary funds earmarked for the purpose of promoting settlement.

In mid-May, the points agreed to were approved by the White House, ending the discussions regarding the Road Map. The determination of the borderlines of where construction would be possible within the settlements, set to be done by a joint Israeli American team, never happened, but the agreement still stands.

Six years later, on June 5 and 17, 2009, President Barack Obama's secretary of state, Hillary Clinton, denied the existing agreements between Israel and the United States on settlements: "There is no memorialization of any informal and oral agreements. If they did occur, which of course, people say they did, they did not become part of the official position of the United States government."

Elliott Abrams, special assistant to President Bush and the National Security Council's senior director for Near East and North African affairs, responded to Clinton's assertions in a June 26, 2009, column in the *Wall Street Journal*:

Despite fervent denials by Obama administration officials, there were indeed agreements between Israel and the United States regarding the growth of Israeli settlements on the West Bank . . .

In the spring of 2003, U.S. officials (including me) held wide-ranging discussions with then Prime Minister Ariel Sharon in Jerusalem. The "Roadmap for Peace" between Israel and the Palestinians had been written. President George W. Bush had endorsed Palestinian statehood, but only if the Palestinians eliminated terror. He had broken with Yasser Arafat, but Arafat still ruled in the Palestinian territories. Israel had defeated the intifada, so what was next?

We asked Mr. Sharon about freezing the West Bank settlements. I recall him asking, by way of reply, what did that mean for the settlers? They live there, he said, they serve in elite army units, and they marry. Should he tell them to have no more children, or move?

We discussed some approaches: Could he agree there would be no additional settlements? New construction only inside settlements, without expanding them physically? Could he agree there would be no additional land taken for settlements?

As we talked several principles emerged. The father of the settlements now agreed that limits must be placed on the settlements; more fundamentally, the old foe of the Palestinians could—under certain conditions—now agree to Palestinian statehood . . .

In recent weeks, American officials have denied that any agreement on settlements existed. Secretary of State Hillary Clinton stated on June 17 that "in looking at the history of the Bush administration, there were no informal or oral enforceable agreements. That has been verified by the official record of the administration and by the personnel in the positions of responsibility."

These statements are incorrect. Not only were there agreements, but the prime minister of Israel relied on them in undertaking a wrenching political reorientation . . .

Clinton's statement was, to put it delicately, inaccurate. It is problematic for the Israeli-American relationship if a new American administration denies the understanding the previous one had with Israel. That is especially the case when it comes to our agreements with the Arabs in which the American commitment is a major factor and influences Israel's ability to compromise. There were those in Israel who were immediately reminded of Hillary Clinton's visit to the PA in November 1999: the then first lady did not protest when Suha Arafat accused the Israelis of poisoning the Palestinians' water wells. Mrs. Clinton listened politely. The two kissed, then Hillary went back to Washington and Suha rushed back to Paris to continue taking care of the needy Palestinians.

Dubi Weissglass returned from the United States on May 16, 2003. My father summarized his meeting with Weissglass in a handwritten note, including the following points:

The Americans are concerned about Mahmoud Abbas's weakness.

On security, we should do whatever we determine is necessary.

The president will request that steps be taken on the matter of settlements.

On the matter of permanent settlement—
Will not discuss now. They want me to say this:

a. That the Palestinian state, if and when it is founded, be a state that has vitality, territorial contiguity and economic logic.

b. That Israel will not use public funds to finance the growth of the Jewish population in Judea and Samaria.

c. That Israel will not build new settlements and will freeze new construction except for that which is done within the already existing constructed areas.

d. That there will be no land appropriation for the purpose of settlement.

e. That several "unauthorized" outposts will be dismantled.

f. That a public statement will be issued saying that the separation fence is for security purposes.

Ten days after the visit four separate declarations will be published. An Israeli one (the six points aforementioned). An American one: making a commitment to Israel's security, a demand to cease terror, allowing for American overseers, and allowing Israel to be freed of responsibility if the process falls apart. A Palestinian one: declaring that Israel has the right to exist as a Jewish state. Terror is to be condemned and abandoned as a means of attaining political goals. There will be a complete cessation of incitement. From the Arab states: recognizing Israel's right to exist and [recognizing] it as a state (diplomats, trade etc.). Arab states will also have to announce their support for a complete cessation of terror.

—The Quartet will not have a role of its own within this arrangement, aside from welcoming each of the announcements.

———

My father's past experience made him skeptical that the Palestinians would fulfill the obligations as outlined in the Road Map. He therefore saw it as a great advantage that the president's plan was contingent upon the Palestinians' ability to fulfill each stage of the agreement. If they truly fought terror and dismantled the terror organizations, then progress was possible. Under the new plan, if the Palestinians failed to do their part, then Israel would be absolved of its responsibility, and no attempts would be made to compel her to accept other solutions such as the Saudi initiative.

Still, the restrictions on settlement were hard for him to swallow. He was indeed, as Abrams would later call him, the father of the settlement movement. But the agreement he had reached allowed for continued growth in the places Israel most needed to keep. He knew that if there were ever to be a peace deal with the Palestinians, then not all of the settlements could be maintained. At the same time, he felt strongly that Israel could simply not afford to return to the borders of 1967, because they were indefensible.

In earlier governmental roles, before he became prime minister, my father made sure that the settlements he considered critical would have ample space to grow. The understandings with the Americans about the settlements was the proper balance between Israel's internal needs and its diplomatic ties, primarily its special relationship with the United States.

———

Once the Road Map had been shown to both parties, as if to emphasize what they thought about the first stage of the Road Map—fighting terror—the Palestinians responded with a series of terror attacks. On May 5, 2003, a Jew was shot near the settlement of Shvut Rachel; his six-year-old daughter was severely wounded, as was another man. On May 9 a six-year-old girl in Sderot was injured by one of six rockets fired toward the city from Gaza. Two days later a Jewish father of six was murdered in an ambush outside Ofra. On Saturday night, May 17, in Hebron, a suicide bomber killed a couple on their way back from prayer.

Late that night my father met with Mahmoud Abbas, the newly appointed prime minister of the PA, and two of his men in Jerusalem.

Abbas spoke first. "You are a historic leader," he said to my father. "You can make decisions; the decision is in your hands. We have suffered, us and you, over the past two years. Now we have the opportunity to better the situation for us and for you."

"Two hours ago," my father responded, "terrorists killed a Jewish couple in Hebron. How can we make progress with that kind of terror?"

The Palestinians asked that "I not give up on the peace process [with them]," my father wrote in his notes from the meeting. "This was the first meeting with [Mahmoud Abbas] since he assembled his government, and he asked to be given a chance. The conversation was open."

My father made the Palestinians an offer that, based on their demands and comments, they should have happily embraced— that they assume responsibility for security in their areas and, to make things easier, do it in stages. In Gaza, for instance, the IDF had not been regularly active, as in Judea and Samaria, and the PA's forces there, tens of thousands strong, had not been harmed, so there was nothing barring them from taking action. In the West Bank, too, certain limited areas that were not problematic could serve as a starting point. Typically, Mahmoud Abbas explained how weak he was. "They are not yet ready to take responsibility for security matters in certain parts of Gaza, Judea, and Samaria," my father wrote in his notes.

In stark contrast, the Palestinians publicly demanded that the IDF withdraw to the pre–Second Intifada lines of October 2000.

When the Americans were told about my father's offer and the Palestinian refusal, they expressed interest in the details. On May 18, my father received a note from Dubi Weissglass.

Elliott Abrams reports:

1. The president was consulted and updated about your offer to the Palestinians regarding the new deployment of IDF troops.

2. The president instructed them to urgently ask us for the exact wording of the offer and the refusal, among other things, in order to brief the press.

3. Elliott asked whether I can be present on Wednesday for a consultation regarding the reality that has been created.

4. There is the feeling (!) that they understand that they must be tougher with the Palestinians and/or Arafat.

The cabinet's acceptance of the Road Map, along with the fourteen Israeli reservations, paved the way for a summit meeting. President Bush was coming to the region, and the Egyptians wanted the summit—with my father, Mahmoud Abbas, and Bush—to take place in Sharm el-Sheikh. My father was unwilling to go to Egypt as long as Azam Azam remained in an Egyptian jail, so, much to the Egyptians' regret, the summit was slated for Aqaba, Jordan.

One day before the summit was to start, leaders of the Arab world convened in Sharm el-Sheikh with President Bush. "All progress towards peace requires the rejection of terror," Bush said in his closing remarks. "The leaders here today have declared their firm rejection of terror, regardless of its justifications or motives. They've also committed to practical actions to use all means to cut off assistance, including arms and financing, to any terror group, and to aid the Palestinian Authority in their own fight against terror." Obviously, one can draw from this that at least until this commitment, they had financed terror. And I wouldn't bet that they stopped after this "commitment."

The Aqaba summit, held on June 4, 2003, was a festive affair. My father spoke of the land of Israel as "the cradle of the Jewish people . . . My paramount responsibility," he said, "is the security of the people of Israel and of the State of Israel. There can be no compromise with terror, and Israel, together with all free nations, will continue fighting terrorism until its final defeat . . . It is in Israel's interest not to govern the Palestinians but for the Palestinians to govern themselves in their own state."

Mahmoud Abbas spoke as well. "Our goal is two states, Israel and Palestine, living side by side in peace and security," he said. He described the suffering of the Palestinians in the years since 1967. "At the same time," he said, "we do not ignore the suffering of the Jews throughout history.

". . . We repeat the renunciation of terrorism against the Israelis wherever they might be. Such methods are inconsistent with our religious and moral traditions and are a dangerous obstacle to the achievement of an independent sovereign state we seek. These methods also conflict with the kinds of state we wish to build based on human rights and the rule of law . . .

"Our goal is clear . . . and we will implement it firmly and without compromise: a complete end to violence and terrorism. And we will be full partners in the international war against terrorism."

Undeniably, the man could write a good speech. As for implementing these points, that was another matter.

President Bush spoke of peace, the war on terror, and the establishment of a Palestinian state. "America is strongly committed, and I am strongly committed," he said, "to Israel's security as a vibrant Jewish state." This seemingly simple sentence stands out in contrast to the ongoing unwillingness on the part of the Palestinians to recognize Israel as a Jewish state. They demand a Palestinian state and, instead of Israel, an additional binational state formed by the free entry of Palestinian refugees to the state from across the Arab world.

The Palestinians, as usual, said one thing and did another. Four days after the summit, on June 8, three Palestinian gunmen joined a group of Gazan workmen and infiltrated an Israeli army base near the Erez checkpoint and opened fire, killing four soldiers before they themselves were killed. The three terrorists were from three separate terror organizations—Fatah, Islamic Jihad, and Hamas. That same day, a soldier was shot and killed in Hebron.

Four days later, on June 12, a terrorist dressed as an ultra-Orthodox Jew blew himself up on the Number 14 bus in Jerusalem. Sixteen people were killed, and over one hundred were injured.

My father convened the cabinet, telling the ministers that the operations against the heads of the terror organizations would continue. "Of course we have no intention of waiting for the Palestinian security apparatus to successfully stop terror attacks," he said. He called the heads of the Palestinian Authority "crybabies" who allowed the terror to run rampant, and referred to Mahmoud Abbas as "a chick that had not yet grown its feathers."

"What's he whining about all the time?" I heard my father say on several occasions. "He has tens of thousands of men, in Gaza we didn't touch any of his people, go right ahead, by all means, who's stopping him from acting?" And yet the central thrust of Abbas's conversations with my father was how weak he was and how he needed time to build up strength so that he could take action.

The Palestinians wanted to use a temporary cease-fire as a means of sidestepping their commitment to dismantling the terror organizations. They clearly used this to avoid the unpleasantness of the task and rid themselves of Israeli and American pressure. (At the same time, Abbas tried to bring Hamas into his government, offering them a national unity government, but the initiative failed.) My father did not fall for that trick, of collaborating with rather than defeating terror.

But there was no real issue in dealing with the evasive Palestinian maneuver; the terror machine never stopped working.

On June 29 the Fatah, and Hamas and Islamic Jihad, declared a three-month cease-fire—as they put it, "the suspension of military operations against the Zionist enemy for the period of three months, beginning today."

"Anything that reduces violence is a step in the right direction," White House spokeswoman Ashley Snee said in response, adding, "Under the road map, parties have an obligation to dismantle terrorist infrastructures. There is more work to be done."

On July 8, the Palestinians illustrated how they interpret a cease-fire. A suicide bomber blew himself up in a house in the Village of Kfar Yavetz, just north of Kfar Malal in the Sharon region. Mazal Ofri was killed in the blast; her three grandchildren were injured.

On July 13 my father flew to the United Kingdom. Prime Minister Blair and Foreign Secretary Straw did everything in their power to make the visit a success. "We are Israel's best friends in Europe," Blair said during their meeting at 10 Downing Street.

But they did not see eye to eye on everything. My father wanted Arafat completely isolated in his headquarters in Ramallah. "Today it is clear to all that Arafat is the main obstacle in the path of progress between us and the Palestinians. He tries to subvert every one of Mahmoud Abbas's initiatives." Every meeting with European representatives, my father said, only served "to weaken Abbas."

The British, however, refused to boycott Arafat.

My father spoke of the illusory nature of the cease-fire: "If it [the dismantling of the terror organizations] doesn't happen, then it endangers the existence of the state of Israel but is an even greater danger to the Palestinian Authority." He saw the cessation of hostilities as a charade: "It is a so-called cease-fire that will allow them to re-arm themselves and prepare for more terror attacks. Actual Palestinian [government] operations against these groups [have] yet to begin."

To both European and Palestinian leaders, my father's position that the terror organizations threatened the existence of the Palestinian Authority seemed to be deeply hypothetical in the summer of 2003. The Palestinians preferred to avoid dealing with Hamas and Islamic Jihad and their like, but four years later PA troops were beaten by Hamas and shamefully run out of Gaza.

At that time, my father wanted Hamas, Islamic Jihad, and the rest to be placed on the list of terror organizations and for their funds to be frozen, but the EU was unwilling to make that move. Britain was the only nation in Europe willing to label Hezbollah a terrorist organization.

The British were worried about the separation fence that was being built.

"I am not a lawyer," my father said in response. "I am just a farmer, but I can tell you in no uncertain terms that the fence is not a political border but merely a means of preventing terror attacks."

The visit ended on a note of appreciation. Israel's diplomatic situation was vastly different from the one my father had inherited two and a half years earlier when he was elected prime minister, and the British no doubt recognized the change.

"Your visit is important," Jack Straw said. "We know the huge amount of work you have been doing to help, in very great difficulties, the peace process between the Israelis and the Palestinians and we would like to commend you for that."

From London my father set off for a short visit to Molde, the hometown of the prime minister of Norway. The pastoral views, the deep blue of the large fjord, the traditional dance the children performed—all reinforced the feeling of tranquillity of the small northern town.

In Norway with Prime Minister Kjell Magne Bondevik, July 2003. (© Amos Ben, Israeli Government Press Office)

Of course, the peaceful atmosphere did not deter Bondevik. In their talks, he insisted on a complete freeze on settlements, opposed the isolation of Arafat, called for the evacuation of settlements, complained about the separation fence, and called for an improvement in conditions for Palestinians.

"If you are concerned about the welfare of the Palestinians," my father said, "you are a wealthy country—perhaps you could arrange for Palestinian refugees to work in Norway and earn money. We would be happy to assist you."

As my father's aides later reported, the already pale-skinned Bondevik blanched.

───────

On August 12 at nine in the morning, a suicide bomber—a Fatah member—blew himself up in a supermarket in Rosh HaAyin in the center of the country. One Israeli was killed. Roughly one hour later another suicide bomber blew himself up alongside a bus stop near the city of Ariel. An eighteen-year-old boy was killed.

On the evening of August 23, a suicide bomber—a Hamas member—blew himself up on the Number 2 bus in Jerusalem. The bus, on the way from the Western Wall to the religious neighborhood of Har Nof, was completely full. Twenty-three people were killed in the blast, and over one hundred wounded, including forty infants and children. This was how the Palestinians continued to make clear the extent to which they intended to uphold their commitments as outlined in the Road Map.

In terms of the numbers of wounded, the national mood, the economy, foreign investments, the disappearance of tourism, Israel was at war. What was unique about this war, though, was that its victims were civilians. There was no front line. All of Israel was a front line. People were afraid to go to crowded places, afraid to go to malls, restaurants, cafés, afraid to ride the bus, afraid even to stand next to a bus at a red light in case it blew up.

Although European leaders may have been right from a public relations point that Israel would look better if it moderated its reaction to terror, when the bodies started to pile up, concern with public relations had to be moved aside. The Europeans seemed to have a hard time understanding what we were facing.

As Tony Blair told me during our meeting in Jerusalem several years later, "I think that at that time it was hard for people from the outside to understand what it was like in Israel at that moment. So, one thing that I understand now far better than I did then was the impact the Second Intifada had on the whole psychology of the Israeli people towards the peace process."

Late in the same week as the bus bombing in Jerusalem, Israel's ambassador to France, Nissim Zvilli, met with President Chirac's diplomatic adviser Maurice Gourdault-Montagne. The Israeli ambassador spoke of the importance of including Hamas in the EU's list of terror organizations and asked for France's assistance in the matter.

"If we find that Hamas and Islamic Jihad are indeed terror groups opposed to peace, we may have to change the EU's stand," Chirac's adviser said. The French needed to make certain that Hamas was in fact a terror organization and not the Palestinian chapter of the Salvation Army. "The current meeting of the European foreign ministers is too early a date to form a European position on this matter."

This stubbornness on the part of the French was not new. A year prior, in a meeting with then foreign minister Shimon Peres, Chirac had objected to the inclusion of Hezbollah in the EU terror list.

International pressure began bearing down, and Arafat published a statement on August 27: Yasser Arafat "calls on all groups and parties to commit themselves to the cease-fire to give a chance to all peaceful international efforts for the road map, which the Israeli government resists."

On August 28 the White House responded, "Arafat has once again shown himself to be part of the problem. He is not part of the solution."

On September 2, six days after his call for the reinstatement of the cease-fire, Arafat told CNN, "The road map is dead, but only because of the Israeli violence over the past few weeks." This type of statement was not unusual for Arafat. Even a polygraph would have had a hard time pinning him down; after all, the device requires some utterance of truth so that it can be compared to the lies.

On September 3, Secretary of State Colin Powell spoke for the administration. "I believe that the road map is the way towards progress," he said. ". . . We have started and there is progress, slow progress, but progress. We did not deal with Arafat when we initiated the road map, so his comments are insignificant to me. I am not responding to them. He did not help and did not want to help, he had to support Prime Minister Abbas and not undermine his efforts."

The next day, September 5, my father called Javier Solana, the EU's foreign policy chief, in advance of the following day's European Union foreign minister's meeting in Lake Garda, Italy.

"We are encouraged by the steps taken by Europe recently to improve relations with Israel," my father said.

The conversation focused on cooperation, improving the atmosphere between the EU and Israel, and Solana's letter concerning a strategic dialogue between Israel and Europe. "We are looking over the proposal," my father said regarding the letter, "and our professional teams will be in touch on this matter."

His main message to Europe, however, was an urgent one:

"I would like to emphasize a few key subjects," he said, "in order to push the process forward. Hamas is a terror organization that is interested in stopping any type of diplomatic progress. They intentionally murder the innocent. Hamas took responsibility for the awful terror attack in Jerusalem last month [the Number 2 bus] where more than twenty people were murdered, including many children. It is critical that the European Union reach the right conclusions at the meeting. Hamas must be included in the EU's list of terror organizations, and its financial assets in Europe must be frozen.

"I know that you were recently in Iran; it represents a danger to Israel and to the rest of the world. It is important that the European Union send a clear message [to Iran] that the continuation of the Iranian nuclear program and its efforts to attain weapons of mass destruction will have a negative effect on their financial ties with Europe. Syria, along with Iran, continues to finance Hezbollah's terror attacks. Europe must declare Hezbollah a terror organization and insist that the organization be demilitarized. These are steps that must be taken, and the European Union [must] be loud and clear."

Everything was laid out clearly.

"Thank you," said Solana. "I have listened to what you said, and I expect to meet with you. We will continue to work and to show our good intentions so that there will be good ties with Israel and the peace process."

Solana had not responded to any of the subjects my father raised, but my father did not give up. "Thank you," he said. "Make an effort that Hezbollah be recognized as a terror organization—same for Hamas. In terms of Hamas there can be no distinction between their humanitarian side [and their military side], it's all under the same organization."

———

We have reached a consensus regarding the inclusion of Hamas on the list of terror organizations," said French foreign minister Dominique de Villepin on September 6. This was a refreshing departure for France, which had previously led resistance to such a move, along with Greece and Ireland. At the time they had contended that "such a move would not contribute to the stabilization of the Middle East"—a very problematic contention in light of the explosions on civilian buses.

Foreign Minister Jack Straw of Britain said, "There was complete agreement that, given the outrage perpetrated by Hamas which killed so many innocent people and for which there was no conceivable justification, we've taken a political decision to freeze the assets of Hamas." Senior Hamas leader Abdul Aziz Rantissi responded by saying, "This is a decision that reflects the Crusader

mentality and the American Zionistic approach." He called on
Muslims to boycott European products.

That same day, Mahmoud Abbas resigned as prime minister of
the Palestinian Authority. Aside from the claims he made against
Israel, he said, "Arafat undermined and obstructed my work."

On September 10, 2003, Ahmed Qurei—one of the heads of
the Fatah and a permanent member of the Palestinian diplomatic
negotiating team with Israel—was appointed prime minister of
the PA.

During the 1998 Wye Plantation negotiations, my father, who
was foreign minister at the time, had proposed building a rail
line connecting Gaza and the West Bank to ease travel for the
Palestinians without harming Israeli security concerns or Israeli
sovereignty. He discussed the issue with Mahmoud Abbas and
Ahmed Qurei. While my father went to look at the cows grazing near the River Wye, one of them quickly hurried after him
and suggested that his son be the one to implement the project.
Palestinian leaders always knew how to mix the personal with the
public. The project never got off the ground.

In terms of combating terror, though, there was no discernible
difference between Abu-Alaa and Mahmoud Abbas. On September 27, the evening of Rosh Hashanah, during the holiday meal,
someone knocked on the caravan door of the Yeberbaum family
in the settlement of Negohot in Judea. The owner of the house,
Eyal, opened the door and was shot by a terrorist, who managed
to kill Shaked Avraham, a seven-month-old baby, before he himself was killed.

S ome Palestinians did understand that the path of terror was
a disaster for both their national aspirations and their private
lives. In an interview with the Associated Press, Muhammad
Dahlan, the minister of internal security under Mahmoud Abbas, said, "Nine-eleven was the turning point of everything. We

did not understand 9-11 in a correct and substantial way in order to . . . bring back international legitimacy for our [Palestinian] Authority and for our president [Arafat]."

Those Palestinians who opposed the doctrines of terror never lamented the intentional murder of civilians but rather the fact that the tactic hurt their image. Regardless of the reason behind their remorse, it did not cause the Palestinians to change their ways.

Preparation for the disengagement from Gaza.
(© Avi Ohayon, Israeli Government Press Office)

THIRTY-FOUR

On Saturday, October 4, 2003, a female terrorist from the Islamic Jihad organization blew herself up in the filled-to-capacity Maxim restaurant in the southern part of Haifa. Twenty-one people were killed, including children and babies. Dozens were injured. The forty-year-old-restaurant was co-owned by Arabs and Jews and had been a model of coexistence.

The following day Israeli planes struck the Islamic Jihad training base in Ein Tsaheb, Syria, twenty miles northwest of Damascus. Syria filed a formal complaint to the UN Security Council and asked for Israel to be condemned and for such actions to be prevented in the future. At the time, the United States' representative to the UN, John Negroponte, was serving as the temporary president of the council. He refused Syria's request, stating that Damascus had "positioned itself on the wrong side of the war on terror." In response to a question regarding the strike, President Bush said, "We would be doing the same thing."

On October 15, an American convoy was attacked near Beit Hanoun in Gaza. The members of the convoy were on their way to interview Palestinians eligible for an American university scholarship. A roadside bomb ripped through one of the vehicles, killing three American security personnel and wounding another man, who was treated in Israel. "We call on all our citizens to leave Gaza for their own personal safety," said U.S. Embassy spokesman Paul Patin. American investigators arrived on the scene and were forced to flee after being attacked with stones.

"The Palestinians should long ago have acted forcefully against terror," the White House said in a statement. The State Depart-

ment offered a $5 million reward to anyone who provided infor-
mation leading to the arrest of the terrorists.

I had long been harboring the feeling that nearly all of the efforts
toward finding a diplomatic solution to the conflict were useless, a feeling I expressed on more than one occasion in meetings
with my father and his advisers. There was too much talk about
negotiations, while the result was always the same—another terror attack, the reshuffling of the deck, the abolishing of all understandings. It was clear that nothing could be worked out with
the Palestinians. I thought that instead the focus should be on the
things that were under our control. Of course, I was not the one
who had to face international pressure.

During our meetings on the farm, another topic, raised mostly
by my brother Omri, was the lack of hope. The people could endure the brutal terror campaigns, he said, but they had to know
that there was something on the horizon—that there was hope for
better times.

In October 2003, during one of those meetings around the
dining room table on the farm, I sat beside my father, with Reuven Adler, Dubi Weissglass, Eyal Arad, Yoram Raved, and
Omri. I turned to my father and said, "I don't understand your
position on the Gaza Strip; it's unclear, not fully formed. Living there together is impossible; the daily reality is unbearable, incessant terror—rockets, mortars, roadside bombs, and
sniper fire. Many settlers have been injured and killed, including
women and children, as have many of the armed forces—the
situation lacks an endgame. In the long term either we won't be
there or they won't be there. We can't live there together." We all
knew that the Israeli public hated Gaza, didn't want to be there,
didn't want to serve in reserves there, and didn't want to pay the
price to keep it, as opposed to Judea and Samaria or other areas
within Israel.

"Since there's no one to talk to on the other side—agreements have never been honored, and as it looks now [they] will never be honored in the future either—still, we want to improve our situation," I continued, "to provide a better and more quiet life to the citizens of Israel. Therefore the way forward is via a unilateral approach, actions that will better our position. It doesn't matter if it's seen as something good or bad for the Palestinians; what's important is that it's good for us. And for that reason our unilateral approach shouldn't be dependent upon any Palestinian actions."

My father listened—it's possible he said something like, "Interesting"—and then he said, "Put it down on paper." That was the first time my father seriously considered the evacuation of the Israeli communities from Gaza.

————————

As difficult as the risk of terror attacks made life in the rest of Israel, life for Jews and Palestinians living side by side in Gaza was intolerable. On a strip of land covering 140 square miles, just 25 miles long and between 3 and 8 miles in width, and populated by a million and a half Palestinians—perhaps the most densely populated place on Earth—it was also untenable. There were two options: to evacuate approximately one and a half million Palestinians to some other place or to evacuate approximately eight thousand Jews.

There had long been a consensus that, if a peace deal with the Palestinians was to be reached, then Israel would have to withdraw from Gaza. The only difference was that some contended that Israel could withdraw only within the framework of an agreement with the Palestinians. However, others thought Israel could do it unilaterally. Although the taking of unilateral steps had not been discussed as a policy before this, Israel did take such steps. Operation Defensive Shield, for instance, was a unilateral step; the separation fence was another unilateral step.

The citizens of Israel were thoroughly fed up with the porous border, which allowed terrorists into the country, affording them access to civilian population centers, and to a lesser extent with the incessant theft of their belongings and the way they kept winding up in the Palestinians' hands.

─────────────

During the aforementioned meeting in our dining room, Dubi Weissglass, my father's chief of staff and the man in charge of all day-to-day ties with the American administration, was intrigued by the idea of a unilateral withdrawal. Eyal Arad said, "It's possible you could do it, but not with this [Israeli] government."

Over the years I had heard my father say that back in the early 1980s, when he was the minister of agriculture and the chairman of the Ministerial Committee on Settlements, he had given the settlements of the Gaza Strip large areas of land in the Negev and millions of cubic meters of water in case, in the future, a situation arose whereby they will no longer remain in Gaza. "There's no way to know how things will develop," he would say. So the only part of the plan that was new was the unilateral aspect—the understanding that what was necessary was action in Israel's best interest, without waiting for an agreement that might or might not materialize.

I wrote a position paper on October 16 and submitted it to my father. I did not explicitly mention a withdrawal or specify Gaza in particular, but rather expounded on the principle of unilateral action.

My father considered the matter and accepted the idea. He knew that the international powers would never tolerate a diplomatic vacuum vis-à-vis the Israeli-Palestinian conflict, and he feared the introduction of new plans, the kind that might endanger Israel. He wanted Israel to decide its future, for us to lead rather than be led, and if there was to be a peace process, for it to proceed according to the Road Map, abiding by each and every step, beginning with the war on terror and the dismantling of the terror organizations.

Less than a month later, on November 14, I wrote another, more detailed position paper for him. He commented on it and wrote questions in the margins.

On November 17, my father went to Italy for three days. There he met secretly with Elliott Abrams at his hotel. It was at this meeting that my father directly discussed a unilateral withdrawal in Gaza for the first time.

Coordinating the move with the Americans was crucial to my father. He sent Dubi Weissglass to Washington, D.C., where Weissglass met with Condoleezza Rice and Elliott Abrams. On November 26, Dubi gave my father a report of these meetings.

The Americans were primarily concerned that the unilateral move not be seen as in conflict or opposition to the Road Map, or that it in any way jeopardize future peace prospects. As time passed, though, the administration came around to the Israeli position: they realized that a withdrawal from Gaza would be difficult for my father on the domestic front, but that, on the other hand, the plan had inherent advantages. They also realized that, were there ever to be serious peace talks with the Palestinians, the unilateral measures would not in any way hinder them. The plan would simply improve Israel's position.

The next stage was the unveiling of the elements of the plan to the public. That took place at the Herzliya Conference. There, after explaining the importance of the Road Map, my father raised the possibility of unilateral steps:

> *If the Palestinians do not make a similar effort toward a solution of the conflict—I do not intend to wait for them indefinitely.*
> *. . . We wish to speedily advance implementation of the Road Map toward quiet and a genuine peace . . . However, if in a few*

months the Palestinians still continue to disregard their part in im-
plementing the Road Map, then Israel will initiate the unilateral
security step of disengagement from the Palestinians.

The purpose of the Disengagement Plan is to reduce terrorism
as much as possible, and grant Israeli citizens the maximum level
of security . . . and minimize friction between Israelis and Pales-
tinians.

We are interested in conducting direct negotiations, but do not
intend to hold Israeli society hostage in the hands of the Palestinians.
I have already said—we will not wait for them indefinitely.

The Disengagement Plan . . . will include the redeployment of
IDF units along new security lines and a change in the deployment
of settlements, which will reduce as much as possible the number of
Israelis located in the heart of the Palestinian population. We will
draw provisional security lines and the IDF will be deployed along
them. Security will be provided by IDF deployment, the security
fence, and other physical obstacles. The Disengagement Plan will
reduce friction between us and the Palestinians.

This reduction of friction will require the extremely difficult step
of changing the deployment of some of the settlements. I would like
to repeat what I have said in the past: In the framework of a future
agreement, Israel will not remain in all the places where it is today.

The relocation of settlements will be made, first and foremost, in
order to draw the most efficient security line possible, thereby creat-
ing this disengagement between Israel and the Palestinians. This
security line will not constitute the permanent border of the state of
Israel. However, as long as implementation of the Road Map is not
resumed, the IDF will be deployed along that line.

Settlements that will be relocated are those which will not be
included in the territory of the state of Israel in the framework of
any possible future permanent agreement. At the same time, in
the framework of the Disengagement Plan, Israel will strengthen
its control over those same areas in the land of Israel that will
constitute an inseparable part of the state of Israel in any future
agreement.

My father characterized the security fence as a measure of safety. Its construction would be accelerated, he said. With its completion, roadblocks could be removed, making life easier for the Palestinian people.

> *In order to enable the Palestinians to develop their economic and trade sectors, and to ensure that they will not be exclusively dependent on Israel, we will consider, in the framework of the Disengagement Plan, enabling—in coordination with Jordan and Egypt—the freer passage of people and goods through international border crossings, while taking the necessary security precautions.*
>
> *I would like to emphasize: the Disengagement Plan is a security measure and not a political one. The steps that will be taken will not change the political reality between Israel and the Palestinians, and will not prevent the possibility of returning to the implementation of the Road Map. Obviously this disengagement plan provides much less to the Palestinians than they could get in a direct negotiation according to the Road Map.*

While the idea of withdrawing from Gaza was evolving into the Disengagement Plan, terrorists had struck in Turkey. In Istanbul, two car bombs exploded outside two different synagogues on November 15, 2003, killing twenty-three people and wounding over three hundred. Several days after the attacks, my father spoke with the Turkish prime minister, Tayyip Erdoğan.

"We have a common problem," said Erdoğan. "Terror has no time or place, and it is impossible to know what its goals are or where it will strike. The Jewish and Turkish citizens were victims last week. It will not stop us from conducting a shared struggle against terror. I want to thank you for your solidarity and to express my condolences for the tragedies that befall you."

". . . I share in your sorrow. We are familiar with this feeling. International terror is the main danger and the most palpable threat to the well-being and stability of the free world, and it must be fought without compromise. I want to stress that it is not a battle against Islam—it is a battle against those who cause terror and are led by extremist Islamic elements."

"In the name of the Turkish people and the government of Turkey I thank you for your sensitivity," said Erdoğan. "We will not succumb to the evil and the malice of terror. We must maintain our solidarity and intensify the cooperation between us."

On December 22, 2003, Foreign Minister Ahmed Maher of Egypt arrived in Israel and met with my father and Foreign Minister Silvan Shalom. Maher conveyed formal greetings from Mubarak, as well as the fact that he saw the meeting as a prelude to a summit between the two leaders.

After the cordial and pleasant meeting, Maher went to pray at the al-Aqsa Mosque on the Temple Mount. The greeting he received there was one he was unfamiliar with: dozens of Palestinians threw shoes at him and yelled, "Traitor, traitor, Allahu Akhbar, traitor, traitor." The Egyptian minister fled the mosque. As he was being rescued by Israeli security forces, he was heard mumbling, "I'm choking, I'm choking." He was evacuated in an ambulance and taken to the Hadassah Medical Center in Ein Kerem.

My father called him in the hospital.

"How are you?"

"I am being treated well," Maher said. "It is just a small thing. People here are very cautious and concerned that I am okay before I leave."

"Stay with us," my father said.

"I have to go home," said Maher. "Thank you for showing interest in my well-being and for the care I am receiving here."

"The visit began successfully. We thought it would continue to be successful," my father said.

"It was a successful day. I am being treated in an Israeli hospital, and that, too, is positive."

"Feel well. You are always welcome in Israel."

Five weeks later, Israel took part in a prisoner for hostage exchange.

Three months before my father was elected, three IDF soldiers had been killed during a cross-border Hezbollah raid on October 7, 2000, in the Har Dov region. One year after the Hezbollah attack, the IDF had formally changed the status of the soldiers from Unknown to Killed in Action.

Also during October 2000, Elhanan Tannenbaum, an Israeli citizen, had been kidnapped in Abu Dhabi—where he had gone to conclude a drug deal with an Israeli Arab dealer who turned out to be a Hezbollah operative—and taken to Beirut.

On January 29, 2004—three years and three months later— the exchange was made. Four hundred Palestinian prisoners were freed, along with thirty-six prisoners of different nationalities— Lebanese, Syrian, Moroccan, Sudanese, Libyan, and German— and the bodies of fifty-nine Lebanese terrorists were exhumed and handed over. Tannenbaum and the bodies of the three soldiers were returned to Israel.

There were those who felt that Tannenbaum should have been left to lie in the bed he had made. But not my father: as far as he was concerned the man could sit in an Israeli prison, if that was what he had coming to him, but not in captivity in the hands of terrorists. Without doubt Israel paid dearly to make this deal, but the arrangement was made in accordance with the policy my father had adhered to since his days as a young officer: we do not leave our people in enemy hands.

Two days later my father spoke with Chancellor Gerhard Schröder and thanked him for his involvement in the exchange:

"I'd like to tell you that we were happy to help," said Schröder. "I think we will continue to work on this project together; we will manage it successfully in the future, too."

"I also wanted to thank the government of Germany and those who aided us in finalizing this important deal, especially Ernst

Uhrlau, whom I met personally on Thursday and Mr. Henning and the impressive team [of German intelligence personnel] that was involved in this important project."

"Mr. Prime Minister, I will pass on your gratitude to Uhrlau and Henning, and will ask them to carry on with the same spirit. We are honored.

". . . Our next mission is to return the abducted pilot Ron Arad. I hope that through joint efforts we will be able to attain information regarding the pilot and afterwards return him home."

"I hope so. I see that as the next stage, and we will take measures so that it will happen."

On the morning of the exchange a terrorist blew himself up on the Number 19 bus in Jerusalem, in Rehavia, not far from the prime minister's residence. The terrorist, a Palestinian policeman and Fatah member, killed seven women and four men and wounded more than forty people.

Kofi Annan published the following statement after the attack: "Once again violence and terror have claimed the lives of innocent citizens in the Middle East. Once again I condemn those who find refuge in those two paths." The statement whitewashed the incident, placing a collective guilt on both parties. In contrast, State Department spokesman Richard Boucher said, "There is no comparing the attack in Jerusalem and Israel's actions."

A week after the attack, on February 6, my father had a phone conversation with Kofi Annan.

Annan opened with a discussion of the Disengagement Plan, which my father had made public on February 3. In an interview with Yoel Marcus of *Haaretz*, he had announced his intention to withdraw from the Jewish settlements in Gaza and from a few settlements in Samaria.

(During the months of December and January, Dubi Weissglass, together with Shalom Turgeman, had conducted talks with senior administration officials Steven Hadley, Elliott Abrams, and occasionally Condoleezza Rice. What had become clear from those talks was that if the Disengagement Plan did not include parts of Judea and Samaria, the Americans would not offer any type of reward for the initiative, nor would it receive their backing. Therefore, already at that point my father had realized that the Disengagement Plan would have to include four isolated settlements in Samaria.)

"I wanted to tell you that I very much appreciated your declarations regarding Gaza," Annan said. After all, at the UN, they do like Israeli withdrawals. "I know that this is a controversial issue but I believe it is an initiative that will help break the stalemate in the peace process and will encourage all parties to return to the negotiating table. It is clear that it is a first step, and there are other issues that must be dealt with in the West Bank. I wanted to encourage you and to ask: When the time comes, how can the Quartet be of assistance in implementation and in harnessing the other side to work with you and carrying out what is required from them?"

"Firstly, Mr. Secretary, thank you for the call. You asked what the Quartet can do. It can pressure the Palestinian Authority to take measures against terror and to dismantle the terror organizations . . .

"Allow me to say that we were very disappointed with the statement the UN published after the dreadful bombing in Jerusalem," my father continued. "Those types of messages send a negative message to the Palestinian Authority. It is important to firmly condemn terror. What was said there encourages terror and does not make them think about [taking] the smallest step to fight terror."

"You are referring to a single statement," said Annan. "I am speaking of a series of statements not only to the media but also to Arab leaders, where I have been very firm. It is not fair to take a single statement that Israel does not find favorable."

"I am sorry that I am repeating this, Mr. Secretary," my father said, "but the soft response was published while all of the victims were in an awful state; some had not yet been identified, and some had not yet been buried. It was a disappointment to us and it pained us, and for the Palestinians it was a sign—Go and kill, because no one blames you."

"That is not at all the intention," said Annan. "The Quartet will continue to put pressure on the Palestinian Authority."

"It must be clear that if the terror continues even if Israel takes the measures I announced [the withdrawal from Gaza], we will not stop acting against terror. Peace and terror cannot go hand in hand. That is the Palestinian strategy, and we will not accept it."

"I agree with that; steps must be taken to break the circle of terror and violence and revenge. Mr. Prime Minister, even if Israel takes steps in the name of self-defense, they must be proportionate, without the use of undue force. That weakens the traditional standing of Israel across the world."

Evidently the formula calculating the correct amount of force changes when it is not one's own countrymen being blown up on buses.

"The world watches the terror attacks, and we are unable to identify all of the bodies—if the world thinks that in order to get more sympathy, we are going to sit with our hands in our laps and be killed every day, that is not going to happen," my father said. "It is not a matter of revenge but of self-defense, and Israel has that right just as all the other countries of the world do."

"We will continue with the pressure," said Annan. "I think that my envoy to the region, [Terje] Rød-Larsen, has been working well with [then head of the National Security Council Giora] Eiland, and we would like to continue with this cooperation."

"Thank you for the call. I know you called several days ago; I apologize that we only speak now," my father said. "We have problems here every day. It took some time."

"No," said Annan. "I am happy we had the opportunity to talk."

"We want to stick to the Road Map—it is the best plan—but if there is no partner, we will have to take unilateral steps, which will create more security and less friction."

———————————

From the moment of its announcement, the Disengagement Plan became the first order of business in Israel until its completion in the summer of 2005 and onward to the next general election on March 28, 2006, in which my father was unable to participate. At the outset it was clear that the majority of the population supported my father and his plan. However, there was a problem in his own party, the Likud, and some of the other parties in the coalition.

On March 12, my father spoke again with Prime Minister Bondevik of Norway:

"I wanted to update you about the meeting with Abu-Alaa," Bondevik said. "I would like to bless you," Bondevik added, "for the decision to withdraw from Gaza and to continue afterwards with a plan along with the Palestinians. Abu-Alaa asked that I pass on to you that he is willing, he is interested in a meeting with positive results. He asked that I pass on that Israel refrain from military action before the meeting . . ."

"Thank you. Insofar as the meeting is concerned," my father said, "I told [Abu-Alaa] long ago that it is only on the basis [of the Road Map] that we are willing to negotiate with the Palestinians . . . They have not done a thing. There has been no effort on the part of the Palestinians, in the places where they have the manpower. Regarding what he said about it being difficult for him to act because we are acting in the area, what does he want? That we sacrifice our lives until they take action?

"Regarding the relocation of the villages in Gaza or the Disengagement Plan—it is not a matter to be negotiated with the Palestinian Authority, and I will tell you why. They want to skip past the first stage of the Road Map, including the steps to be taken against terror, and to address the matters to be touched upon in the third stage of the Road Map. That is why we will not negotiate with them about the Disengagement Plan.

"You should know that I preferred the Road Map. The fact that I have chosen the Disengagement Plan is because we have no partner to negotiate with . . . The Palestinians should stop whining and start preventing terror. We cannot live with terror; everyone sees what has been happening with terror in the world."

"We saw what happened yesterday in Spain," Bondevik said, referring to the four blasts at Madrid subway stations on March 11, 2004, that had claimed 192 lives and injured some 1,400 people.

"When one sees that," my father said, "then it should be understood that there is no compromising with terror. I thank you that from where you are, you still find interest in our problems." It was a delicate hint that was probably not understood.

"I hope that you appreciate it," Bondevik said.

"I very much appreciate it. It is hard to understand that people can live a quiet life and interest themselves in what happens in other places."

"I am happy that you want to move forward, and I understand the difficulties you face on security matters. Regarding control of the security forces, there is the newly formatted National Security Council, where Arafat is the chairman and the prime minister is a member," Bondevik observed.

"They took the murderer and made him a chairman."

The Disengagement Plan stirred considerable international interest. On April 7, 2004, British Prime Minister Tony Blair called my father:

"I thought it might be good to talk," Blair said. "I know you are going to Washington."

"I want to apologize for being unable to meet with you in the United States," my father said. "I want to continue the dialogue with you and hope to meet with you soon. I have to be back here early because of the [party] referendum. It won't be easy."

"Will you succeed?" Blair asked.

"I hope," my father replied.

"So do we. I wanted to ask whether you will announce the plan in Washington."

"I've spoken about it with members of the administration in Rome," my father said. "I'll announce it in Washington after I speak with President Bush."

"I want there to be a positive reaction to the plan from other locations as well. We will react positively," Blair told him. "I spoke with President Bush this morning. Once you introduce the plan, everyone else will back it and demand that the Palestinians take action on the security front. What's important is that there not be simply an Israeli offer but also that they react positively as well and that there be a true and positive security plan so that the situation improves."

"There are daily attempts to send suicide bombers. We stop many of them; otherwise we would be in a grave situation."

"That has to change if we want to return to the Road Map," Blair said.

"Otherwise the Palestinians will not be able to realize their hopes and dreams. If the terror continues after the measures that we will take," my father said, "I do not foresee a Palestinian state as part of the Road Map. We will not be able to live with terror. What you and President Bush have been doing against terror is important."

"Undoubtedly," said Blair, "and now all of Europe is experiencing it."

With President Bush at the White House, announcing
details of the Disengagement Plan, April 14, 2004.
(© Avi Ohayon, Israeli Government Press Office)

THIRTY-FIVE

The time came for my father to make a formal announcement to the international community about the Disengagement Plan. This was to occur in a meeting with President Bush at the White House, where my father was to present Bush with a letter explaining the plan, and Bush was to present a letter of U.S. support and commitment.

Almost until the last moment there were negotiations as to what the Americans would do in exchange for the withdrawal. In fact, my father was so concerned about several unresolved issues that immediately before his departure he considered canceling his trip, and it was in fact delayed by five hours.

He would not accept a situation whereby the president's letter would include the matter of exchanging territory for settlement blocs, and he wanted more clarity on the subject of negating the Palestinian claim of return to Israel and the matter of keeping the settlement blocs in Israeli hands. He was willing to fly only once he had received reliable assurances.

On April 14, 2004, in the conference room alongside the Oval Office, President Bush handed my father what has become known as the Bush-Sharon letter. In return, my father handed him a letter with details of the Disengagement Plan. Bush's letter, which he read, began with a strong statement against terrorism.

The United States will do its utmost to prevent any attempt by anyone to impose any other plan. Under the roadmap, Palestinians must undertake an immediate cessation of armed activity and all acts of

violence against Israelis anywhere, and all official Palestinian insti-
tutions must end incitement against Israel . . .

. . . There will be no security for Israelis or Palestinians until
they and all states, in the region and beyond, join together to fight
terrorism and dismantle terrorist organizations. The United States
reiterates its steadfast commitment to Israel's security . . .

Israel will retain its right to defend itself against terrorism, in-
cluding to take actions against terrorist organizations . . .

The United States is strongly committed to Israel's security and
well-being as a Jewish state. It seems clear that an agreed, just, fair
and realistic framework for a solution to the Palestinian refugee issue
as part of any final status agreement will need to be found through the
establishment of a Palestinian state, and the settling of Palestinian
refugees there, rather than in Israel.

. . . In light of new realities on the ground, including already ex-
isting major Israeli population centers, it is unrealistic to expect that
the outcome of final status negotiations will be a full and complete
return to the armistice lines of 1949.

Aside from the supportive words, the letter reflected several unprecedented changes in the official American position:

The dismissal of the notion of settling Palestinian refugees in Israel;

A de facto recognition of the settlement blocs in Judea and Samaria; and

A rejection of the idea of returning to the 1949 Armistice Lines.

There was no mention of a territorial swap.

President Bush's letter was supported by an overwhelming
majority in both houses of Congress. In the House, 407 supported
and nine were against. In the Senate, 95 voted for it; 3 against—
a firmly bipartisan decision. America's acceptance of the Disengagement Plan constituted an historic accomplishment.

The Palestinians will never forgo the right of return . . . the fanatical Israeli leaders are wrong, and so are those who support
them," Arafat said.

Prime Minister Abu-Alaa said, "We are stunned and very disappointed by the declarations," adding, "Bush is the first president to recognize the settlements as legitimate."

Reaction to the letter was less than enthusiastic in Europe and
in the halls of the UN. The official responses were carefully
worded, so as to avoid a direct confrontation with the U.S. administration.

None of this seemed to matter to President Bush. On April 21,
one week after handing his letter to my father, President Bush
said, "Ariel Sharon came to America, stood beside me, and said,
'We will withdraw from the Gaza Strip and parts of the West
Bank.' As far as I can tell the entire world should say, Thank you,
Ariel." He added, "Iran has set a goal of destroying Israel, and
therefore its plan of acquiring nuclear weapons is intolerable."

On May 4, Quartet representatives met in New York and announced their support for the Disengagement Plan as part of the
Road Map, demonstrating, in effect, the support of the international community.

In Israel, the Disengagement Plan had yet to be approved in
its final form. In the Likud Party's referendum, 59,000 party
members opposed the plan and 40,000 supported it.

My father realized that he had erred in putting a national matter in the hands of a single party. In a May 30 cabinet meeting,
he said, "It's possible I was mistaken in going for the referendum,
and I take responsibility for that mistake. But the results of the
referendum are not a commandment. There is without a doubt a
significant suggestion that is to be weighed carefully, along with,

and not in isolation from, other weighty considerations. Likud Party members are an important part of Israeli society, but they are only a part. We are the government of all of Israel, and we must be attuned to the positions of the public at large, which overwhelmingly wants with all its might to see the plan enacted."

On the day of the referendum, on the road leading from Gush Katif in Gaza to the Kissufim Crossing, a terror attack robbed David Hatuel of his entire family—four daughters and his pregnant wife. My father made a condolence visit. These visits to the families of terror victims among the settlers in Judea, Samaria, and the Gaza Strip were difficult for him. Family members would assail him verbally, claiming he wasn't doing enough to fight terror, and criticized his diplomatic moves. He didn't have to be there but he felt the need to be there. At the home of the Hatuel family he said, "I've endured tragedies myself, sorrow and pain, so I understand yours." When family members and residents of Gush Katif started to complain to him about the Disengagement Plan, my father's response was, "the Disengagement Plan has to go on in order for things to get better in other places." It wasn't easy to say this to them but he spoke honestly.

———

By this time, the number of terror attacks in Israel was on the wane. In 2002, 452 Israelis had been murdered in terror attacks. In 2003, that number was cut to less than half, with 208 murdered. In 2004, 117 Israelis lost their lives to terror, and in 2005 the number of Israelis killed in terror attacks was down to 56.

Nonetheless, there were still attacks. On March 14 two terrorists from Gaza, one from Fatah and one from Hamas, carried out a suicide bombing in the port of Ashdod. Exploiting the fact that Israel allowed the transfer of shipping containers from Gaza to Ashdod for export, they hid inside one of the incoming containers. The two blasts killed ten people and wounded thirteen. Subsequently it was discovered that a Palestinian officer from the PA's

Preventative Security Service, stationed at the Karni Crossing, had been directly responsible for facilitating the attack.

On March 22, 2004, the Israeli Air Force liquidated the head of Hamas, Ahmed Yassin. There was much innocent blood on the hands of Yassin and his replacement, Abdul Aziz Rantissi. By that point Hamas had carried out 425 terror attacks, killing 377 Israelis and wounding 2,076.

Shimon Peres, at the time the head of the opposition in the Knesset, said of the strike, "Were I in government I would have voted against the liquidation. I think it is a mistake. I think it has worsened Israel's situation. I believe that negotiation is the way to stop terror."

Condoleezza Rice expressed a different opinion. In an interview after the strike she said, "Hamas is a terror organization," and "Yassin was personally involved in terror."

After Rantissi, too, was killed by Israeli forces on April 17, 2004, Peres changed his tune. "He who deals in murder pays the price," he said, "and his blood is on his own head." This change of position was perhaps influenced by his aspirations to join the government.

Once the cabinet had passed the Disengagement Plan on June 6, my father embarked on a round of talks with world leaders. The idea was to present to the world Israel's willingness to make painful concessions for peace. Until this point my father was known more for his military acumen and success, but he performed excellently on the diplomatic front as well.

On June 8, my father called the new prime minister of Spain, José Luis Rodriguez Zapatero, to congratulate him on his appointment and bring him up to date on Israel's plan:

"Our intention is to withdraw from military installations and all of the Israeli settlements in the Gaza Strip, in addition to the military installations and small number of settlements in Samaria," my father said. "This will reduce the friction between Israelis and Palestinians . . ."

Zapatero replied, "I would like, as part of the international community, to contribute so that peace will be attained . . . I would like to work together and would like for you to see in Spain a friendly nation that wants to contribute to peace."

"Thank you, Prime Minister. We have been experiencing terror for one hundred and twenty years. My grandfather, my father, my children—all of them have experienced terror."

O n June 11, my father spoke with Kofi Annan:
"Mr. Prime Minister, how are you?" Annan said.

"Okay, how are you? Good morning, I hope I am not calling too early."

"It's fine. I wanted to speak with you. I wanted to congratulate you on the decision in the cabinet. It is a brave step. The withdrawal will not be easy because of the emotions involved and the determination of those who will oppose it. I wanted to tell you that I've followed the moves closely, and I believe that a full withdrawal from Gaza could also revive the [peace] process in the West Bank. I hope this will be a new beginning, and we must find a way to use this to help the peace process in general. The UN and the Quartet are willing to play a role with the Palestinian Authority to ensure that they fulfill their part. My feeling is that at some point on the route you will need an international force, a third party, which will manage the situation in terms of possible points of friction and will work together with the Palestinian Authority to ensure that the process is well managed and that there not be any additional problems, as we did in South Lebanon. I would like to congratulate you on your courage."

My father had no intention of allowing international forces into the area. South Lebanon was an excellent example of why not. Those types of forces do not prevent terror; they merely hinder the fight against it. But that was not the topic of discussion, and my father had already made the point clear to Annan on many previous occasions.

"Regarding South Lebanon—I regret that there have been several incidents there along the Blue Line. I have asked Terje Rød-Larsen to organize our troops in the area. He, Hansen [a UN official in the area], and the troop commanders would like to invite General Yaalon [then IDF chief of staff] to see what the problems are and how they might be solved."

"That type of cooperation is important. Any such meeting might help. I want there to be quiet on the northern border. We are making an effort not to worsen the situation. We must take certain measures in self-defense. We want the front there to be quiet. Pressure must be brought to bear on the Iranians and the Syrians, who support Hezbollah. We suffer from terror based in Lebanon, as executed by the Iranian Revolutionary Guards and Hezbollah. "Back to the Palestinian issue—it was not easy, but I have decided to see the plan through, and that is what I intend to do.

"I understand you are traveling today."

"I am going to Brazil."

"I have been there several times. A beautiful country. All the best."

———————————

On July 6, 2004, my father called the prime minister of Canada, Paul Martin:

"I'm calling to congratulate you on your victory in the elections, and I wish you good luck in assembling a coalition and putting together a government in government," my father said.

"That's nice of you. The important part of the process is putting together the coalition and joining all of the elements together. I have to ask for your advice on the matter. You have experience."

"Yes," my father said, "we have experience. I'd also like to congratulate you on the occasion of Canada Day . . . I would like to express my appreciation for your friendship and friendliness towards Israel and the Jewish people. I am familiar with your relationship to the Jewish community of Canada and appreciate your position regarding the growing and spreading expressions of anti-Semitism. On the same note, I would like to express my appreciation for the Canadian government's friendship and for understanding of Israel's troubles and needs and for the solidarity you have shown us. We know your position on terror and appreciate the fact that you included Hamas and Hezbollah on your terror list . . . Iran is making efforts to arm itself with nuclear weapons. It is the greatest danger to Israel and the region."

". . . The Canadian people feel tremendous appreciation for Israel . . . You are the only democracy in the region and [we] will support [you] in any way we can.

"I would like to address an issue you raised earlier, the matter of Iran and nuclear capability. Can you be more specific: is it serious, and what needs to be done?"

"Iran is a large country and they take pains to hide [their actions]," my father said. "The oversight has to be more serious, and the matter will have to be brought before the UN Security Council. Iran is a grave threat. They received help from Russia, China, and North Korea and managed to develop ballistic missiles—the Shahab-3, with a range of thirteen hundred kilometers, covering all of Israel and other states in the region, is now operational. There must be economic and political pressure applied regarding several different developments, including nuclear weapons. I'd be happy to carry on and send someone, or you could send someone from your office whom we could update."

"I would appreciate that . . . ," Martin said.

"We will be in touch with your office so that you can receive information. Iran is the center of global terror. They arm Hezbollah in Lebanon, and they [now] have over thirteen thousand rockets, a dangerous development. They are active among the Israeli Arabs, and although the majority of them [the Israeli Arabs] want to blend in to Israeli society, a growing minority is involved in terror. They are acting via Islamic organizations.

"I'd like to congratulate you on the occasion of your son's marriage, set for the coming weekend."

"Your intelligence is excellent."

"If we were able to learn about the wedding, then our intelligence services will be able to get other information as well."

"I'll tell my son. That's very nice of you."

On July 9 the International Court of Justice in The Hague ruled, "The construction of the wall being built by Israel, the occupying Power, in the Occupied Palestinian Territory, including in and around East Jerusalem, and its associated régime, are contrary to international law."

Two days later the Palestinians expressed their support for the ruling in the usual way: a suicide bomber from the Fatah organization blew himself up alongside a bus on Har-Zion Avenue near the old bus station in Tel Aviv. A young woman was killed, and forty people were injured.

Sami Masraweh, an Arab from Jaffa, was lightly wounded in the attack. It was the second time he had narrowly escaped death at the hands of Palestinian terror. Ten years earlier he had nearly been killed in the deadly Number 5 bus blast in central Tel Aviv. He said, "I am an Arab from Jaffa, and I want to emphasize that. I am left wing, and I opposed the separation fence till now. After what I saw today during the attack, I hope to establish an organization supporting the fence."

At the beginning of the cabinet meeting held on July 9, my father said, "What the judges of the Court refused to see, the Palestinians were quick to demonstrate . . . On Friday, the right of the war against terror received a slap in the face from the International Court of Justice at The Hague . . . The State of Israel rejects outright the International Court of Justice at The Hague's opinion. This is a biased opinion, which is supported solely by political considerations. The opinion completely ignores the reason behind the construction of the security fence—which is murderous Palestinian terror."

Palestinian inaction on the front against terror was so glaring that even Terje Rød-Larsen sharply rebuked Arafat and the Palestinian Authority in a report given to the Security Council on July 13, saying that "although the leader of the Palestinians remains confined to his headquarters in Ramallah . . . this is not an excuse for passivity and inaction of the Palestinians in fighting terror. The PA had," he said, "made no progress on its core obligation to take immediate action on the ground to end violence, combat terror, and to reform and reorganize." He called on Arafat to take immediate steps to restore confidence in the Palestinian Authority and for the Palestinian prime minister to be given authority so that he could not only make decisions but also execute them. As of now, he said, "there is so far no sign of any of these measures being taken."

On August 17, 2004, President Mubarak of Egypt called my father:

"Hello, my friend, how are you?" Mubarak began.

"How are you, Mr. President?"

"Very well. Since it has been a long time since I have heard from you, I decided to call and see what was happening."

"I am happy to speak with you," my father said. "How is your back feeling?"

"I feel very well," replied Mubarak. "I called to let you know that I will be sending the foreign minister to you at the end of the month and also Omar Suleiman to see you, in order to see what is required of us vis-à-vis the Palestinians. We are convening a conference with Hamas and the other groups, and are trying to reduce the tension, in an attempt to see how it will be possible to move forward with the process."

"In the meantime the terror continues," my father said. "[They] are making many attempts, and we have been able to either catch them or kill them."

"This is predictable. I must tell you in all honesty," Mubarak said, "that we found a large stockpile of weapons along the border several days ago.

"We passed it on to the armed forces and are trying to find the people responsible for it. We found it in the area of el-Arish."

"That's very important. Thank you. We must continue with this effort."

My father was well aware that the Egyptians' efforts were mere theater. If they didn't want arms smuggling at the isolated border crossing of Rafah, there would be no smuggling. For that matter, the senior officer from the IDF's Intelligence Corps had recently told the Knesset Foreign Affairs and Defense Committee that "RPG rockets found in Gaza were made by the national Egyptian weapons manufactory."

"This must be continued. Our decision is to continue with this. We want to cooperate in all sectors, without restrictions," Mubarak said amiably.

"It is very important. I want to congratulate you on the occasion of Revolution Day, July 23."

"Thank you."

"In addition I would like to congratulate you on the establishment of your new government." My father said this ironically, for there were no obstacles in Mubarak's path, no real elections or coalition government negotiations. His election was a mere technicality in Egypt.

"We are working hard," Mubarak said.

"I hope to tighten the cooperation between us and to continue the dialogue regarding security issues in Gaza. We have to work close together," my father replied.

"Be certain that I am willing to work with you even closer."

"Mr. President, I am determined to implement the Disengagement Plan as quicky as possible. I have encountered political difficulties, but I believe I will be able to overcome them."

"I am sure you will overcome them," Mubarak said, and both of them laughed. Mubarak, the commander of the Egyptian air force in the Yom Kippur War, was speaking from firsthand experience.

"I told my people not to worry," Mubarak continued, "because we need Mr. Sharon; he is the only man who can do what he has committed to do."

"Thank you for calling. Together we can do things that will push the peace process forward."

"Absolutely. I am willing to proceed," Mubarak said.

On September 2 the UN Security Council passed Resolution 1559, which called for the withdrawal of foreign troops from Lebanon and the dismantling of the armed militias, both Lebanese and foreign. The American and French ambassadors to the UN, the primary backers of the resolution, said explicitly that the intent was to pressure Syria. Hezbollah and the other terror organizations that had found a home in Syria were similarly targeted. On occasion the political pressure produced results, and the Syrians, who had occupied Lebanon for years without much of a squeak from the international community, were pushed hard against the wall.

On September 6, my father called President Putin. The call was made in the wake of a series of terror attacks in Russia that had claimed hundreds of lives, culminating with the siege on the school in Beslan. The attack began on September 1 and only ended three days later, by which point 322 people had been killed and 700 wounded. Among the dead were nearly 200 children.

"I've called to express my profound shock and the shock of the government of Israel at the series of terror attacks that struck Russia over the last two weeks," my father said.

"Thank you, Mr. Prime Minister."

"Especially the last terror attack, which took place in the school, where hundreds of people lost their lives, including many students. I ask that you pass my condolences on to the government of Russia, the people of Russia, and the families of the victims.

"We must link hands to fight this dangerous phenomenon of terror, which makes no distinctions and also attacks women and children. We must make a diplomatic and intelligence-based effort to eradicate this phenomenon."

"Thank you, Prime Minister. The Jews have for years been the victims of terror; we stood by you and showed sympathy, and so will we continue to do . . ."

"Mr. President, I would like to offer our assistance, in terms of security, in terms of intelligence, and in terms of humanitarian aid. We have experience with the rehabilitation of those wounded from terror attacks," my father said, "and we would be happy to put that experience at your service."

"Thank you. I will instruct our minister of health to contact his Israeli counterpart. I would like to let you know that after the terrorists were neutralized, we found out that half of the terrorists in this attack were of Arab origin. We continue to believe that terror has no nationality, and that terror cannot be compromised with.

"On May 9, we will be celebrating the sixtieth anniversary of the end of World War II and we intend to build a memorial for the Jews in the center of Moscow."

"Thank you," my father said. "We always remember the courage of the Red Army and the fact that it saved what was left of the Jews of Europe. And we will never forget it."

While my father was working on the implementation of the Disengagement Plan, terrorists coming from Syria were launching attacks on American troops in Iraq. Apart from these attacks, the identification of Syria as a state that sponsors terrorism, along with the Security Council's resolution calling for the withdrawal of all foreign troops from Lebanon—Syria being the only country with troops based there—put pressure on Syria's president, Bashar al-Assad. His country's actions risked isolating it from the international community. In an attempt to escape this isolation, Syria tried to initiate so-called peace talks with Israel.

My father had no intention of handing over the Golan Heights. He was well aware of the situation in Syria, where a small Alawite minority repressively rules over a vast Sunni majority, and all state policies, both domestic and foreign, are made with a view toward maintaining this dictatorship. (It is worth noting that Islamic doctrine regards those of the Alawite faith as worse infidels than Jews or Christians, for those are, at least, monotheistic, while the Alawites are considered to be idol worshippers.) Peace talks with Syria would not have brought a resolution. They would simply have bolstered those who were striking our American allies in Iraq and given Syria's regime a way to avoid international isolation. For those reasons, my father rejected the Syrian attempts at initiating peace talks. The Turks were keen to serve as the brokers, but Israel politely declined. The February 2005 murder of former Lebanese prime minister Rafik al-Hariri—an assassination whose trail seemed to lead to Damascus and Hezbollah—and the ensuing UN investigation further isolated Syria.

After he became prime minister, Ehud Olmert, in an act of folly, paved the way for the Syrians' release from their international isolation (at least until the Syrian massacres in 2011) by consenting to Turkish-brokered peace talks. But as soon as the noose had been loosened, the regime in Damascus turned a cold shoulder to Olmert. The Syrians' goal had been not peace but a way out of their quandary.

But while my father was prime minister, Assad continued to make a show of his peaceful intentions. "I am willing to renew peace talks with Israel," he said, "if Sharon is willing."

———————

On the night of October 7, suicide bombers struck the sleepy beaches of the Sinai Peninsula. The two attacks, at the Taba Hilton and the Ras el-Shatan beach, frequented by many Israelis, killed over thirty people, including twelve Israelis. The next day my father spoke with Mubarak:

"I want to thank you for your help and efforts," began my father. "There is no compromising with terror. It must be fought. It strikes now in Egypt, but it does so all over."

"All over," Mubarak agreed.

"This has to be the goal of the free world," my father said, "a war on terror."

Perhaps Mubarak was not the ideal audience for a discussion about the goals of the free world.

"We took the greatest possible precautions, yet it happened, and I do not know how," Mubarak said.

"We must fight and join ranks."

"Yes," Mubarak said.

"We have many missing, and we are making an effort," my father said. "We have a problem, though: we need to send heavy equipment in, but there is a problem getting that in through Taba. There may be many people buried there in the ruins."

"Our people are working together to find people among the ruins."

"Not all of the equipment has entered," my father noted.

"We have no objection to the admission of any type of equipment," Mubarak replied.

"I would like to thank you for your efforts," my father said. "And let us join ranks in the fight against terror."

"Thank you very much for that," Mubarak said.

———————————

On October 15, my father called Prime Minister John Howard of Australia:

"Mr. Prime Minister," my father began. "This is Arik Sharon. I called to congratulate you on your reelection."

"That's nice of you to call," said Howard. "Very generous of you."

"You've been reelected for the fourth time," my father observed; "that is definitely an impressive political accomplishment."

"Mr. Prime Minister, thank you. I am a friend and an admirer of Israel and the Jewish people. Australia is a friend of Israel's and I hope there will be peace in the region, with Israel protected and acknowledged and with a Palestinian state. That is not easy—you live constantly under threat. We made our positions clear in the UN regarding the security fence and the legal issues because we felt the attitude towards Israel was insincere. We support your steps in Gaza; we know how difficult it is and how serious is the threat of terror."

"I admire your determination on the matters of national security and the war against terror. I am referring to the terror attacks against the Australian embassy in Jakarta and the awful attack in Bali that claimed the lives of eighty-eight Australians. We see eye to eye on the need for a united front against terror in all its forms. Terror is terror," my father said, "and all attempts to differentiate one kind from the other based on its motives only weaken those who are trying to eradicate it."

"That's true," said Howard. "We need a united front. That has been our position throughout. That is the stance of the majority of Australians. We have difficulties in our region, but not like the problems you have in your region, where your country is under the constant threat of terror and you are forced to take steps to defend yourselves. I don't understand why other countries don't understand this."

"I also wanted to express my gratitude towards Australia for its position against the Iranian nuclear threat."

"We share the same positions you do in terms of Iran. We do not think Iran meets the world's standards," Howard stated. "We will continue in the same vein."

"Now that the elections are over, I would like to invite you to come visit Israel. You will find yourself among friends."

"Thank you, Mr. Prime Minister," Howard said. "I would like to make that visit. I visited Israel in the sixties and the eighties, and would like to do so again as prime minister. I wish you and the state I admire all the best."

L ate in October 2004, reports started coming out of the Muqata in Ramallah that Arafat was not well and that his condition was worsening. On October 29, he boarded a Jordanian helicopter that took him to Amman, where a French jet sent especially for him awaited. He was admitted to a French military hospital outside of Paris. (Suha, his wife, managed to find the time to ensure that no matter what happened to her husband, she would continue to receive financial support from the coffers of the Palestinian people.)

On November 9, 2004, Arafat died. In France, the terrorist received a full military funeral. After a stop and another ceremony in Cairo, the body arrived in Ramallah for burial.

Arafat's burial was a reflection of his life—all chaos and violence. The masses circled around the landing helicopter: the Palestinian police officers opened fire; the armed men in the crowd followed suit; pandemonium ensued. People were injured by the gunfire, and amid the bedlam, the coffin remained in the helicopter for half an hour. Hundreds of people were trampled and bruised. Many fainted. Armed men climbed atop the coffin before it was finally put into the ground.

Arafat had embezzled from his people and brought them nothing but pain and suffering, and yet they clung to him. The story of the Palestinians' life—bad choices.

O n October 25, 2004, a day before the vote on the Disengagement Plan in the Knesset, my father addressed the Israeli Parliament:

> *You know that I do not say these things with a light heart . . . This is a people that has courageously faced and still faces the burden and terror of the ongoing war, which has continued from generation to generation; in which, as in a relay race, fathers pass the guns to their*

*sons; in which the boundary between the front line and the home
front has long been erased; in which schools and hotels, restaurants
and marketplaces, cafes and buses, have also become targets for cruel
terrorism and premeditated murder.*

*. . . I know the implications and impact of the Knesset's deci-
sion on the lives of thousands of Israelis who have lived in the
Gaza Strip for many years, who were sent there on behalf of the
governments of Israel, and who built homes there, planted trees,
and grew flowers, and who gave birth to sons and daughters, who
have not known any other home. I am well aware of the fact that I
sent them and took part in this enterprise, and many of these people
are my personal friends. I am well aware of their pain, rage, and
despair. However, as much as I understand everything they are
going through during these days and everything they will face as
a result of the necessary decision to be made in the Knesset today,
I also believe in the necessity of taking the step of disengagement
in these areas, with all the pain it entails, and I am determined
to complete this mission. I am firmly convinced and truly believe
that this disengagement will strengthen Israel's hold over territory
which is essential to our existence . . .*

*. . . Israel has many hopes and faces extreme dangers. The most
prominent danger is Iran, which is making every effort to acquire
nuclear weapons and ballistic missiles, and establishing an enormous terror network together with Syria in Lebanon.*

*The Disengagement Plan does not replace negotiations and is not
meant to permanently freeze the situation that will be created. It is
an essential and necessary step in a situation that currently does not
enable genuine negotiations for peace.*

The following day the plan was put to a vote.

Netanyahu, who had helped draft the plan and voted in favor
of it in the cabinet meeting, along with three other ministers—
Limor Livnat, Yisrael Katz, and Dani Naveh—suddenly added
a condition: they called for a national referendum, without which

they would not support the plan. They asked for an urgent meeting with my father in order to discuss their ultimatum.

My father refused to talk to them.

When the time came for a vote, he left his Knesset office and walked out onto the floor. I watched him on TV. He sat alone at the head of the government's table and listened to various speakers. Again, he was approached regarding a meeting with Netanyahu and his colleagues.

"After the vote," he said calmly.

Alone, at the head of the government table in the Knesset.
(© Amit Shabi, Yedioth Ahronoth)

Meanwhile, Netanyahu stood in a corridor, surrounded by reporters. Clearly anxious, breaking out in a sweat, he frantically voiced his demands. On TV, the picture kept changing between my father, who sat quietly and self-confidently inside the plenum, and Netanyahu, who was squirming in the corridor with his allies. The vote began, and the roll was called. Netanyahu was not present. When the second roll call began, he and three colleagues hurried onto the floor and cast their votes in favor of the plan.

This was a true manifestation of Netanyahu's character. Not only was he subversive, but he was also a coward.

The victory was clear and absolute. Sixty-seven members of Knesset supported the plan, forty-five opposed it, and seven abstained, most of whom were members of the various Arab parties in Parliament. After the vote, Netanyahu announced that if within two weeks his demand for a referendum was not met, he would resign his post. The allotted time passed; he took no action and stayed in the government.

Israel does not have a process for conducting referendums and would have had to enact a new law to do so. In addition, the plan received an unequivocal majority and was similar to the public response to the plan as reflected in published polls.

On November 3, 2004, the Knesset passed an aspect of the Disengagement Plan law, which settled the format for compensation of the settlers slated for evacuation. Sixty-four MKs were in favor, forty-four opposed it, and nine abstained. It's worth noting that my father had the bill amended before its submission so as to increase the compensation for the evacuees.

THIRTY-SIX

At President Bush's ranch in Crawford, Texas, April 2005.
(© Avi Ohayon, Israeli Government Press Office)

O n November 24, 2004, a momentous event occurred at the UN: the annual resolution condemning Israel's action in the territories finally included a clause that expressed "deep concern about the suicidal terror attacks that cause the loss of life of Israeli civilians."

On December 2, President Hosni Mubarak of Egypt said in a speech in Port Said, Egypt, "I believe that if they [the Palestinians] do not make diplomatic progress during the tenure of the current prime minister, it will be very difficult to make progress towards peace at a later date. Sharon is capable of making peace and he can reach solutions. He will help, he will take down roadblocks, but he asks for one thing—that the Palestinians stop the attacks." He referred to my father as "the best chance for peace."

That same day, Prime Minister Jean-Pierre Raffarin of France announced that Hezbollah's television station, al-Manar, which was broadcast in Europe via French satellite, was "incompatible with the values of France . . . and I therefore intend to take legislative action shortly in order to acquire the means to shut down such programs immediately, which espouse hate, violence and human disgrace." The prime minister was referring specifically to an al-Manar program that claimed that Zionists had been spreading AIDS and other diseases among the Arabs. (The fact that al-Manar was a terrorist organization station had not been sufficient to shut it down before this repugnant accusation.)

On the same day, Foreign Minister Miguel Moratinos of Spain, speaking at a forum at Tel Aviv University, said, "Israel must not feel distanced from the Union, and the new Europe must offer Israel some type of status beyond what it has today. Spain, like Israel, has suffered terrible terror attacks. Terror is the cancer of the modern age, and there is no excuse for the targeting of the innocent. The Second Intifada and the terror were a mistake and they must be stopped here and now. The struggle against terror must be waged with unlimited cooperation. The European Union, Spain, and Israel are committed to working together in order to slay terror once and for all."

This was quite a departure from the statements made by the Spanish in conversations with my father two years earlier. With that speech, with a little extra effort, Moratinos could run for a spot on the Likud's Central Committee.

Mahmoud Abbas also spoke out against the use of arms in the service of the Palestinian cause. Speaking to the Arabic newspaper *Asharq al-Awsat* in December 2004, he said that it was "a legitimate right of the people to express their rejection of the occupation by popular and social means," but that "the use of arms has been damaging and should end."

Robert Heinsch, the Dutch ambassador to Israel, gave an interview on December 24, just before the end of the Netherlands' turn as president of the European Union. "It's true," Heinsch said, "it was hard for us to accept it; we have been focused on the negotiations track for 40 years and then suddenly along comes Mr. Sharon and presents a unilateral plan. As soon as we understood its significance, and we reached the conclusion that this prime minister is very serious—and this took some time, I admit—then we have supported it with all our heart[s]. We disagree on quite a few issues—settlements, illegal outposts, the positioning of the fence and more—but that does not detract from the respect I have for him.

"We changed our policy," Heinsch continued, "and in November we announced that we view Hezbollah as a terror organization, as was the case with Hamas, and we will push for it to be included in the European list." In terms of Israel's ties to the EU, the ambassador said, "there is a plan of action for the next three years and that is the crown achievement of our turn at the presidency. That will strengthen ties between Israel and Europe. We go farther with Israel than we do with other countries."

Prime Minister Anders Rasmussen of Denmark also voiced pro-Israel comments on the campaign trail while meeting with students. "Israel is surrounded by enemies that want to throw it into the sea and we must recognize that it has a unique history. Israel must use a somewhat firm manner in order to defend itself."

Clearly the EU was smiling on my father; perhaps once the most defamed man in Europe, now he was dramatically raising

Israel's standing while still landing harsh and painful blows on Palestinian terror. The Belgians, though, remained loyal to their earlier convictions: their ambassador to Lebanon met with the secretary general of Hezbollah, Hassan Nasarallah.

———

On January 27, 2005, my father spoke with President Mubarak:

"Mr. President," my father said.

"Good evening, my friend," Mubarak responded. "I want to send Suleiman over to you on February 1 with some ideas."

"I'll be happy to meet with him," my father said.

"Thank you for that."

"Thank you for the assistance," my father said. "Together we will strengthen our ties, and we will be able to do things."

"Suleiman will pass several ideas on to you. How many hot dogs will you give him?"

"Maybe we will be able to add one more," my father said, and they both laughed.

———

On February 8, 2005, my father participated in the summit in Sharm el-Sheikh, along with the newly elected chairman of the Palestinian Authority, Mahmoud Abbas; King Abdullah of Jordan; and the host, Hosni Mubarak. He agreed to go there after the Egyptians finally released Azam Azam on December 5, 2004.

The Egyptians treated my father and the rest of the Israeli delegation with enormous respect, in pointed contrast to the way they had treated Ehud Barak in the same locale toward the end of his term as prime minister. This time no one was locked in his room, phones were not confiscated, Israeli journalists were not boycotted, and Israel's flag was flown. After a friendly meeting between my father, Abdullah, Mubarak, and Mahmoud Abbas, the ceremony began.

My father spoke in Hebrew. He began by greeting the host, Mubarak, and thanking him for his hospitality.

I also wish to convey special congratulations to Your Majesty, King Abdullah, on the birth of your son Hashem and on the occasion of your birthday. May you live a long life filled with joy, and be able to lead your people to tranquility and prosperity, and hopefully we can, together, strengthen the relations between us.

Congratulations are also due to you, the chairman of the Palestinian Authority, Mr. Mahmoud Abbas, on your impressive victory in the Palestinian Authority elections. We have an opportunity to break off from the path of blood which has been forced on us over the past four years. We have an opportunity to start on a new path. For the first time in a long time, there exists in our region hope for a better future for our children and grandchildren.

. . . I am determined to carry out the Disengagement Plan which I initiated. The Disengagement Plan was initiated by a unilateral decision. Now, if new change does emerge on the Palestinian side, the disengagement can bring hope and become the new starting point for a coordinated, successful process.

To our Palestinian neighbors, I assure you that we have a genuine intention to respect your right to live independently and in dignity. I have already said that Israel has no desire to continue to govern over you and control your fate. We in Israel have had to painfully wake up from our dreams, and we are determined to overcome all the obstacles which might stand in our path in order to realize the new chance which has been created.

You, too, must prove that you have the strength and the courage to compromise, abandon unrealistic dreams, subdue the forces which oppose peace and live in peace and mutual respect side-by-side with us.

To the citizens of Israel, I say: we have passed difficult years, faced the most painful experiences and [overcome] them. The future lies before us. We are required to take difficult and controversial steps, but we must not miss the opportunity to try to achieve what we have wished for, for so many years: security, tranquillity and peace.

Then it was Mahmoud Abbas's turn to speak: "We have differences on several issues,

which may include the settlements, prisoner releases, the wall, the closure of Jerusalem institutions and other issues. We shall not be able to settle all these issues today, but our stand on them remains clear and firm. Intensifying efforts to fulfill our obligations will lead us to another commitment from the Road Map . . . I stress our eagerness to honor and implement all our obligations. We hope that our brothers in Egypt and Jordan will continue their good efforts. We are also expecting the Quartet Committee to perform its tasks to ensure speedy progress along the Palestinian-Israeli track, together with efforts to reinvigorate the peace process on the Syrian and Lebanese tracks . . . It is time for our people to enjoy peace and the right to lead a normal life like the rest of the world under rule of law . . . for the language of dialogue to replace that of bullets.

*Shaking hands with Abbas at the
Sharm el-Sheikh Summit, February 8, 2005.
(© Avi Ohayon, Israeli Government Press Office)*

The Egyptians and the Jordanians announced that they would, after a four-year hiatus, return their ambassadors to Israel, who had been withdrawn in late 2000 as an act of protest against Israel's military actions.

———

One day after the summit, King Abdullah called my father: "Mr. Prime Minister, how are you? . . . I wanted to congratulate you on the meeting yesterday; it was held under very positive circumstances. My wife was very moved by your words; they touched her heart deeply. I would also like to thank you for yesterday's speech and the meeting we had."

"How is Hashem doing?" my father asked.

"Thank you for the things you said about him yesterday. We will meet soon. I am very proud of what transpired yesterday."

"I enjoyed our meeting," my father responded.

"Thank you, it was wonderful meeting with you."

———

Despite all these nice words and promises of peace, there was another terror attack in Tel Aviv on Sabbath eve, February 25, 2005. A Palestinian suicide bomber from the village of Dir al-Ruson near Tulkarm blew himself up at the entrance to the Stage, a club on the city's beachside promenade. Five people were killed, and fifty were injured.

After years of resistance, on March 10, 2005, the European Parliament finally agreed that Hezbollah was a terror organization. The parliament also called on Syria to withdraw its forces from Lebanon.

On March 15, the Yad Vashem Holocaust Museum in Jerusalem celebrated the opening of its new wing. More than forty world leaders attended the ceremony. The Holocaust has always been regarded as separate from political matters, but this degree of attendance was unusual. My father met with the leaders of Poland,

Switzerland, Belgium, France, Sweden, Denmark, the Netherlands, and Romania, among others. Due to time constraints, the
representatives had to be moved in and out of the office at half-
hour intervals.

The following week, my father flew to the United States to
meet President Bush. In an unusual gesture that illustrated
the personal friendship and camaraderie between the two men,
Bush invited my father to his ranch in Crawford, Texas. On
April 10, my father's plane touched down at Andrews Air Force
Base near Washington, D.C. From there he continued on to
Waco, Texas, where he had dinner with Condoleezza Rice,
Steve Hadley, Elliott Abrams, and David Welsh, along with his
own advisers.

The next day, my father and his team rode down to Crawford, passing through many beautiful tracts of land. The ranch
itself looks simple. They went into the guest quarters—a humble
structure.

Only a small staff was on duty at the house. The coffee was
served in mugs, and the president himself served marzipan
cookies in the shape and colors of the Israeli flag. Not only did
President Bush wait on my father, but he continued on to Dubi
(who even got a kiss) and the rest of the Israeli delegation. The
meeting was informal, and the conversation was friendly. Afterward, they piled into Bush's double-cab pickup truck. The
president drove. My father sat by his side, and Dubi and Dick
Cheney shared the backseat. The bodyguards rode in the back
along with Avi Ohayon, the Israeli Government Press photographer.

With President Bush on a tour of his ranch, April 2005.
(© Avi Ohayon, Israeli Government Press Office)

When they settled down in the truck, the president said to Dubi, "You see this microphone here—don't tell any of your dirty jokes, 'cause the whole convoy back there will be able to hear you." They drove the length of the ranch, which is roughly the size of ours. Bush drove and talked about the ranch. (Speaking at an American Israel Public Affairs Committee [AIPAC] meeting a month and a half later, he related a discovery: "Last month I met with President Bush on his ranch in Crawford, Texas. A beautiful place, I enjoyed being there. As an agriculturist I found that the prime minister of our tiny state, Israel, has more cattle than the president of the world's greatest superpower." The crowd laughed hardily.)

The conversation revolved around landscape, nature, trees, and childhood memories. They saw a giant tree that had been felled by lightning. Bush talked about his neighbor, the owner of the nearby ranch, who had asked the real estate agent the name of the new landowner. When the agent replied that it would be the president of the United States, the neighbor became angry. He didn't like being toyed with.

They returned to the ranch's simple dining room and had lunch. There was no diplomatic agenda. The visit was simply an expression of respect, of the strength of the bond between Bush and my father.

———

On April 27, Vladimir Putin arrived in Israel. It was the first visit by a sitting Russian president. Putin stayed in Jerusalem and visited the Western Wall and the Church of the Holy Sepulcher. The visit was a clear indication of the excellent personal relations between my father and Putin and of Israel's world standing. "We are strategic allies in all matters concerning terrorism," Putin said when he met with my father on April 29. "A nuclear Iran worries us no less than it worries you. So long as I am president, Russia will not take any action that could harm Israel."

The next day, April 30, my father spoke with the Canadian prime minister, Paul Martin:

"I know you are having some internal troubles," my father said. "I ran a minority government here for a few months and understand the problem."

"I could learn from you; you are handling it excellently."

"You are handling it excellently," my father countered. "I hope you overcome these troubles."

"I will be meeting Mahmoud Abbas in two weeks," said Martin. "How can I be of help? He's also got some difficult problems and must be helped. I know that you released prisoners and handed cities over [to Palestinian security control]; can more be done?"

"I appreciate the fact that you want to help. On the matter of prisoners—we released five hundred, and we are checking another group of four hundred. The main problem is that the terror continues, albeit to a lesser extent, but there still is terror, not only [within] the PA's territory but also within Israel. Rockets are fired at Israeli towns."

"Are you concerned about the expansion of the settlements at this time?" Martin asked.

"There is no expansion," said my father. "There are no new settlements, only construction along the already existing line. A quarter of a million people live there, there are big families there and people get married young, they need to live a normal life. The matter of Jewish settlement will only be decided in the final negotiations."

"I'd like to raise two matters. [Mahmoud Abbas] is coming to Canada in two weeks, and I will tell him to take steps towards reform. Are there messages you would like me to pass on to him?"

"We will provide you within a few days a list of steps he must take so that we can move forward," my father said.

"That would be very helpful . . . [Mahmoud Abbas's] son spoke with us. They are concerned for his personal safety. They are not worried about you [Israel] but internal elements. Mahmoud Abbas's son wanted me to pass this on to you, and I promised to do so. I would like to come and visit Israel."

When it comes to their own personal safety, it is only Israel that they trust. But they would never admit it publicly.

"We would be happy for you to come, the earlier the better. You will find yourself among friends. I know you are planning to come in September."

"True . . . Maybe you could send me some members of Knesset to vote for me, and that way we will pass through the votes. I'll send you in return some of my parliament members," Martin said, laughing. Then he added, "I will report to you after Mahmoud Abbas's visit."

AIPAC, a lobbying group that advocates for tightening the relationship between the US and Israel on Capitol Hill and within the executive branch, has always been dear to my father.

The organization is composed of Americans from all across the United States who believe this relationship is in America's best interest. My father appreciated the group's businesslike conduct as well as the fact that it is not affiliated with the right or left in Israel or the United States. Each spring the organization holds its annual conference, generally attended by high-ranking U.S. government officials and a representative of Israel, as well as its own members.

In May 2005, shortly before the conference, the organization found itself in a tight spot. Two of its members, Steve Rosen and Keith Weissman, were accused of receiving classified information from Lawrence Franklin, a Pentagon official. They were charged with conspiracy to disclose national security secrets for allegedly passing the information on to an Israeli embassy official in Washington. Four years later, in May 2009, the Justice Department dropped all charges against the two men. In 2005, however, with the FBI searching AIPAC offices and the media in the midst of a feeding frenzy, the organization found itself in a crisis.

Several weeks before the 2005 conference, AIPAC executive director Howard Kohr called my father's chief of staff, Dubi Weissglass, and relayed that because of the ongoing investigation and the media reports, none of the American government officials who had been invited had confirmed their attendance. At AIPAC the thinking was that the officials were waiting to see which way the investigation seemed to be heading, and that their attendance was therefore very much in doubt.

Dubi was well aware of my father's affection for the organization. Whenever the heads of AIPAC came to Israel or he flew to the United States, my father made a point of meeting with these officials.

When my father heard about the organization's predicament, he made an immediate decision. "We're going," he ruled. Dubi called Howard Kohr, who was silent for several seconds. A few minutes later Dubi received a phone call from

Amy Friedkin, the former president of the organization. "Is it true?" she asked.

Several minutes later Bernice Manocherian, who followed Friedkin as president, called with the same question.

"Look," Dubi said, "you can't keep calling me every minute to ask me if it's true." This was on Thursday.

"On Monday we're sending out the invitations," Bernice said, noting that of course "[my father] would be the keynote speaker but if in the end he is unable to come people will attribute it to the investigation and the damage to the organization will be enormous."

Weissglass said, "If he said he's coming, he's coming."

My father asked not to have any other formal meeting scheduled while he was in Washington so that no one would think he had come to the conference on the way to another meeting.

On May 24 he arrived at the convention. The police had closed off the area, and limousines lined the sidewalks. Eight thousand people filled the convention center, which was alive with the sounds of music and the sights of flag-waving. A giant hologram on the wall depicted the intertwined Israeli and American flags.

The master of ceremonies said, "Ladies and gentlemen, the prime minister of Israel," and the crowd got to its feet and applauded hard for several long minutes. The MC asked them to be seated, but the crowd continued with the applause. The crowd refused to settle down. They knew the event had been saved, and they appreciated the gesture. My father was embarrassed. It's not that he is a shy man, but public expressions of endearment embarrass him. His reactions, however, are authentic.

Howard Kohr told Dubi that twenty-four hours after the invitations had been made public on that Monday after they spoke, all of the invitees had confirmed their attendance.

Sharon speaking at the UN General Assembly, September 2005. (© Avi Ohayon, Israeli Government Press Office)

THIRTY-SEVEN

On the morning of July 7, 2005, one day after London had been chosen as host of the 2012 Olympics, four Muslim suicide bombers blew themselves up in a synchronized attack in three subway stations and one bus. Dozens were killed, and hundreds were injured. My father called Tony Blair the next day. This time, in a departure from the usual routine, the condolences came from the other direction:

"Mr. Prime Minister," my father said.

"Hi, Arik."

"We were shocked to hear about the tragedy and the terrible terror attacks," my father said. "I hope these kinds of things do not occur again. If we can help in any way, we are willing to provide any assistance that we can. I would like to express my condolences to the people of Britain from the people of Israel and from myself personally. It is an awful and cruel thing. It only underscores that terror cannot be compromised with, that today it is the central threat."

"It could have been a day of celebration, as Britain was picked to host the Olympic Games," said Blair, "but instead it was turned into a tragedy."

"We are willing to provide any assistance," my father said. "We are open to anything."

"That is very kind of you, I appreciate it, and very nice of you to call. You have suffered from terror, and there is a strong tie between us."

"I know that you are going through a tough time," my father said. "We are familiar with it from our own experience. I am sure that you and the British people are strong enough to perse-vere through this tragedy that you are going through. Our con-dolences."

"Thank you. Thank you for the call. Thanks, Arik."

On July 27, my father flew to France. As the media reported, the "French were going out of their way to warmly welcome Prime Minister Ariel Sharon who was until recently considered to be virtually a pariah in their eyes."

Chirac truly went out of his way to make the visit a success and to convey his warmth, after all his former friends in the region, including Saddam Hussein and Yasser Arafat, had left the stage. *Le Figaro* wrote, "France realizes that it risks isolation for a long period of time in all matters pertaining to the Middle East if it does not listen to Israel." As far as Israel was concerned, the paper noted that without stable ties with France, it would have a hard time getting anything accomplished in the European Union.

My father is not naïve. He did not believe that the French had undergone a change of heart. They had simply fallen into line with the rest of the Western nations, and they very much liked the idea of an Israeli withdrawal from Gaza. Nonetheless, the visit was defi-nitely more pleasant this way, and the French know how to make a guest feel wonderful when they want to. My father viewed the me-dia circus realistically. He had seen everything by then. He recalled that the newspapers that now called for reconciliation with Israel were the very same ones that had fired darts at him in the past.

The implementation of the Disengagement Plan—August 15—drew near, and as during other difficult periods, my father intensified his focus.

Two or three days before the plan went into effect, the secre-tary of the prime minister's office, Marit Danon, went into my father's office. "I had a very bad dream last night," he told her. "I dreamed that I was hanging from a rope above a well, and that the rope snapped." He did not elaborate, and she did not ask.

Never before had he revealed anything so intimate to her—nor would he ever do so again. Marit had worked closely with my father for five years, and she thought that his dream was an expression of the responsibility he felt and his concerns about Israelis being wounded during the upcoming evacuation.

Just as always before military operations, his office tolerated no banter, just focusing on preparation and an attention to detail during the weekly and twice-weekly meetings with the general director of the prime minister's office, Ilan Cohen, and others. When certain measures were caught up in the inevitable bureaucracy, he would raise his voice—it was simply a means of cutting through the red tape—but other than that he was quiet.

Withdrawing from Gaza was difficult and painful, and extraordinary leadership was needed to implement the plan. It was done with great sensitivity and determination, with orderliness, and precisely on schedule. There was no one who was not moved by the pain of the relocated families and the tears of their children, forced from their homes—homes and farms that had been cultivated for years.

The disengagement proved that Israel can be sensitive, that the evacuators can cry with the evacuees, but that it is a country that knows how to follow the rule of law—a strong democracy with a wide base and a deep will to survive.

There are those who argue that the disengagement precipitated the onslaught of rocket fire on Israel. The reality is that these rocket attacks had begun several years before the disengagement. Indeed, there were more attacks from Gaza in 2004 alone than after the disengagement in 2005 and 2006 combined. The very fact of the disengagement enabled Israel to respond to the attacks, as it was no longer governing and responsible for the inhabitants.

S everal days after the completion of the Disengagement Plan,
my father received a call from Kofi Annan:

"Mr. Prime Minister, how are you?" Annan said.

"Hello, Mr. Secretary-General, how are you?"

"Fine. I've called to thank you and have published a statement
saluting the manner, the courage, and the determination with
which you carried out the disengagement."

With Kofi Annan at the UN, September 2005.
(© Avi Ohayon, Israeli Press Office)

"Thank you," my father said. "It was not easy, but I managed
to do it, and I carried it out exactly as I had committed to."

"Many people around the world appreciate the bold move you
made," Annan said. "The disengagement is a testament to your
leadership, and the people in Israel showed maturity and their
willingness to sacrifice for peace. The army's professionalism and
restraint when dealing with civilians is worthy of appreciation
and admiration."

"Thank you very much."

"It is an historic event that offers a true chance for the promotion of peace. I spoke with Abbas and told him he has a chance to
instill law and order. I read in the paper that the two of you may
soon meet, and that would be good."

"It's important. Thank you, I'm grateful for the call. I want to move forward. The problem is terror. Yesterday there was a suicide attack. Fifteen people were injured, two of them critically."

(The morning before their conversation, a suicide bomber had blown himself up in Beersheba's central bus station.)

"My condolences and my sympathies are extended to the government and to those injured in the brutal attack."

"Thank you," my father said. "I look forward to seeing you."

"Are you coming to the General Assembly?" Annan asked.

"Yes."

"It's an important mission. The Quartet would like to work with you. And we want to work along with you. Congratulations," Annan said, "and see you in New York."

—————

The UN General Assembly convened in September 2005. It was the world's biggest diplomatic event. Of the 192 UN members, a record 177 nations sent their heads of state to the assembly. According to Danny Gillerman, then Israel's ambassador to the UN, there has been no such event before or since.

My father was staying at the time in the Palace Hotel in New York City. My wife, Inbal, and I went to visit him there. We passed through the many layers of security in the lobby, went up in the elevator, continued on up the stairs, and finally arrived at the room—a lovely suite. The back of St. Patrick's Cathedral on Madison Avenue should have been visible from the window. I looked and instead found a drawn curtain and a bulletproof barrier. I checked all of the windows and found them to be similarly blocked.

"Dad," I asked, "how do you know you're in New York? Maybe they took you to Netanya or Petach Tikva [cities in Israel]. You can't see out the window, can't walk down the block, can't go to a good restaurant or out to a show. How do you know?" It was not a fair question, because he really appreciates all those things, but he accepted my remarks good-naturedly, even though they did not make him laugh. He knew I had just articulated what he had been feeling. After all, he used to call the prime minister's residence in Jerusalem a comfortable jail. He hated its closed windows and the dark curtains that stood between him and the view; he hated the tall wall surrounding the yard; and he missed the open landscape of our farm more than anything else.

In a conversation I had with Danny Gillerman, he said of my father's visit to the General Assembly, "Even though everyone was there, the star of the event was Arik Sharon—the one who was considered a radical for years and had been demonized. He was greeted like an international hero, the superstar of the universe."

Gillerman said that people requested, even pleaded, for a meeting with my father, and that he had never seen anything like it. As someone who represented Israel at the UN both before and after this assembly, Gillerman was accustomed to asking for meetings for visiting Israeli dignitaries, but at this event, there were so many requests to meet my father that he had to apologetically reject many of them.

The only one of the 192 heads of state to have an office at the UN is the president of the United States of America. The other 191 do not. The area is partitioned with screens and outfitted with couches, and the leaders bounce from station to station to meet with one another. My father was allotted a conference room—something even the Russians were not given.

Kofi Annan's welcoming ceremony on the fourth floor of the UN building was the occasion for a surreal sight: heads of state pushing and elbowing each other in order to get to my father, shake his hand, and express their high regard for him.

"I admire your courage. I know it was hard to carry out the disengagement, and you are a brave leader," President Bush said when the two met.

Gillerman revealed that my father had a condition that had not been made before: no secret meetings; whoever wanted to meet with him would do so in the full light of day. And they did—

Pervez Musharraf, the president of Pakistan, who brought his wife for the meeting; the prime minister of Turkey, Tayyip Erdoğan; and the leaders of Indonesia, Qatar, Oman, Morocco, and Australia, among others.

Most interesting was my father's meeting with Vladimir Putin, with whom he had a special relationship. According to Gillerman, the Russians had taken over the entire eighteenth floor of the Waldorf-Astoria, and once they had given it the treatment, it resembled a wing of the Kremlin, with carpets on the floor and decorations on the walls.

The meeting was scheduled to take half an hour. At first Putin said, "I miss you. Come and visit again."

"I will," my father said, "but I want to see more villages and cows and less people."

They discussed the dangers of Iran and Syria and the nature of the Russian vote in the UN regarding Israel. My father had a habit: before any meeting with a nation's leader, he would ask to have that nation's voting record in the UN regarding Israel.

The meeting with Putin went on and on. At around the forty-minute mark Russian foreign minister Sergey Lavrov started coming in and out of the room, clearly agitated. He sent a note in to Putin, who waved it off. After an hour and ten minutes Putin said, "I want to tell you a joke—a Jew and a Chinese man are riding on a train. The Chinese man asks the Jew, How many of you are there? Thirteen million, the Jew tells him. After a while the Jew turns to the Chinese man and asks, How many of you are there? One and a quarter billion, the Chinese man says. You've got one and a quarter billion people and we've got thirteen million people, the Jew says, so why are people always talking about us and never about you?" The Russian delegation laughed hard at the joke, and then Putin said, "The reason the foreign minister has been running around is because the president and premier of the billion and a quarter people have been waiting outside for fifty minutes for the leader of seven million people."

With President Putin.
(© Amos Ben Gershom, Israeli Government Press Office)

The next day, September 15, 2005, my father, along with his small entourage and an army of Israeli, American, and UN bodyguards, returned to the UN building. They made their way down a long, wide hall lined with UN workers' cubicles. UN employees from all over, of every color and ethnicity, came pouring into the hall, cameras flashing incessantly, some trying to extend hands that the bodyguards were trying to push away and some waving.

My father sat with his small team in a room adjacent to the General Assembly Hall. Dubi Weissglass described the scene to me: every few minutes the commander of the bodyguard squad would come in and ask Dubi Weissglass to please come out and authorize this or that world leader who had asked to shake my father's hand and exchange pleasantries with him. There were dozens of such requests—from Asia, Europe, Africa. After a while Dubi decided to stay outside so that he would not have to go in and out every few minutes. As Dubi said, "An entourage approaches—they see a commotion, they stop, find out what's happening, find out that it's Prime Minister Sharon, come over, introduce themselves, ask to shake his hand. That's how they came in, one after another, handshake, a few words, some compliments and then the next delegation."

Then my father entered the General Assembly Hall and began speaking. Ambassador Danny Gillerman had tried in vain to convince my father to speak in English, or at least to say a few words in English in the beginning, but he wouldn't have it.

"I arrived here from Jerusalem, the capital of the Jewish people for over three thousand years and the undivided and eternal capital of the State of Israel." With those words, my father began his speech.

I stand before you at the gate of nations as a Jew and as a citizen of the democratic, free, and sovereign State of Israel, a proud representative of an ancient people, whose numbers are few, but whose contribution to civilization and to the values of ethics, justice, and faith surrounds the world and encompasses history. I was born in the Land of Israel, the son of pioneers—people who tilled the land and sought no fights—who did not come to Israel to dispossess its residents. If the circumstances had not demanded it, I would not have become a soldier, but rather a farmer and agriculturist. My first love was, and remains, manual labor; sowing and harvesting, the pastures, the flock and the cattle.

I, as someone whose path of life led him to be a fighter and commander in all Israel's wars, reach out today to our Palestinian neighbors in a call for reconciliation and compromise to end the bloody conflict, and embark on the path that leads to peace and understanding between our peoples. I view this as my calling and my primary mission for the coming years. The Land of Israel is precious to me, precious to us, the Jewish people, more than anything. Relinquishing any part of our forefathers' land is heartbreaking, as difficult as the parting of the Red Sea. Every inch of land, every hill and valley, every stream and rock, is saturated with Jewish history, replete with memories. The continuity of Jewish presence in the Land of Israel never ceased. Even those of us who were exiled from our land, against our will, to the ends of the earth—their souls, for all generations, remained connected to their homeland, by thousands of hidden threads of yearning and love, expressed three times a day in prayer and songs of longing.

The Land of Israel is the open Bible, the written testimony, the identity and right of the Jewish people. Under its skies, the prophets of Israel expressed their claims for social justice, and their eternal vision for alliances between peoples, in a world that would know no more war. Its cities, villages, vistas, ridges, deserts, and plains serve as loyal witnesses to its ancient Hebrew names. Page after page, our unique land is unfurled, and at its heart is united Jerusalem, the city of the Temple upon Mount Moriah, the axis of the life of the Jewish people throughout all generations, and the seat of its yearnings and prayers for three thousand years. The city to which we pledged eternal vows of faithfulness, which forever beats in every Jewish heart: "If I forget thee, O Jerusalem, may my right hand forget its cunning!"

I say these things to you because they are the essence of my Jewish consciousness, and of my belief in the eternal and unimpeachable right of the people of Israel to the Land of Israel. However, I say this here also to emphasize the immensity of the pain I feel deep in my heart at the recognition that we have to make concessions for the sake of peace between us and our Palestinian neighbors.

This week, the last Israeli soldier left the Gaza Strip, and military law there was ended. The State of Israel proved that it is ready to make painful concessions in order to resolve the conflict with the Palestinians. The decision to disengage was very difficult for me, and involves a heavy personal price. However, it is the absolute recognition that it is the right path for the future of Israel that guided me. Israeli society is undergoing a difficult crisis as a result of the disengagement, and now needs to heal the rifts.

Now it is the Palestinians' turn to prove their desire for peace. The end of Israeli control over and responsibility for the Gaza Strip allows the Palestinians, if they so wish, to develop their economy and build a peace-seeking society, which is developed, free, law-abiding, and transparent, and which adheres to democratic principles. The most important test the Palestinian leadership will face is in fulfilling their commitment to put an end to terrorism and its infrastructures, eliminate the anarchic regime of armed gangs, and cease the incitement and indoctrination of hatred towards Israel and the Jews.

Until they do so—Israel will know how to defend itself from the horrors of terrorism. This is why we built the security fence, and we will continue to build it until it is completed, as would any other country defending its citizens. The security fence prevents terrorists and murderers from arriving in city centers on a daily basis and targeting citizens on their way to work, children on their way to school, and families sitting together in restaurants. This fence is vitally indispensable. This fence saves lives!

. . . I am among those who believe that it is possible to reach a fair compromise and coexistence in good neighborly relations between Jews and Arabs. However, I must emphasize one fact: there will be no compromise on the right of the State of Israel to exist as a Jewish state, with defensible borders, in full security and without threats and terrorism.

. . . The Jewish people have a long memory. We remember events that took place thousands of years ago, and certainly remember events that took place in this hall during the last sixty years. The Jewish people remember the dramatic vote in the UN General Assembly on November 29, 1947, when representatives of the nations recognized our right to national revival in our historic homeland. However, we also remember dozens of harsh and unjust decisions made by the United Nations over the years. And we know that, even today, there are those who sit here as representatives of a country whose leadership calls to wipe Israel off the face of the earth—and no one speaks out. The attempts of that country to arm itself with nuclear weapons must disturb the sleep of anyone who desires peace and stability in the Middle East and the entire world. The combination of dark fundamentalism and support of terrorist organizations creates a serious threat that every member nation in the UN must stand against.

. . . Peace is a supreme value in the Jewish legacy, and is the desired goal of our policy. After the long journey of wanderings and the hardships of the Jewish people; after the Holocaust that obliterated one third of our people; after the long and arduous struggle for revival; after more than fifty-seven consecutive years of war and terrorism that did not stop the development of the State of Israel; after all this—our heart's desire was and remains to achieve peace with our neighbors. Our desire for peace is strong enough to ensure that we will achieve it, only if our neighbors are genuine partners in this longed-for goal. If we succeed in working together, we can transform our plot of land, which is dear to both peoples, from a land of contention to a land of peace—for our children and grandchildren.

Before my father began speaking, there had been a constant flow of delegates returning to their seats in the hall, according to Gillerman. When he reached the last line, there was thunderous applause, a rare tribute at the UN. Then he stepped down and was mobbed by hundreds of people, heads of state, ministers, ambassadors, who wanted to shake his hand. He spent over an hour standing there.

The world loves Israeli withdrawals—there is no doubt about that. That said, the members of the assembly clapped for a speech that spoke about Israel as the land of our bible and of Jerusalem as the eternal and indivisible capital of Israel—in which my father said that Israel would continue to fight terror, continue with the construction of the security fence. He talked about the anti-Israel bias at the UN, and the deadly intentions of the terror-supporting regime in Iran that has called for the annihilation of Israel.

Apart from the members' satisfaction with Israel's decision to leave Gaza, the welcome he received at the UN demonstrated respect and appreciation for my father's courage and determination. It was the dramatic conclusion to four and a half years as prime minister, a complete turnaround from open hostility, even before he launched the war on terror, to fame and recognition after all that he had accomplished—the war on terror, the construction of the fence, and the withdrawal from Gaza.

———

Assuredly, the determined fight against terror and the reestablishment of military activity over all of Judea and Samaria, along with the construction of the separation fence, brought results. My father's directives, combined with his policies, broke the spirit of the Palestinian terror campaign. This was accomplished without any assistance from the Palestinian Authority.

———

Public servants, certainly those who have served as prime minister, must return the country that they were entrusted with in better shape than when they received it. That is their duty. If they do not fulfill that basic condition then what exactly were they doing during their term of office and to what aim?

My father received a country in a chaotic state—terror raged, lives were lost, fear reigned; the power of deterrence had been lost; the economy was in crisis; the country's international standing at a nadir; Israel was disrespected abroad and the government and the man who headed it were disrespected at home, deemed illegitimate by an untrusting public.

Even my father's adversaries do not dispute the fact that from the moment he sat down in the prime minister's chair it was clear that there was someone in charge. Honor and respect were restored to the office and the man who occupied it.

At the end of his five-year tenure, the country was in a far different state, much improved from its condition at the beginning of his first term: terror had been routed, personal safety had been restored, along with the public's faith in the country and its ability to withstand adversity. Israel's economy had been placed on a track of rapid growth, and the country enjoyed a privileged place among nations and was held in high esteem by the international community.

The question "What would your father do?" comes up again and again whenever Israel reaches a crossroads. The answer, I believe, lies in his judgment and in the principles that guided him throughout his adult life—the safeguarding of Israeli citizens and Jews the world over, an unflagging war against our enemies, and, at the same time, a hand extended toward peace. Not a hand that hangs in the air, divorced from reality. Rather, a solid policy: this and this we are able to do, and beyond that we are unable, because it will threaten Israel, because the state cannot afford to take that risk. Everything clear and out in the open.

He had many plans for his third term in office. With terror suppressed, he intended on focusing on internal affairs—changing Israel's impossible electoral system, pushing hard for the advancement of education, improving personal security, and advocating strongly for mass immigration—an issue always close to his heart. All his attention was given to guarding the citizens of Israel and bettering their lives. It is sorrowful that he was unable to serve a third term in office.

EPILOGUE

I t is July 2006, and I am sitting beside my father in the intensive care unit in Tel HaShomer Hospital. I've turned the radio on. They don't work well in hospitals. The medical equipment interferes with the reception. I find a local station. The song "Yesh Li Ahuv b'Sayeret Haruv" (I Have a Beloved in the Haruv Recon Unit) comes on. It takes me back to when I was young, a kid in a family whose father is a career officer in the army. The vibe is of the early seventies. We live in Beer Sheva, my father is commander of the southern front, the Command has its own musical troupe, my mother looks out for them, brings the girls makeup kits, etc.

The next song on the radio is "B'aviv at Tashuvi Hazara" (In the Spring You'll Be Back), and we're in my father's army-issue Plymouth, driving north out of Beer Sheva. This was in the early seventies, I was around six, my brother two years older than me. There was no Tel Aviv–Beer Sheva road back then and we drove on the local road toward Netivot. Omri, my mother, my father, and I rode in the car and we tried to teach my father the song that is now playing on the radio, which we'd learned in school. He kept bungling the words and there was nothing we could do to make him get it right. We laughed, tried again, "Autumn winds, golden leaves, scattered all around," and my father, "Autumn winds scatter the golden leaves," etc.

The radio plays on. The Southern Command's troupe is singing "Eretz Yisrael Yaffa" (The Land of Israel Is Beautiful) in the background. I am scheduled to go to the Ukraine for an agricultural project, bio-diesel, environmentally friendly energy production, but there has been an emergency call-up of the reservists. Yesterday I called the entire team and told them that they shouldn't be shocked if we were all on our way toward Lebanon (the second Lebanon war) on Sunday. Tomorrow, I will join my reserve unit.

My fellow reservists aren't kids anymore. Most are forty-five years old. One is closer to fifty, which doesn't stop him from calling every few minutes from abroad and searching for the next flight to Israel, despite my attempts to calm him. All the guys are closer to retirement age than draft age and yet they come immediately.

"Yom Yavo" (The Day Will Come), from the album *Don't Call Me Black,* is on the radio. I like this song.

On December 18, 2005, I received a call from Marit Danon. She was in the office and felt my father might not be okay. She wasn't sure what to do. My father's personal assistant, Lior Shilat, said, "Call Gilad, if anybody'll know it's him." She made the call and put him on. We exchanged two words and I could tell something was wrong. I believed it was a stroke. During the seconds it took before I got her back on the line, he had headed down to the car. "He has to be rushed to the hospital," I told her. I reached him in the car via the secret service. Go straight to the hospital, I told the bodyguards. Inbal and I raced to the hospital and spent the night with him. Omri was there, too, along with all of my father's aides and office staff. In the morning, I lay down in his bed and he sat on the couch and handled affairs.

The next time was far worse. On the fourth of January in the evening, Inbal and I were sitting in his room, the kids with him on the bed. We were watching something funny on TV. Then we went to our rooms. At around nine, I felt that something wasn't right. I came to his room. He was speaking to the IDF chief of staff, instructing him to fire from navy ships in the event of rocket fire on Israel from Gaza. "Why, Dad?" I asked him. "Apply pressure that way, too," he explained. He would not accept fire on Israel. A few days later, amid the madness of surgery and intensive care in the hospital, I heard the news on the radio, navy ships had returned fire on targets in Gaza. I smiled to myself: even with his life in danger, he was still protecting the Jews; he knows no other way.

―――――――――――――――

I t was Saturday morning, on the seventh floor of the Hadassah University Hospital in Ein Kerem, the neurosurgical intensive care unit. I was in my father's room and a tall, thin man walked in, young-looking, dressed casually, leather straps on his wrist for a bracelet. I realized he was a doctor, mostly on account of his self-confidence, but I checked to be sure. He noticed and introduced himself, Dr. Pikarsky. He was the surgeon who was called when my father began having stomach problems. "Fight, fight for my father's life," I said. This was not the first time I had asked this of the doctors. He had already had three brain surgeries. He turned to me with a look that said, you do not understand. "Come," he said, "we have to talk." I followed him out. "Come to my office." "Where is your office?" It was on another floor. "I'm not moving from here," I said. "Okay, then we'll talk here in the ward."

We sat down in department head Professor Felix Umansky's office. Umansky, who had done the three previous surgeries, was there along with Dr. Yoram Weiss, who was taking care of my father (these are the two who orchestrated the treatment that saved his life). Pikarsky, a new addition to the team, told me in not so many words that based on the CT scan, the game was over. I said, I understand, but I want you to keep on battling.

Surgeons are a different breed of doctor. In many ways they perform the classic tasks of the physician. There's a problem— they open up, intervene, and fix it.

They share many characteristics, the most prominent of which is their speech pattern, more accurately referred to as a growl. They do not seem to speak; they issue rapid and impatient monosyllabic directives. At times you can identify familiar words amid the chain of growls. The language is often accompanied by declarative hand gestures.

Their familiarity with the subcutaneous regions of man can be likened to the knowledge of a butcher, who knows the insides but not the souls of the animals. The same is true of their handiness with the knife. They are a tough bunch, shaped by their training to be what they are, an intervention force. Abdomens cannot be opened hesitantly. Carefully, yes, but with decisiveness, with self-confidence and assured movements.

We were joined by Professor Avi Rivkind, a surgeon, head of the hospital's trauma unit. He tried to tell me the same thing: there's no chance whatsoever.

I remember one of the doctors who treated my mother, one of the less affable ones, saying that if it were his mother he'd allow her to be released from her suffering. Back then, his sentiment had made an impression on my father, who hadn't considered the option that perhaps this particular doctor wasn't crazy about his mother.

The line is an old doctors' trick: if this were my mother or father, this is what I would do, or, in our case, what I would not do. I started to think that maybe not all doctors love their mothers and fathers.

When they started telling me "if it were my father . . ." I suddenly remembered my mother's doctor saying the same thing. They tried to explain to me that there was nothing more they could do, I told them about a dream I had had many years ago. In that dream I was with my father in the hospital. He was lying in bed, surrounded by medical staff, and they had all either given up or lost hope and were about to leave, and my father didn't say a thing, but he stared at me with this look, with those green-gray eyes of his, and I knew I would never give up, and that I simply would not leave him. This was a dream I had when my father was healthy and strong and the scenario was completely divorced from reality. I did not tell a soul about the dream at the time. During my father's previous hospitalization, in the morning after he was admitted, Inbal and I sat with him in his hospital room. It was then that I told him about my dream, but I did not think that it was going to come true. But now I shared it with the doctors and my fear that it was happening now and that I would never be able to forgive myself if we did not fight to the end. Dr. Weiss suggested that they could take him down to the OR and assess the extent of the damage. If it seemed catastrophic, they'd come and talk to me. Pikarsky snapped, "I don't step out in the middle of a surgery to consult with the family. If I go in, I go in."

He was not overly sensitive, but I liked that, liked the spirit. In general I was fond of Pikarsky. Yoram Weiss, my father's devoted doctor, did not use halfway measures. If it's a battle, then let's do battle. He brought my father to the OR in excellent condition, under the circumstances.

Inbal and Omri arrived quickly from the farm. She knew that we were on our way down to the operating room, and in order not to miss us, she ran up the seven flights of stairs. "Wait a second," I told Dr. Weiss. Inbal reached us and gave my father a kiss. "A kiss from Inbalika is a blessing," I told Dr. Weiss. We went down.

We sat outside the operating room, just as we had done the three times before. The mood was grave. Everyone thought it was over. Friends gathered in the waiting area. Dr. Segev, our family doctor, told them that based on the information he had it was a one in a thousand chance. I was later told that our friend Adler said, "I'm going with the one."

After a while, Dr. Weiss came out and said that his condition was a lot better than the CT scan had indicated. It looked all right. Waiting outside of the OR is not an easy experience. Every time a green-clad shape in a disposable mask and hat came through the doors, I scrambled to my feet to see whether there was any news.

Later, in the post-operation meeting, Rivkind used a blade a lot like his scalpel to excoriate the person who had read the CT.

When the surgeons sat down with us after the operation, their clothes still soaked in blood, Pikarsky gave me the thumbs-up, letting me know that it had gone okay.

After several days in the intensive care unit, I asked the doctor, Dr. Maroz, "What are the chances."

"I can't say," he responded. "No one's ever survived this sort of thing."

———

One day, when we were still in the ward, on Hadassah's seventh floor, a woman in a nurse's uniform entered. She had black hair and black eyes, a penetrating stare, and to me she looked like a witch. She said, "You can release and let him go, it's only in your hands."

"I'll never let go," I told her.

I don't know who she was, and I never saw her again.

For perhaps a month I did not leave the hospital. I stayed by his side. Omri and Inbal would go home for a few hours and then come back to the hospital. Later, Omri and I started to take turns, working in shifts.

We visit every day, haven't missed once. He lies in bed, looking like the lord of the manor, sleeping tranquilly. Large, strong, self-assured. His cheeks are a healthy shade of red. When he's awake, he looks out with a penetrating stare. He hasn't lost a single pound; on the contrary, he's gained some.

———

If ever I don't come to," he said to us once in jest, "all you have to do is let me smell a salami omelet and I'll get right up." He said this in the kitchen of our house on the farm, while we were eating just that with some sharp mustard. Anyone who thinks I didn't bring that exact dish to my father in the hospital simply doesn't know me.

The farm is developing; we've built greenhouses and more new sheepfolds, and we've increased the size of the flock. We've built more sheds and yards for the cattle. Everything is in order, set up, waiting for him, just like us, patiently and expectantly.

———————————

During his second stroke, before we got into the ambulance and headed to the hospital, I lay down beside him in his bed on the farm. We were alone. "You know that we love you," I said to him, hiding behind the "we." A moment later I overcame my shyness and said, "You know that I love you."

He responded, "I love all of you," still one bashful stage behind me, but I knew what he meant.

NOTES

49 "An olive grove near Hulda": Ariel Sharon, *Yedioth Ahronoth*, article from April 19, 1998.

52 "I owe him my life": Ariel Sharon, newspaper interview, summer 1973.

57 My father's next position: The IDF's Northern Command is the regional command along the northern front with Lebanon and Syria, with units stretching from Mount Hermon in the north to Netanya in the south.

63 Rabbi Zalman Tzoref's story stuck: Cabinet decision, July 21, 2002.

64 Article 4 of the agreement states: Weizmann-Faisal Agreement, January 3, 1919.

65 In a 2007 article: Interview with the Park Hotel terrorist, April 7, 2007, Yoram Schweitzer, Maariv, April 6, 2007.

66 Churchill refused, saying, "It is manifestly right": Martin Gilbert, *Churchill: A Life* (New York: Henry Holt, 1991), 435.

73 The IDF attempted several: Ariel Sharon, *Warrior* (New York: Simon & Schuster, 2001), 80.

88 On October 19, 1953: Michael Bar-Zohar and Eitan Haber, *The Paratroopers Book* (A. Levin Epstein Press Inc., 1969), 78.

89 "Arik never sent us": Shimon "Katcha" Kahaner, in conversation with the author, December 27, 2010.

90 "Our first law is to execute": David Ben-Uziel, "Tarzan," *Bamachaneh* magazine, December 1980.

91 Yitzhak Rabin: Yitzhak Rabin: *A Soldier's Diary* (Maariv Library, 1979), 95.

91 Zvi Yanai described: Zvi Yanai, warrior in the paratroopers, "Becoming Israeli," *Haaretz*, May 2000.

100 According to Moshe Dayan: *Ben-Gurion: Moshe Dayan, Milestones* (Jerusalem: Eidanim, 1976), 641.

100 "Every time Arik arrived": Dalia Goren, conversation with the author, December 23, 2010.

101 Yaakobi's lawyer: Amiram Harlaf, *Be'saarat Regashot* (Ozar HaMishpat, 2006), 100.

107 France and Great Britain sought to reestablish: Egypt nationalized the Suez Canal Company on July 26, 1956.

110 As prime minister and defense minister: Transcription of Ben Gurion's speech, on his visit to the paratrooper division, April 25, 1957.

112 Ben-Gurion had intended: I heard this from my father himself and in Shlomo Nakdimon, "The men who would be chief," *Haaretz*, February 18, 2001.

123 The five: Yael Dayan, *Israel Journal: June 1967* (New York: McGraw-Hill, 1967), 19.

125 "Moshe, it seems to me": Sharon, *Warrior*, 186.

127 "I wasn't opposed to close air support": Ibid., 191.

129 "It was like watching a snake": Ibid., 193.

130 Another prisoner of war: Dayan, *Israel Journal*, 71.

131 Yael Dayan described his actions: Ibid., 91.

132 While eating his C-rations: Sharon, *Warrior*, 198–201; Dayan, *Israel Journal*, 110.

132 "Arik, the legendary warrior": Dayan, *Israel Journal*, 118.

139 "Our goal is to position": Lieutenant-General Bar-Lev, cited in Dov Tamari, *The Reserves: A Unique Israeli Phenomenon*, 288.

140 "You cannot win a defensive": Sharon, *Warrior*, 220.

141 On December 19, 1968: Sharon note for stronghold discussion December 19, 1968.

142 The officer on the phone wanted to know: Sharon, *Warrior*, 222.

151 Tzafi was very popular with the terrorists: At that time, the Saudis kept forces in the area to help the Jordanians.

153 In 2007, during the skirmishes: Ali Waked, "Gaza: 17 were killed during Fatah-Hamas confrontations," *Ynet*, June 12, 2007.

159 In September 1972, after the murder: Zeev Shif, *Haaretz*, September 1972.

163 "The mere thought of the Egyptian army": Sharon, *Warrior*, 25.

164 Dayan said, "Arik, we aren't": Ibid., 271.

167 "Who is this Abu Rashid": Ibid., 282.

171 Not southern front commander: Eitan Haber, "The Hero Who Failed," *Yedioth Ahronoth*, October 1, 1991.

172 "What are you going to tell the journalists": Uri Dan, *Bridgehead* (E. L. Special Edition, 1975), 104.

175 "What could have been more logical": Uri Ben-Ari, *Yedioth Ahronoth*, December 13, 1974.

177 This is what Chief of Staff David Elazar said: Elazar quote from a chief of staff briefing for Prime Minister Meir in April 1973, cited in Dov Tamari, *The Reservers: A Unique Israeli Phenomenon* (PhD thesis, Haifa University, The Class For Israeli Studies, August 2007).

179 "Arik's here," they said: Dan, *Bridgehead*, 45.

179 He had seen warriors retreating: Ariel Sharon lecture in front of the paratrooper officers on the twentieth anniversary of the Canal crossing, October 17, 1993.

179 Speaking earlier in the day: Ibid.

180 "We have a large force in hand": Dan, *Bridgehead*, 28–29.

181 "Please send us help": Sharon, *Warrior*, 297.

181 "The plan was based on": Sharon testimony before the Agranat Commission, July 29, 1974, page 13.

181 "I tried to convince": Ibid., 14.

182 It was Avi Yaffe, radio operator: Avi Yaffe, *Yedioth Ahronoth*, August 15, 2003.

183 Rabin, in an uncharacteristic gesture: Sharon, *Warrior*, 299.

185 special commando force: Dan, *Bridgehead*, 69.

185 A quick look at the maps: Amatzia "Patzi" Chen, conversation with the author, January 24, 2011.

187 "I will dismiss you right now!": Sharon, *Warrior*, 301–302.

188 Several months after the war: Dan, *Bridgehead*, 278.

189 "Arik's division": Dayan, *Milestones* (Eidanim, 1976), 606–607.

189 "More than a few of the orders": Sharon testimony before the Agranat Commission, July 29, 1974, 24; 61.

189 "After that I scrutinized": Sharon, *Warrior*, 304.

191 We'll think about it: Ibid., 308.

192 On October 14, at 6:20 a.m.: Ibid., 311.

193 The Egyptians, in their arrogance: Ibid., 310.

196 "That left me speechless": Yossi Regev, *Hashirion*, October 19, 2003.

197 In the yard, my father personally directed the bulldozer: Lieutenant-Colonel Amos's description in Dan, *Bridgehead*, 182.

197 My father described the scene: Ariel Sharon, *Yedioth Ahronoth*, September 24, 1993.

199 "You are cut off!": Sharon, *Warrior*, 318.

199 "But it was certainly open": Ibid., 318.

199 Bar-Lev didn't make it down: Dan, *Bridgehead*, 163.

200 "Look at it for yourselves": Sharon, *Warrior*, 319.

201 Shalom had a message as well: Dan, *Bridgehead*, 205.

201 He heard someone say, "Our friend just bought it": Sharon, *Warrior*, 323.

201 "Does anyone have a bandage?": Dan, *Bridgehead*, 200.

202 "To the left": Sharon, *Warrior*, 324.

202 "How had I left that place unguarded": Ibid., 324.

204 The APCs drove: Dan, *Bridgehead*, 206.

204 Speaking in his slow: Sharon, *Warrior*, 326.

207 There is only one name missing: Haim Hefer, "The Commanders," *Yedioth Ahronoth*, October 12, 1973.

208 "This I can say to you": Dayan, *Milestones*, 655–56.

209 Aside from a description: Translated from the Arabic by Ovadia Danon.

212 "After midnight I received": Dayan, *Milestones*, 661.

213 "From my observation post": Amiram Azov, *The Crossing: 60 hours in October 1973* (Dvir, 2011), 142.

212 This struggle is evidenced: Ibid., 266–67.

214 And Gorodish was barred: Gorodish died on September 30, 1991, in Milan, Italy.

218 Rabin recalled in his memoir: Rabin, *Rabin*, 488–90.

219 Dayan crossed the lines: "The Political Prostitute," *HaOlam HaZeh*, May 22, 1977.

223 In the decade following: Israel's Central Bureau of Statistics, "Table C/4, Population and Population Growth in Jerusalem, 1967–2008."

223 As Albeck explained: Plia Albeck, *Haaretz*, April 4, 2004.

223 My father would laugh: Ibid.

224 During the following four: Sharon, *Warrior*, 366.

224 Alongside his efforts in Judea and Samaria: Ibid., 366.

227 In early 1981: Ibid., 374.

227 This fell in line with Sadat's vision: Ibid., 376.

228 When one of Sadat's ministers: Ibid., 376.

230 According to Shlomo Nakdimon: Shlomo Nakdimon, *Tamuz in Flames* (Yedioth Aharanoth, 1986), 90.

231 Sitting in the Knesset cabinet room: Sharon, *Warrior*, 382.

232 On May 9, Peres wrote to Begin: Nakdimon, *Tamuz in Flames*, 213.

232 During the meal Ali leaned in: Sharon, *Warrior*, 380.

232 As he got ready to go: Ibid., 383.

233 "We were wrong": James Baker, *This Week*, February 13, 2005.

233 In a pre-election ad: Labor Party advertisement in *Yedioth Ahronoth*, June 12, 1981.

234 "Would I send Jewish boys": Shlomo Nakdimon, *First Strike: The Exclusive Story of How Israel Foiled Iraq's Attempt to Get the Bomb* (Hebrew edition), 364.

256 During the course of the fighting: IDF—Israel Defense Forces, http://dover.idf.il/IDF/About/history/70s/1978/031501.htm.

258 In late July 1981: Sharon, *Warrior*, 433.

258　The majority of the stockpiling: Ariel Sharon, "Facts of the Lebanon War," lecture presented at Tel Aviv University, Israel, August 11, 1987.

258　In a meeting my father had: Ibid.

258　During the eleven-month cease-fire: Sharon, *Warrior*, 433.

259　Begin addressed the cabinet: Prime Minister Begin, in a cabinet meeting, June 4, 1982, from Sharon, "Facts of the Lebanon War."

260　At that June 5 meeting: Prime Minister Begin, cabinet meeting, June 5, 1982.

260　The resolution that was passed: Sharon, "Facts of the Lebanon War."

261　On June 6: Ibid.

261　"Yes, we will cross": Ibid.

262　Earlier that day, speaking before the Knesset: Ibid.

262　The cabinet decided: Ibid.

262　During the June 8 cabinet meeting: Ibid.

264　An attack on: Sharon, *Warrior*, 429.

264　At the June 9 cabinet meeting: Prime Minister Menachem Begin, cabinet meeting, June 9, from Sharon, "Facts on the Lebanon War."

265　As my father pointed out: Ibid.

268　Rabin, in his memoirs, explained: Rabin, *Rabin*, 502–503.

268　"If the Christians in the north collapse": Chaim Bar-Lev, April 13, 1981, in Sharon, *Warrior*, 428.

269　"Yes," Gemayel said: Ibid., 438.

270　"If we do come in": Ibid., 439, 441–42.

271　Yitzhak Rabin voiced unequivocal support: Sharon, "Facts of the Lebanon War."

271　They stood on a lookout point: Uri Dan, *Ariel Sharon: An Intimate Portrait* (New York: Palgrave MacMillan, 2007), 123.

271　To the Brother Hadj Ismail: Sharon, *Warrior*, 430–31; R. Avi-Ram, *Arab Documents and Sources, Volume One: The Path to Peace for the Galilee War* (Maarchot, 1987), 107.

272　This is Shimon Peres: Yossi Sarid, "From Resistance to Support, and from Support to Resistance," *Haaretz*, August 21, 1987.

273　Rabin and Dayan fully understood Peres's agenda: Rabin, *Rabin*, 534.

273　In addition, Dayan wrote this: Note from Moshe Dayan, February 14, 1979.

274　"They have to leave": Sharon, *Warrior*, 478–79.

277　The Syrians did not want: Yossi Melman, "Waltz without Bashir," *Haaretz*, September 22, 2010.

277　Victor Shem-Tov: Victor Shem-Tov, during the Knesset session, September 22, 1982, as quoted in Sharon, *Warrior*, 472.

278　As far back as June 15 the cabinet decided: Sharon, *Warrior*, 501.

278　Pierre Gemayel emotionally thanked: Yossi Melman, "Waltz Without Bashir," *Haaretz*, September 22, 2010.

279　This was a development: Kahan Commission of Inquiry, testimony of the defense minister, unclassified section, October 25, 1982, pages 49–56.

281　Atrocities such as the massacre: Kahan Commission, Sharon testimony, unclassified section, page 16.

283　At the time, the media: Uri Porat, media adviser to Prime Minister Begin, in an affidavit, February 11, 1997.

283　Begin decided to go on air: IBA interview, June 15, 1982, transcript from *Moked* program, interview conducted by Yaakov Achiemeir and Elimelech Ram.

285　At the cabinet meeting on Saturday: Ibid.

285　In the June 27 cabinet meeting: Ibid.

285 In the cabinet meeting of August 1: Ibid.
286 After the terror attack: Cabinet meeting, April 8, 1980.
286 In 1980 my father: Sharon, "The Facts of the Lebanon War."
286 Yechiel Kadishai: Yechiel Kadishai, affidavit, June 12, 1996.
286 On February 10, 1997: Yechiel Kadishai, affidavit, February 10, 1997.
287 In a Channel 2 interview: Sivan Dornon and Ben Caspit interviewing Yechiel
     Kadishai, *Reshet al ha-Boker*, Channel 2, November 7, 1997.
287 Begin's military liaison officer: Ben Caspit interviews Ozriel Nevo, *Maariv*,
     May 13, 2011.
287 "There is no country that runs its wars like this": Begin, August 5, 1982, as
     quoted in Sharon, "The Facts of the Lebanon War."
287 This was in sharp contrast: Sharon, *Warrior*, 463.
288 Defense Minister Ehud Barak boasted: Ahiya Raved, "Barak: The Operation
     Has Changed Reality Fundamentally," *Ynet*, January, 21, 1991
288 On July 17, 2006: Prime Minister's Office archive.
290 The bodies of the soldiers: Yoni Schenfeld, Army Radio web site, http://glz.co.il.
290 At the beginning of the government meeting: Maya Bengel, "Report: Hamas
     agreed for One Year Long Appeasement, Starting from Thursday," nrg.co.il,
     February 1, 2009.
294 But at the July 18 cabinet meetings: Sharon, "The Facts of the Lebanon War."
295 After the funeral: Knesset general assembly memorial service for the fallen of
     the Tyre accident, November 15, 1982.
295 "And I will not hide": Dubi Weissglass, in a conversation with the author,
     February 28, 2010.
296 This is the one and only instance: Dr. Avigdor Klagsbald, "The Kahan
     Commission of Inquiry: Jurisdiction and Standards in Determining Questions
     of Ministerial Responsibility," *Public Law Journal* (UK), Autumn 1983.
296 Nor has it happened: Dr. Avigdor Klagsbald, interview with the author, Feb-
     ruary 8, 2011.
303 All told, after two years: Ministry of Housing and Sector of Information and
     Economic Analysis.
304 Rabin and my father spoke: Dan, *Ariel Sharon*, 222.
305 In 1994 Arafat arrived: Barry M. Rubin and Judith Colp Rubin, *Yasir Arafat:
     A Political Biography* (New York: Oxford University Press, 2003), 125.
305 In an article published: Ariel Sharon, "Oslo Accords—Seed of War," *Yedioth
     Ahronoth*, September 4, 1994.
308 He of course supported: Peace treaty between Israel and Jordan, October 26,
     1994, Article 9, Clause 2.
308 As my father told Uri Dan: Dan, *Ariel Sharon*, 222.
309 "True, no one would have gotten": Dan, *Ariel Sharon*, 224.
313 the Mossad tried to assassinate: My father's notes in real time; my conversa-
     tions with Majalli Wahabi; and Shlaim, *The Lion of Jordan*; and my own recol-
     lections.
313 This contradicted the notion: Lieutenant-Colonel Majalli Wahabi, interview
     with the author.
313 Although the Israel-Jordan peace treaty: Israel-Jordan Peace Treaty, July 25,
     1994, Annex 2, Article 1, 2d, 3.
313 water would be given to them: Ibid.
322 Majalli hung up the phone: Majalli Wahabi, interview with the author.
323 Several years later: Avi Shlaim, *The Lion of Jordan: The Life of King Hussein in
     War and Peace*, 499 (Hebrew edition).

345 It seems that Dayan's position: Uzi Narkiss, *The Liberation of Jerusalem: The Battle of 1967* (Am Oved, 1975), 213–14 (Hebrew edition).

346 The visit to the Temple Mount: "However, we were provided with no persuasive evidence that the Sharon visit was anything other than an internal political act. . . . The Sharon visit did not cause the "'Al-Aqsa Intifada.'" *The Mitchell Report*, 2001.

346 President Clinton blamed: Rubin and Rubin, *Yasir Arafat*, 196–99.

347 One year later, on October 4, 2001: Ali Waked and Efrat Weiss, "Darama in The Territories," *Ynet*, October 5, 2001.

347 This decision contradicted: Prime Minister's Office, the completion of the IDF's withdrawal from Lebanon, May 25, 2000.

348 It was published: Al-Hayat Al-Jadida, "The Press Pass of the Journalist Who Apologised for Photographing the Lynching Was Taken from Him," *Ynet*, October 18, 2001.

350 It seemed like: "Peres and Arafat meet in Gaza," *Ynet*, November 2, 2000.

350 On November 18: "Perpetrator of Attack Promoted in Rank," *Ynet*, November 19, 2000.

350 On November 20: Avi Mosesko, "Barak to the Wounded: 'We Are on the Path to Making Peace,'" *Ynet*, November 23, 2000.

351 He thought he was: Ibid.

351 There were also calls: "Boycott Ariel," *Ynet*, December 2, 2000.

356 And yet Marit Danon: Marit Danon, interview with the author.

366 "But this peace": Handwritten note from when Sharon was prime minister, Sharon Archives.

368 The Clinton proposal at Camp David: President Clinton's outline for a final resolution to the conflict, as issued in 2000. The outline includes a Palestinian relinquishment of the right of return; an Israeli relinquishment of sovereignty on the Temple Mount; the division of Jerusalem and the founding of a Palestinian state on the majority of the West Bank, aside from the large settlement blocs, which will be annexed by Israel.

370 Secretary of State Colin Powell's: Amir Rapaport, "Tough Criticism from Powell on IDF Actions in Gaza," *Yedioth Ahronoth*, April 18, 2001.

371 Israel accepted: On March 22, 2001, Prime Minister Sharon convened a press conference and declared a unilateral cease-fire. Linoy Bar-Geffen and Ali Waked, "Sharon Ordered the IDF to Cease Fire," *Ynet*, May 23, 2001.

373 This underscored the truly absurd: U.S. Department of State, "Country Reports on Terrorism," April 30, 2008.

374 In the wake of: Felix Frisch and Ali Waked, "CIA Chief in Israel Issues Timetable for Mitchell Report Implementation," *Ynet*, June 9, 2001.

378 In June 2001: "Panorama," *BBC News*, June 17, 2001.

379 The Belgians were also very concerned: "Lawsuit Filed against American General Tommy Franks in a Belgian Court," Aljazeera.info.news, May 14, 2003.

379 On February 18, 2003: "Weapon of Mass Destruction," Globalsecurity.org, March 18, 2003.

380 The Belgians realized: Sharon Sadeh, "Belgium: Scope of Law, Enabling Sharon Trial, Has Been Reduced," *Haaretz*, April 3, 2003.

381 Oren, who supported: B. Siglovitz, "Oren to *Davar*: My pleasure. Espionage Charge—False," *Davar*, May 14, 1956.

386 Personal ties were forged: Sharon, *Warrior*, 134.

387 "You are my personal friend": Anne Appelbaum, "Farewell, Jacques Chirac," *Washington Post*, May 8, 2007.

388 That attitude was fanned: Dan, *Ariel Sharon*.

390 Five years later: Israel fought the Second Lebanon War against Hezbollah in the summer of 2006, during which thousands of rockets were fired from Lebanon into northern Israel, and the Israel Defense Forces invaded Lebanon. The war began on July 12, 2006, with the capture of two IDF soldiers.

391 In Poland: For example, the Kielce Pogrom, perpetrated on July 4, 1946, roughly one year after the Holocaust, targeting the small, tattered Jewish community in the city. Some forty Jews were killed in the pogrom.

391 My father gave a tally: Felix Frisch and Ali Waked, "CIA Chief in Israel Issues Timetable for Mitchell Report Implementation," *Ynet*, June 9, 2001.

392 "Why not use the Security Commission": A meeting of Israeli and Palestinian security personnel, coordinating security matters and held in the presence of American observers.

392 "But on the other hand": A closure is the blockading of a Palestinian area for security purposes. The goal is to prevent the flow of terrorists to and from the area.

399 Even after admitting: Shlomo Shamir, "The UN Still Refuses to Hand Over Uncensored Tape," *Haaretz*, July 9, 2001.

400 We expect to receive everything: Nakura is a UN base on the shores of the Mediterranean Sea, not far from Israel's border.

403 Among other things: Anat Roeh and Felix Frisch, "Arab-Israeli Kidnapped by Palestinian Security," *Yedioth Ahronoth*, June 7, 2001.

408 The Jewish population: Yehuda Slutsky, *The Haganah*, vol. III (Am-Oved: 1973), 617–18, 699.

408 By contrast, the most visible Palestinian: The Rashid Ali rebellion was an armed uprising from April to June 1941 in which pro-Nazi Iraqi rebels battled the Iraqi monarchy and the British Army. During the course of the uprising there were riots against Jews. Zvi Elpeleg, *The Grand Mufti: Haj Amin al-Hussaini, Founder of the Palestinian National Movement* (Defense Ministry Press, 1989), 67–75.

410 Not only did the British: "The Army Prevented Rescue of Convoy," *Davar*, April 15, 1948.

410 On May 13, 1948: Nearly 130 residents of Kfar Etzion were slaughtered after the battle, once they had put down their weapons and surrendered. Yitzhak Levi, *Nine Measures: Jerusalem during the War of Independence* (Defense Ministry Press, 1986).

410 Once the legionnaires: Website of the Jewish Quarter of the Old City of Jerusalem, http://www.myrova.com.

412 More than sixty years after: Tamara Traubman, "New U.K. Bid to Boycott Israeli Universities, Profs," *Haaretz*, May 9, 2006.

413 The murder of Israeli citizens: "Hamas 'Exhibit' in Nablus: Body Parts and Slices of Pizza," Associated Press, September 24, 2001.

417 Throughout the Palestinian Authority's realm: Ali Waked, "Dancing in the Streets in the Territories," *Yedioth Ahronoth*, September 12, 2001.

417 The footage of these: Linoy Bar-Geffen, "AP and Reuters Threatened, Not Showing Celebrations," *Ynet*, September 12, 2001.

420 "We were disappointed": The decision to create a list of terrorist organizations came in the wake of the 9/11 terrorist attacks. All assets belonging to groups or individual terrorists on this list were frozen by the European Union authorities in fifteen European states.

420 "Hezbollah and Iran were involved": March 17, 1992, a Hezbollah suicide

bomber detonated himself and a car full of explosives alongside the Israeli embassy in Argentina. The attack claimed the lives of 29 people and 220 were injured. On July 18, 1994, a Hezbollah terrorist detonated a car bomb outside the Jewish community's AMIA building in Buenos Aires, killing 86 people and wounding 300.

431 As far as I know: Peres and Arafat met at a conference for Mediterranean states held on the Spanish island of Majorca, where they had lunch along with Egyptian president Mubarak and Spanish prime minister Aznar.

443 On January 3, 2002: "Cheney: Arafat Involvement in the Weapons Ship Clear," *Ynet*, AP, January 27, 2002. "Powell: Arafat Took Partial Responsibility for Weapons Ship Affair," *Ynet*, AP, February 14, 2002.

443 Days later he sent a letter to President Bush: Ibid.

447 The Saudi Initiative: Clause 11 in UN Resolution 194, issued on December 11, 1948, relates to the return of refugees.

455 Two weeks later, the Saudi Arabian ambassador: "Diplomat Censured over Bomb Poem," http://bbc.news.co.uk, April 18, 2002.

457 Only after the IDF's failures: War waged in July/August 2006 during Prime Minister Olmert's term in office.

461 The same day, April 8, 2002: Itim "Danish Foreign Minister: 'Israel Has Crossed Every Line,'" *Ynet*, April 8, 2002.

462 Continuing the UN's long-standing tradition: "UN Secretary General: The Entire World Demands that Israel Withdraw," *Ynet*, April 8, 2002.

463 Terje Rod-Larsen was awarded: Jonathan Tisdoll, "Foreign Ministry to Investigate Money to Rød-Larsen," *Aftenposten* (Norway), April 24, 2002; "Norway: The Larsens Were Scolded for Accepting Money from the Peres Center," *Ynet*, April 29, 2002; "Terje Larsen's Wife to Return Funds Received from Peres Center," *Ynet*, May 24, 2002.

463 Apparently, there is no business: Ibid, *Ynet*, April 29, 2002; *Ynet*, May 24, 2002.

464 "Powell is a neo-Nazi agent": Al Jazeera, April 7, 2002.

466 Rajoub himself left the site: Felix Frisch and Ali Waked, "The Surrender at Rajub Headquarters," *Ynet*, April 2, 2002.

474 The committee's findings: Felix Frisch, "Amnesty Report," Ynet.com, November 4, 2002; "Human Rights Organization: No Massacre in Jenin," *Yedioth Ahronoth*, May 3, 2002; "UN Report: No Massacre, Israel Prevented Disaster," *Yedioth Ahronoth*, August 1, 2002.

476 "Sign already, you dog": Rubin and Rubin, *Yasir Arafat*, 145.

478 Javier Solana, in response to the attacks: "Condemnations of Attack in Europe and U.S.A.,". *Ynet*, June 18, 2002.

479 The Spanish foreign minister: Ibid.

479 "Today it is hard": Ibid.

481 "Europe must play": Orly Azoulay-Katz, "Arafat Must Go," *Yedioth Ahronoth*, June 25, 2002.

481 He noted that the United States: Richard Boucher, June 27, 2002, cited in "State Department Spokesperson Clarifies," *Ynet*, June 28, 2002.

482 Arafat, Secretary of State Powell said: News Agencies, "Spokesman for The US Department of State Has Cleared," *Ynet*, June 28, 2002; Diana Bahur-Nir, and the News Agencies, "Powell: There Is a Price to Not Fighting against Terror," *Ynet*, June 29, 2002.

496 Hussein-McMahon correspondence: An exchange of letters taking place from July 1915 to January 1916 between Hussein bin Ali and Sir Henry McMahon, the British High Commissioner in Egypt, concerning the political boundaries of land under the Ottoman Empire.

485 The EU confirmed: Ali Waked, "EU Representative Met with Yassin," *Ynet*, July 28, 2002; Roee Nachmias, "This is How EU Representative Flattered Ahmad Yassin," *Ynet*, December 3, 2004.

485 Jacques Chirac, in a conversation: Diana Bachur, "Chirac: Do Not Include Hezbollah in List of Terror Groups," *Ynet*, July 29, 2002.

485 Five of the dead: "FBI to investigate Hebrew University Terror Attack," *Ynet*, August 2, 2002.

488 In August, EU Commissioner: Diana Bachur, "EU to Arafat: Stop the Incitement," *Ynet*, August 9, 2002.

489 The Egyptian army: UN Resolution 181—the November 29, 1947 partition plan.

492 "We were so insulted": Shimon Schiffer, "Sharon, Mubarak and the Frankfurter Crisis," *Yedioth Ahronoth*, August 9, 2002.

492 Mubarak also used to complain: Ibid.

494 The sentiment is expressed: "Anti-Semitic Article in al-Ahram," *Ynet*, August 10, 2002.

494 During that meeting: Conversation with Smadar Perry, correspondent for Arab affairs, in *Yedioth Ahronoth*, May 4, 2011; Roee Nachmias,"The Ambassador to Cairo, the Editor and the Storm over Normalization," *Ynet*, September 17, 2009.

494 She went to Cairo: Roee Nachmias, "The Ambassador to Cairo," *Ynet* September 17, 2009.

494 The Egyptian minister of culture: Nahum Barnea and Smadar Perry, *Yedioth Ahronoth*, June 5, 2009.

494 A year later: Charles Bremner, "'Hebrew book-burning' minister Farouk Hosni is front-runner to head Unesco," *Times*, May 30, 2009.

495 The handshake caused: Nahum Barnea and Smadar Perry, "I Came, I Spoke, I Conquered," *Yedioth Ahronoth*, June 5, 2009.

499 Henry Kissinger: *Henry Kissinger: White House Years* (New York: Little, Brown and Company 1979), 631.

503 "Did we not throw mud": "Senior Palestinian Official: We Should Have Accepted the Clinton Proposal," *Ynet*, September 4, 2002.

503 That same month: Ali Waked, "The PA: Storm Over Fire on House of Senior Palestinian Official," *Ynet*, September 4, 2002.

503 Two years later Arafat: "Unknown Assailants Shoot Palestinian Legislative Council Member Nabil Amar," *Ynet*, July 21, 2004.

503 On September 11, 2002: Ali Waked and Felix Frisch, "Palestinians Expressed No Confidence in Arafat and His Way," *Ynet*, September 11, 2002.

503 Muhammad Dahlan: Ali Waked, "Dahlan: Palestinian Leadership Has Made Every Possible Mistake," *Ynet*, October 16, 2002; Ali Waked, "Dahlan: Palestinians Better Off before Intifada," September 29, 2003.

504 In the wake of Peres's report: Note left on Sharon's desk on October 28, 2002.

505 On November 5, 2002: Stephen Farrell and Robert Thomson, "The Times Interview with Ariel Sharon," *Times*, November 5, 2002.

507 On the night of November 11, 2002: Felix Frisch and Ali Waked, "Metzer Attack Terror Suspect: Sirhan Sirhan Is From Tul-Karm," *Ynet*, November 13, 2002.

508 "Mitzna's plan": Ali Waked, "Abu-Alaa: If Mitzna Elected—Agreement Can Be Reached," *Ynet*, November 20, 2002.

509 "Ma-fish Christmas": "Arafat: No More Christmas," *Ynet*, November 27, 2002.

509 "It is hard to believe": Ibid.

509 The blast: Felix Frisch, "Al-Qaeda has Strong Presence in Kenya," *Ynet*, November 28, 2002.

510 At this time there was some tension: "The Guardian: Britain Imposes Arms Embargo on Israel," *Ynet*, April 13, 2002.

510 The president of the EU: "European Union to Israel: Allow Palestinians to Visit London," *Ynet*, January 8, 2003.

513 As he himself testified: "I am an unpopular prime minister," *Globes*, March 15, 2007.

513 On January 29: "Arafat: Prepared to Meet Sharon Tonight," *Ynet*, January 29, 2003.

514 "As part of Israel's efforts": Prime Minister's Office, press announcement, February 1, 2003.

514 The terrorist, for his part: Israel Ministry of Foreign Affairs: "Suicide Bombing of Egged Number 37 in Haifa," March 5, 2003.

514 "The President stands strongly": "Bush: 'We Stand Alongside Israel in the Fight against Terror,'" *Ynet*, March 6, 2003.

515 Once again I urge: Ibid.

515 EU foreign policy chief: Ibid

515 UN secretary-general: Ibid.

515 Eventually, the efforts: Ali Waked, "Likely: Abu Mazen to Be Prime Minister," *Ynet*, March 4, 2003.

515 In the end, Arafat agreed: *Ynet*, March 4, 2003.

518 In an August 2, 2003: *Herald Tribune*, August 2, 2003 via *Ynet* August 2, 2003, "Sharon: 80 Percent of The Israelis Fell in Love with The Fence."

518 "Israel will stay": Ibid.

518 In mid-October 2002: Diana Bachur, "Sharon: the American Proposal—A Non-binding Draft," *Ynet*, October 18, 2002.

518 In November: Diana Bachur, "New Road Map: Palestinian State in 2003," *Ynet*, November 21, 2002. From that point on, the discussions revolved around Israel's comments on the plan.

521 Six years later: "Clinton: No Understanding Regarding Construction within Settlement Blocs," *Ynet*, June 6, 2009.

526 "All progress towards peace": "Arab Leaders in Sharm Commit to Stop Funding Terror," *Ynet*, June 3, 2003.

528 He called the heads: "Sharon: Abu Mazen Is a Chick Without Feathers," *Haaretz*, June 12, 2003.

529 At the same time, Abbas tried: Ali Waked, "Abu Mazen Offered Hamas: Join Unity Government," *Ynet*, June 18, 2003.

529 On June 29: *Yedioth Ahronoth*, June 29, 2003.

529 "Under the road map": White House statement, June 30, 2003.

529 "We are Israel's": Diana Bachur-Nir, "The British Remain Unconvinced: Arafat Is Elected Leader," *Ynet*, July 15, 2003.

530 Every meeting: Ibid.

530 Britain was the only nation: Diana Bachur-Nir, "Straw to Sharon: We Will Continue to Meet with Arafat," *Ynet*, July 15, 2003.

531 "Your visit is important": Ibid.

531 "We know the huge amount of work": "Sharon, Blair Mend Fences," July 14, 2003, http://edition.cnn.com/2003/WORLD/europe/07/14/blair.sharon/index.html.

532 Bondevik blanched: Sharon in Norway, July 15–17, 2003.

533 "If we find that": Diana Bachur-Nir, "France to Israel: Unproven that Hamas and Islamic Jihad Are Terror Organizations," *Ynet*, August 25, 2003.

533 A year prior: Diana Bachur, *Ynet*, "Chirac: Don't Include Hizbollah in List of Terror Organizations," July 29, 2002.

533 International pressure: Ali Waked, "Arafat Calls for Return to Hudna; White House Is Part of the Problem," *Ynet*, August 28, 2003.

533 On August 28: Ibid.

533 On September 2: "Arafat: 'Road Map Is Dead,'" *Ynet*, September 3, 2003.

534 On September 3: "Powell: Road Map Still Alive," *Ynet*, September 3, 2003.

535 Foreign Minister Jack Straw: "Agreement within European Union: Hamas to Be Declared Terror Organization," *Ynet*, September 6, 2003.

536 He called on Muslims: Ibid.

536 Aside from the claims: Ali Waked, "Abu Mazen Resigns; Arafat Declares Transitional Government," *Ynet*, September 6, 2003.

536 In an interview: Ali Waked,"Dahlan: Palestinians Better Off before Intifada," *Ynet*, September 29, 2003.

539 Syria filed: "Bush on the Strike in Syria: We Would Have Done the Same Thing," *Ynet*, October 7, 2003.

539 In response: Ibid.

539 "We call on": Ali Waked and Efrat Weiss, "3 Americans Killed in Gaza Blast," *Ynet*, October 15, 2003.

539 American investigators: "FBI Sends Team Terror Attack in Gaza Strip," *Ynet*, October 15, 2003.

539 The State Department offered: "American Offer: Five Million Dollars for Gaza Terror Attack Perpetrators," *Ynet*, October 31, 2003.

541 There were two options: Israel's Central Bureau of Statistics, Table 3: Population and Demography 2004, *The World Fact Book*, CIA, http://www.cbs.gov.il/reader.

546 He was evacuated: Efrat Weiss and Ali Waked, "Temple Mount: Egyptian Foreign Minister Attacked and Evacuated to Hospital," *Ynet*, December 22, 2003.

547 On January 29, 2004: "Tannenbaum and Bodies of Soldiers en Route to Israel," *Ynet*, January 29, 2004.

548 "Our next mission": Ron Arad is an Israeli Air Force weapon systems officer who has been missing in action since October 1986. He was lost on a mission over Lebanon, captured by Amal, a Shi'a Muslim militia, and later handed over to Hezbollah.

548 Kofi Annan published: "At UN: Attempt to Condemn Jerusalem Terror Attack Fails," *Ynet*, January 30, 2004.

548 In contrast: Ibid.

548 In an interview: Yoel Marcus, "The Scheduled Evacuation: Twenty Settlements in Gaza Strip and The West Bank within a Year or Two," *Haaretz*, February 3 2004.

551 At the outset: During the run-up to the March 2006 elections, months after the disengagement from Gaza, my father continued to enjoy tremendous public support and high approval ratings, as seen in the pre-election polls. In fact, the March 2006 election results, in which the Kadima Party won handily, are evidence of public support for the disengagement—Kadima was founded on account of the withdrawal from Gaza and was fully identified with the move. It seems fair to say that without the withdrawal there would never have been a Kadima. My father won the individual elections for prime minister in 2001, and the general elections in 2003 and again in 2006, this time without actually participating. All evidence points to the fact that had he run in 2006, Kadima would have achieved an even greater victory at the polls.

557 "Bush is the first": Aluf Benn and Nathan Guttman, "Bush Letter: Israel Will

Not Be Sent Back to Green Line, Refugees Will Not Return to Israel," *Haaretz*, April 14, 2004.

557 The official responses: "UN and Europe Assail 'Washington Understanding,'" *Ynet*, April 15, 2004.

557 On April 21: "Bush: Iran Has Set for Itself the Goal of Destroying Israel," *Ynet*, April 22, 2004.

557 On May 4: Yitzhak Ben-Horin and Diana Bachur, "Quartet Has Adopted Disengagement Plan," *Ynet*, May 4, 2004.

557 In a May 30: Diana Bachur-Nir, "Sharon to Netanyahu: 'Moving to Hear You Speak of Democracy,'" *Ynet*, May 30, 2004.

558 Subsequently it was discovered: Ron Greenberg, "Palestinian Officer in Karni Enabled Attack at Ashdod Port," *Yedioth Ahronoth*, June 22, 2004.

559 By that point: "Hamas Carried Out 425 Attacks during Intifada; 377 Killed," *Yedioth Ahronoth*, March 22, 2004.

559 Condoleezza Rice: "Yassin was personally involved in terror": "Sharon: We Struck Arch Enemy of Israel," *Ynet*, March 22, 2004.

559 "He who deals": "The Assassination: Peres Supports," *Ynet*, April 18, 2004.

563 On July 9: Diana Bachur-Nir "International Criminal Court in The Hague: Separation Fence and Settlements Illegal," *Ynet*, July 10, 2004.

563 "After what I saw today": Ofer Meir, "Injured Man from Jaffa: 'I'm in Favor of the Fence now, too,'" *Ynet*, July 11, 2004.

563 Palestinian inaction: Diana Bachur-Nir, "Larsen: Arafat Doing Nothing to Fight Terror and Violence," *Ynet*, July 13, 2004.

564 As of now: United Nations Security Council Meeting, July 13, 2004, http://unispal.un.org/unispal.nsf/9a798adbf322aff38525617b006d88d7/e698d621 47c5151985256ed2005407be?OpenDocument&Highlight=0,S%2FPV.5002 ,of,13,July,2004.

566 The American and French: Itzhak Ben-Horin, "A Compromising Suggestion: The UN Will Call for Foreigners' Withdrawal from Lebanon," *Ynet*, September 03, 2004.

568 It is worth noting: Ishaan Tharoor, "Syria's Alawites: The Minority Sect In the Halls of Power," Global Spin/Time.com, March 30, 2011, http://globalspin.blogs.time.com/2011/03/30/syrias-alawites-the-minority-sect-in-the-halls-of-power.

568 "I am willing to renew": Sharon Rofe-Ofir, "Assad: I Am Ready for Negotiations—If Sharon Is Ready," *Ynet*, September 7, 2004; Roee Nachmias, "Syria to the World: Apply Pressure on Israel," *Ynet*, December 4, 2004.

571 Suha, his wife: "Suha Arafat, the Senior Palestinian officials and the battle for the money," *Ynet*, November 12, 2004.

571 Armed men: Ali Waked, "Tens of Thousands Accompany Arafat on Final Journey in Ramallah," *Ynet*, November 13, 2004.

573 When the second roll call: Attila Somfalvi, "Game Just Beginning," *Ynet*, October 26, 2004; "Knesset Decides: There Will Be No Settlements in Gaza," *Ynet*, October 26, 2004.

574 It's worth noting: Attila Somfalvi, *Ynet*, "Knesset Passes Evacuation-Compensation Law," November 3, 2004.

575 On November 24, 2004: Yitzhak Ben-Horin, "A First: Condemnation of Attack against Israel," *Ynet*, November 25, 2004.

576 He referred to my father: "Mubarak to Palestinians: Only Sharon Offers Chance for Peace," *Ynet*, December 12, 2004.

576 That same day: Diana Bachur-Nir, "French PM: Al-Manar Broadcasts at

Odds with Our Values," *Ynet*, February 12, 2004.

576 The fact that al-Manar: Ibid.

576 "The European Union, Spain, and Israel": Atilla Shomfalvie and Guy Ronen, "Moratinos: Israel Should Get a Significant Position in The EU," *Ynet*, December 2, 2004.

577 Mahmoud Abbas also spoke out: Roee Nachmias, "Abu Mazen: Use of Arms is Only Damaging," *Ynet*, December 14, 2004.

577 "We changed our policy": Diana Bachur-Nir, "Dutch Ambassador: Europe Learned to Trust Sharon," *Ynet*, December 24, 2004.

577 Prime Minister Anders Rasmussen: Diana Bachur-Nir, "Danish PM: Israel's Enemies Want to Throw It into the Sea," *Ynet*, February 2, 2005.

578 The Belgians: Diana Bachur-Nir, "Belgian Ambassador Reprimanded over Meeting with Nasrallah," *Ynet*, January 24, 2005; Reuters, "Security Council: Shaba Farms are not part of Lebanon," *Ynet*, January 29, 2005.

581 A Palestinian suicide bomber: Ali Waked and Hanan Greenberg, "The Man Responsible: Kais Obeid, Hizbollah Case Officer," *Ynet*, February 26, 2002.

581 The parliament: Attila Somfalvi, "European Parliament Determines: Hizbollah Is Terror Organization," *Ynet*, March 10, 2005.

582 due to time constraints: Doron Sheffer, "Annan at Yad Vashem: We Failed in the Past; Committed to the Jewish People," *Yedioth Ahronoth*, March 15, 2005.

583 Speaking at: Ariel Sharon, AIPAC speech in Washington, May 24, 2005.

586 Four years later: Yitzhak Ben-Horin, "US: Charges against AIPAC's Leaders were Dropped," *Ynet*, May 1, 2009.

590 As the media reported: "Sharon meets with Chirac." The headline in *Le Monde* read: "After Years of Bittersweet Relations, The Time Has Come for Détente"; Yoav Toker, "Love Depends upon Something," *Ynet*, July 27, 2005.

590 As far as Israel: Diana Bahur-Nir and Ronen Bodoni, "Sharon Meets Chirac—The Relationship Defrosting," *Ynet*, May 27, 2005.

592 "I've called to thank you": Kofi Annan published a statement on August 15, 2005, calling the disengagement "a moment of promise and hope": http://unispal.un.org/UNISPAL.NSF/0/3A3AFE483388C8668525705E00608 C2F.

593 According to Danny Gillerman: Danny Gillerman in an interview with the author.

594 My father was allotted: Ibid.

594 Kofi Annan's welcoming: Ibid.

594 "I admire": Ibid.

594 And they did: "UN General Assembly, Sharon and the President of Pakistan," Ynet.com, September 14, 2005.

595 The meeting with Putin: Gillerman interview.

597 Ambassador Danny Gillerman: Gillerman interview.

597 With those words: Ariel Sharon's speech to the UN can be found in English at http://www.un.org/webcast/summit2005/statements15/isr050915eng.pdf and in Hebrew at http://www.haaretz.co.il/hasite/pages/ShArt.jhtml?item No=625556&contrassID=1.

600 When he reached: Ibid.

BIBLIOGRAPHY

Books

Avi-Ran, Reuven, *Lebanon War: Arab Documents and Sources*. Maarachot— The Ministry of Defense Publishing House, 1987.

Ben-Gurion, David. *The New Nation Of Israel*. Am Oved, 1969.

Bergman, Ronen. *Authority Granted*. Yedioth Ahronoth, 2002.

Bush, George W. *Decision Points*. Crown, 2010.

Dan, Uri. *Ariel Sharon: An Intimate Portrait*. Palgrave Macmillan, 2007.

Dan, Uri. *Bridgehead*. E. L. Special Edition, 1975.

Dayan, Moshe. *Milestones*. Eidanim, 1976.

Dayan, Moshe. *Story of My Life*. William Morrow, 1976.

Dayan, Yael. *Israel Journal: June, 1967*. McGraw-Hill, 1967.

Dershowitz, Alan. *The Case for Israel*. John Wiley & Sons, 2004.

Elpeleg, Zvi. *The Grand Mufti*. Taylor & Francis, September 1973.

Ertel, Heinz and Richard Sculze-Kossens. *Europäische Freiwillige*. Munin Verlag, 1986.

Ezov, Amiram. *The Crossing: 60 hours in October 1973*. Dvir, 2011.

Haber, Eitan and Michael Bar-Zohar. *The Paratroopers Book*. A. Levine Epstein Press, 1969.

Harlaf, Amiram. *Emotional Storm: Emotional Chapters in My Life*. Ozar Hamishpat, 2006.

Kahalani, Avigdor. *I Swear To You: Personal Conversation On Soldiers On The Yom Kippur War*. Maarachot— The Ministry of Defense Publishing House, 2004.

Kissinger, Henry. *White House Years*. Little, Brown and Company, 1979.

Klagsbald, Avigdor. "The Kahan Commission of Inquiry: Jurisdiction and Standards in Determining Questions of Ministerial Responsibility." *Public Law*, Autumn 1983.

Lebel, Jennie. *The Mufti of Jerusalem: Haj-Amin el-Husseini and National-Socialism*. Cigoya Stampa, 2007.

Levi, Yitzhak "Levitza." *9 Measures: Jerusalem in the War of Independence*. Maarachot—The Ministry of Defense Publishing House, 1986.

Littlejohn, David. *Foreign Legions of the Third Reich*. R. James Bender, 1981.

Milstein, Uri. *The Lesson of a Collapse*. Yedioth Ahronoth, 1993.

Nakidimon, Shlomo. *First Strike: The Exclusive Story of How Israel Foiled Iraq's Attempt to Get the Bomb*. Summit Books, 1987.

Narkiss, Uzi. *One Jerusalem*. Am Oved, 1975.

Pinsker, Leo. *Auto-Emancipation* (translated from German). Ahad Ha'am,1882.

Rabin, Yitzhak. *A Soldier's Diary*. Maariv Library, 1979.

Rabin, Yitzhak. *The Rabin Memoirs*. University of California Press, 1996.

Rubin, Barry M. and Judith Colp Rubin. *Yasir Arafat: A Political Biography*. Oxford University Press, 2003.

Sharon, Ariel. *Warrior: An Autobiography*. Simon & Schuster, 2001.

Shlaim, Avi. *Lion of Jordan: The Life of King Hussein in War and Peace*. Alfred A. Knopf, 2008.

Slutsky, Yehuda. *History of the Haganah, Volume 3: From Resistance to War*. Am Oved, 1972.

Tamari, Dov. *The Reservers: A Unique Israeli Phenomenon*. Doctorate for

Philosophy Thesis, Haifa University, The Class For Israeli Studies, August 2007.

Internet sites of the following organizations
Central Statistics Office
European Union at United Nations
First Decade Paratroopers
IDF
IDF Radio Station
In Memoriam for Citizens Killed in Terrorist Attacks
Jewish Agency for Israel
Jewish Virtual Library
Knesset website
MidEastWeb
Ministry of Foreign Affairs
Palestinian Media Watch
Prime Minister's Office of the State of Israel
Shabak
U.S. Department of Defense

Journals and newspapers
Aharam Weekly
Dvar
Globus
Haaretz

Haboker
Haolom Hazeh
Hashiriyon-Bytaon Amotat Ma'ariv
Makor Rishon
Yad Lashrion
Yedioth Ahronoth
The Wall Street Journal
The Washington Post

Additional sources
Ariel Sharon, Major General Sharon, testimony before the Agranat Commission, July 29, 1974.
Ariel Sharon, Minister of Defense, testimony before the Kahan Commission, October 25, 1982.
Ariel Sharon, Prime Minister, "Facts of the Lebanon War," Tel Aviv University lecture, August 11, 1987.
Ariel Sharon, "Twenty Years Passing Through the Canal," October 17, 1993.
Ariel Sharon, Prime Minister, speech in the Herzliya Convention, December 18, 2003.
Benjamin Netanyahu, Bar-Eylan University speech, June 14, 2009.

ABOUT THE AUTHOR

Gilad Sharon is the youngest of former Israeli prime minister Ariel Sharon's sons. Gilad holds a master's degree in economics and writes a column for the prominent Israeli newspaper *Yedioth Ahronoth*. A major in the Israel Defense Force reserves, he currently manages his family's farm in Israel.